SUBMERSIBLES AND THEIR USE IN OCEANOGRAPHY AND OCEAN ENGINEERING

FURTHER TITLES IN THIS SERIES

1 J.L. MERO
THE MINERAL RESOURCES OF THE SEA

2 L.M. FOMIN
THE DYNAMIC METHOD IN OCEANOGRAPHY

3 E.J.F. WOOD
MICROBIOLOGY OF OCEANS AND ESTUARIES

4 G. NEUMANN
OCEAN CURRENTS

5 N.G. JERLOV
OPTICAL OCEANOGRAPHY

6 V. VACQUIER
GEOMAGNETISM IN MARINE GEOLOGY

7 W.J. WALLACE
THE DEVELOPMENT OF THE CHLORINITY/SALINITY CONCEPT IN OCEANOGRAPHY

8 E. LISITZIN
SEA-LEVEL CHANGES

9 R.H. PARKER
THE STUDY OF BENTHIC COMMUNITIES

10 J.C.J. NIHOUL
MODELLING OF MARINE SYSTEMS

11 O.I. MAMAYEV
TEMPERATURE—SALINITY ANALYSIS OF WORLD OCEAN WATERS

12 E.J. FERGUSON WOOD and R.E. JOHANNES
TROPICAL MARINE POLLUTION

13 E. STEEMANN NIELSEN
MARINE PHOTOSYNTHESIS

14 N.G. JERLOV
MARINE OPTICS

15 G.P. GLASBY
MARINE MANGANESE DEPOSITS

16 V.M. KAMENKOVICH
FUNDAMENTALS OF OCEAN DYNAMICS

Elsevier Oceanography Series, 17

SUBMERSIBLES AND THEIR USE IN OCEANOGRAPHY AND OCEAN ENGINEERING

edited by

RICHARD A. GEYER

Department of Oceanography
Texas A & M University
College Station, Texas, U.S.A.

ELSEVIER SCIENTIFIC PUBLISHING COMPANY
Amsterdam — Oxford — New York 1977

ELSEVIER SCIENTIFIC PUBLISHING COMPANY
335 Jan van Galenstraat
P.O. Box 211, Amsterdam, The Netherlands

Distributors for the United States and Canada:

ELSEVIER NORTH-HOLLAND INC.
52, Vanderbilt Avenue
New York, N.Y. 10017

Library of Congress Cataloging in Publication Data

Main entry under title:

Submersibles and their use in oceanography and ocean engineering.

(Elsevier oceanography series ; 17)
Includes bibliographical references and index.
1. Oceanographic submersibles. 2. Underwater exploration. 3. Ocean engineering. I. Geyer, Richard A.
GC67.S9 551.4'6'0028 77-4023
ISBN 0-444-41545-9

ISBN 0-444-41623-4 (series)
ISBN 0-444-41545-9 (vol. 17)

Printed in The Netherlands

Dedication

This book is dedicated to Dr. Thomas Bright, Kenneth Bottom and Michael Cooke, who were the first to demonstrate that a submersible operation by an oceanographic institution could be economically viable and self-sustaining, as well as scientifically sound.

PREFACE

The ocean, since mankind first ventured on its surface, has become an ever-increasing means of communication and source of material resources, as well as recreational and esthetic satisfaction. Therefore, it is not surprising that the desire and eventually the ability to successfully venture beneath the ocean in further search of tangible as well as intangible rewards, challenged his mind and occupied his attention for many thousands of years.

The history of man's ability to develop means for directly exploring, scientifically studying and finally exploiting the bottom of the ocean and its superjacent waters falls into four eras. The first era, wherein he can go beneath the ocean only briefly and to shallow depths by himself. The second which can be divided into two phases — phase one, in which he encapsulates himself initially in a rather crude vehicle and again is limited in depth, endurance, range and speed — phase two where a more sophisticated version, known as a submarine became available. It is capable of transporting a large crew long distances, at significant speeds; and if so desired can cause widespread death and destruction. Era three is the one wherein he has designed and used, generally for constructive purposes, a much smaller vehicle which actually is a small submarine. However, it is generally called a submersible, and is the primary subject of this book. The fourth era coincides closely with the time of the writing of this book. It may be defined as the period in which a parallel and highly specialized technological effort to study and harvest the varied resources in and beneath the ocean has been successfully consummated using an unmanned submersible. Its availability and diversified uses are so relatively recent that when it was decided to write this book no thought had been given to including the topic of unmanned submersibles. However, it soon became apparent that the original idea had been overtaken by events to the extent that at least a section of the book should include a discussion of the role and respective merits of manned versus unmanned submersibles. This has been done in the form of several chapters written by acknowledged authorities. In addition, evidence both pro and con with respect to this controversy appears directly as well as indirectly to varying degrees in each of the case histories describing the uses of manned submersibles. It is also pertinent to state that this controversy over the use of manned versus unmanned vehicles to explore and exploit a hostile environment is not confined to the oceans. The same applies to the man-in-space counterpart, where the arguments and attentions have been more widespread as well as acrimonious, and have been underway for a much longer period.

The remainder of this book concerns itself with a series of case histories describing the results of the most recent developments in the use of submersibles to solve diversified problems in the ocean for a wide variety of scientific and engineering disciplines. These have a broad geographic distribution including parts of the Atlantic and Pacific Oceans, as well as the Mediterranean and Caribbean Seas and the Gulf of Mexico. The widespread geographic origins of the contributors also facilitates presenting broader and even differing perspectives. This is especially valuable in the discussions developing criteria for selecting manned or unmanned submersibles to solve specific problems. More specifically, these case histories are taken from such diversified scientific disciplines as geological, geophysical, and biological oceanography, as well as nautical archaeology. Examples of the solution of ocean engineering problems with submersibles include the survey for and construction of underwater pipelines, the evaluation and surveillance of ocean dumpsites and sewage outfalls, and the establishing of ecological base lines prior to offshore oil and gas production. Other major topics include the search for and salvage of an atomic bomb, and other objects lost in the sea, as well as describing methods for improving undersea navigation and communication, and discussing the role with examples of a teaching and research program of a university which owns a submersible.

It is evident from this brief description of the contents as well as the philosophy used in developing criteria for selecting the various elements comprising this book that it contains information of interest to a broad cross-section of readers. These include scientists, engineers, and management personnel in the academic, industrial, and governmental sectors, as well as lawyers, bankers, and even insurance specialists concerned in a variety of ways with solving problems bearing on the exploration and exploitation of the ocean. Students, especially those on a graduate level in the fundamental sciences, as well as of oceanography and ocean engineering, should also find the contents of this book to be a major reference source in these areas. Hopefully it may even stimulate some to make a professional career of some aspects of ocean science or technology.

Acknowledging the help of those involved in the writing of any scientific book is generally difficult because of the wide variety of persons who have played major roles either directly or indirectly from its conception to consummation. It is especially difficult for a book of this type which is composed of contributions by many authors. Although they have broad geographic origins and talents they have one basic objective in common. It is to demonstrate the versatility of the submersible as a means of gaining a better understanding of its successful role in helping to study, explore and exploit the oceans for the benefit of mankind. Obviously, without their cooperation and taking a great deal of time from exceptionally busy schedules this book would never have become available. The editor also acknowledges the help he received from the book, *Men Who Dared the Sea*, written by Gardener Soule in preparing the section on the history of submersibles appearing in the introduction.

The sponsors of the meeting, held in London in June 1975 under the auspices of the Marine Studies Group of the Geological Society of London and the Society of Underwater Technology, must be acknowledged for their role as catalysts in germinating the idea for this book. It was as a result of participating in this meeting on the uses of submersibles that the editor decided such a book should be written.

RICHARD A. GEYER

FOREWORD

Without rehashing the standard arguments about "why man?", the idea of placing the trained mind and eyes at the site of investigation is certainly not novel. In general, the scientist depends on in situ examination of the object of his studies. It is not enough to have some third party take samples from their natural environment and bring them back to the scientist's laboratory. He must study the context from which the samples came and their relationship to their natural environment.

Until relatively recently ocean studies have been frustrated by a denial of this basic principle of direct observation. Surrogate eyes and hands in the form of cameras and relatively crude grabbing devices brought up out-of-context samples from the deep oceans and their floors. From these fragments the scientist was supposed to imply the nature of the processes at work. It has been a tough, and often inexact business. Yet, with so little known about the oceans, much good work was done. Besides, there were no alternative choices.

Science effectively went *into* the sea in the early 1930's with Dr. William Beebe and his BATHYSPHERE. Four years of observational work in Bermuda, culminating with a dive to a depth of a half mile, forever established the principle of direct observation as well as the utility of in situ oceanography.

This is not to say that all of the oceanographic disciplines benefit equally from the use of submersibles. Where the work is largely observational in nature, the submersible's unique qualities make it a powerful tool for the marine scientist and engineer. The utility is a little more uncertain in areas such as physical and chemical oceanography, where only *measurements* are needed. One does not need to see a measurement or be at the site to take certain types of samples. Furthermore, the case in favor of the deep submersible is not strengthened by making blanket claims that all types of ocean science are better served through their use. Nevertheless, there are examples where the use of a submersible has helped the physical oceanographer to directly observe microscale turbulence at the seafloor. Such work was conducted by Dr. Eugene La Fond near San Diego, California using the bathyscaph TRIESTE, in the early 1960's.

Nearly two decades ago I had the good fortune to be personally involved during the early days of deep submergence. As a lieutenant in the Navy, I was the first officer-in-charge of the Navy's bathyscaph TRIESTE from 1959 to 1962. TRIESTE was purchased by the Navy from Professor Auguste Piccard and brought to the United States in 1958. At the time we began our diving operations in December, 1958, TRIESTE was virtually the only *operational* deep submersible in the world. Within the next decade there would be a small armada of these vehicles being developed in many nations. But few survived that decade to become successful deep-submergence platforms.

Much has been written about the "unfortunate situation" in the 1960's when so many submersibles were developed . . . but few survived to become operational successes. There were several factors causing this situation. But, most certainly, the apparent failure of deep submergence to develop rapidly cannot be assigned solely to lack of interst on the part of the government and potential users. Some key factors must be realized in attempting to understand the apparent failure of deep submergence during a period when many others areas of ocean science and engineering flourished.

First, there was the real expectation that manned exploration of the oceans was upon us. The United States had just quickly and successfully entered the space age in the late 1950's. The general enthusiasm was mutually fueled by both those involved in ocean studies and those who undertook to design and build the submersibles. The younger ocean scientists, especially those who had had some SCUBA diving work, were intrigued by the fact that they could *technically* go to any place and any depth in the world ocean for their work. Our dive to the deepest place in the ocean with TRIESTE in 1960 demonstrated this. The submersible builders came almost entirely from the aerospace industry and they were riding the crest of the space program. They had the high technology, the capital to invest and an eye to the future . . . the world of "innerspace". All of this was healthy but there was little moderation. The hard questions were not asked, or if asked, not understood.

A second factor was the increasing U.S. involvement in Vietnam and increased Navy resources began to be diverted to support Vietnam operations. The Navy in the 1960's was still the principal supporter of ocean science and the primary developer/user of advanced ocean engineering systems. Also, both the Congress and Defense Department began to worry about the Navy's "mission relevance" in supporting the majority of the national ocean program. Thus, the primary user-community for the submersibles then under construction began to slip away.

Most of the first generation of submersibles were built on speculation by a variety of industrial organizations. The design philosophies, systems and materials reflected a diversity of engineering approaches to a nearly common mission. There was no "company solution" resulting from government specifications. The result was a very large national base for deep-submergence engineering even though these vehicles were generally one of a kind. Operational teams were formed and most of the vehicles were operated for periods ranging from months to years. We learned how to do a variety of difficult tasks in the deep ocean. All this was done with the expectation that things would turn around and that there would be a profitable business in the building and operation of deep submersibles. It did not happen that way.

And one-by-one the first generation of submersibles was retired, scrapped or put into storage. There were few survivors. Today the Perry submarine series, ALVIN . . . perhaps a few others all modified, represent the first generation. By and large the original (aerospace) companies have left the field.

Was it all a loss as many claim? Most certainly not, if one thinks in terms of national capability development. There was neither instant success nor instant failure. The diversity of present design philosophies and the initial development of deep-ocean systems have their origins in this period. The personnel needed to design, build and operate the submersible fleets of today, largely got their start in the 1960's. The basic foundations for the orderly and evolutionary growth of today were laid during this period.

In the 1970's the manned submersible came into its own, practical and fully competitive with other systems. The principal successful applications have been in support of offshore gas and oil development; however, there also have been significant developments in support of marine science. The important factor is that submersibles have been selected for these jobs because of their unique ability to do the job. Offshore contractors are governed by doing the job at least cost; submersibles have been found to meet this test in numerous applications. In ocean sciences, the highly successful work along the Mid-Atlantid Ridge in Project FAMOUS is an example where this "uniqueness" has been applied to this area. This recognition of utility is a far better condition than in the 1960's when "unique" was often confused with "novelty".

There are still problems in bringing more deep-submersible operations to the support of ocean science and engineering. But many of these lie with a general lack of overall facilities support in the national ocean program, rather than with the deep-submergence technique itself. Federal government support for oceanographic ships and shore facilities has been considerably below expectations and needs for the past several years. It is difficult to see widespread support of scientific submersible platforms if there is inadequate support for oceanography's primary platform, the research vessel. Supplementary platforms such as aircraft and deep submersibles will find increased support only when the research-ship fleet is fully active.

To compound this problem, the submersible assets do not exist in the present ocean-science facilities inventory. In general the oceanographic ships do exist in sufficient numbers, although there is a pressing need to improve and upgrade the national fleet. In the case of submersibles there are only two active in the U.S. research fleet (Woods Hole's ALVIN and Texas A&M's DIAPHUS). The potential need is much greater, but this will require a considerable capital investment to acquire scientific submersibles and their requisite support systems. Clearly, the source must be the Federal government, but understandably any major efforts in this direction will come after the research-ship fleet is first revived.

The absolute costs are not great compared to other Federal expenditures, but the relative costs are considerable when compared to other, immediate needs in the ocean community.

Another recent development that will influence the deep-submersible community is where to put the man in the system. In the 1960's there was little favorable comparison between manned and unmanned systems. Manned submersibles were used where precision and in situ program flexibility were important. Unmanned systems were used for routine, programmed missions or for those tasks where there was considerable risk for a manned system. But all of this has changed in the last seven years. Man is still in the loop, but he is often on board a surface ship operating the submersible through very sophisticated control and sensor systems. This is not to say that the day of the manned submersible is over, but to suggest that we must consider manned and unmanned systems as a complimentary, overlapping set of capabilities. For a model of this relationship consider that this is essentially the way we have looked at diving systems versus submersibles in the past. The interface between systems capabilities is flexible and the choice of systems to be used for a specific task must take this into account. When we were first diving TRIESTE to 4000 feet off San Diego in the late 1950's diving was generally much limited to a maximum of 600 feet; saturation concepts were just then being worked out for practical applications. Today, working divers (of a very special type) are at 1500 feet and deeper, while most contemporary submersibles used for offshore work are not working much deeper (although most are capable of greater depths). This is not a direct equation of capability, of course, but it does illustrate the point of a flexible interface between capabilities. This is the way we must look at unmanned versus manned submersibles systems.

While I have dealt with the question of deep submersibles from the U.S. point of view, the general notions advanced here apply to most of the similar programs in other nations. There has to be an appreciation of what submersibles can do for the support of research in ocean sciences and engineering. This can only be achieved through the conduct of actual operations and the publication of successful results. Operations such as the French—American Project FAMOUS have done a great deal in advancing this area. But in addition to practical results, there has to be some means for introducing researchers to these very specialized techniques. It takes time to make a good in situ scientist. The entire scheme of observation and manipulation tends to be different from past experience, and real productivity often does not come until after several diving experiences. Recall that most marine scientists go on practice cruises during their university training. They get the feel of the shipboard platform and they learn to make the standard measurement manipulations at sea. This applies in most oceanographic institutions for scientific SCUBA diving courses. But for submersible operations this has not been possible. I am not sure what the answer might be; perhaps some sort of basic submersible scientific training program at selected institutions. It will not be possible to economically do this for large groups of studients. The only alternative is to "learn by doing" once the student becomes an oceanographic professional with the opportunity to use submersible platforms.

Demand follows experience, and with submersibles if scientists are not given the chance to see what they can do, then there will be no demand for the addition of these platforms to the national research fleet. Most oceanographic institutions cannot afford to purchase and support a submersible on a full-time basis out of their own funds. Even the outright gift of a vehicle can be a mixed blessing, as the support costs and facilities requirements often exceed the original cost of the submersible. This is especially true for deep-diving probes such as the bathyscaphs. The total annual bill for an active deep-ocean program could run into the millions of dollars. This is not an argument for not having capability (although this has seemed to be the case in the past), but rather it suggests that this is a proper role for the Federal government.

If we think of submersible systems in the same way we think of research vessels, the conceptualization is much easier. They tend to be large, expensive systems; costly to own and to maintain. They can be used for multiple missions simultaneously with careful planning, and with rare exception, they are acquired and supported through government grants. Their sole reason for existence is that they can be used to do certain types of marine research that cannot be done equally well with any other platform.

I am optimistic. The 1970's are showing the real value of deep-submergence systems for both industry and science. Once the present slowdown in the support of marine sciences eases, an orderly, logical expansion of deep-submergence activity should follow. This expansion will be based on need, not novelty, with the government participating in the development of these systems.

It is good to know that our early work with TRIESTE has helped some in bringing us to this point.

DON WALSH
Director of the Institute for
Marine and Coastal Studies
University of Southern California

LIST OF CONTRIBUTORS

Ardus, D.A.
Institute of Geological Sciences
Edinburgh, United Kingdom

Bass, G.F.
American Institute of Nautical Archaeology
Texas A & M University
College Station, Texas, U.S.A.

Bline, D.
Hydrotech Products
Hydrotech Int. Inc.
Houston, Texas, U.S.A.

Bright, T.J.
Department of Oceanography
Texas A & M University
College Station, Texas, U.S.A.

Busby, R.F.
R. Frank Busby and Associates
Arlington, Va., U.S.A.

Eden, R.A.
Institute of Geological Sciences
Edinburgh, United Kingdom

Geyer, R.A.
Department of Oceanography
Texas A & M University
College Station, Texas, U.S.A.

Heezen, B.C.
Lamont-Doherty Geological Observatory of Columbia University
Palisades, N.Y., U.S.A.

Keller, G.H.
School of Oceanography
Oregon State University
Corvallis, Oreg., U.S.A.

McQuillin, R.
Institute of Geological Sciences
Edinburgh, United Kingdom

Palmer, H.D.
Dames and Moore
Washington, D.C., U.S.A.

Rezak, R.
Department of Oceanography
Texas A & M University
College Station, Texas, U.S.A.

Rosencrantz, D.M.
American Insitute of Nautical Archaeology
Texas A & M University
College Station, Texas, U.S.A.

Talkington, H.
Ocean Technology Department
Naval Undersea Center
San Diego, Calif., U.S.A.

Treadwell, T.K. (USN, Ret.)
Department of Oceanography
Texas A & M University
College Station, Texas, U.S.A.

Vadus, J.R.
Office of Ocean Engineering
NOAA
Rockville, Md., U.S.A.

Van den Berg, L.
Shell U.K. Exploration and Production
Aberdeen, United Kingdom

Wilson, J.B.
Institute of Oceanographic Sciences
Wormley, Godalming, United Kingdom

CONTENTS

CHAPTER 10. U.K. EXPERIENCE OF THE USE OF SUBMERSIBLES IN THE GEOLOGICAL SURVEY OF CONTINENTAL SHELVES by R.A. Eden, R. McQuillin and D.A. Ardus

CHAPTER 11. SUBMERSIBLES: GEOLOGICAL TOOLS IN THE STUDY OF SUBMARINE CANYONS by Harold D. Palmer

CHAPTER 12. USE OF SUBMERSIBLES IN THE CONSTRUCTION OF SUBMARINE PIPELINES by Dolf van den Berg

INTRODUCTION

DEFINITION

The origin of the term "submersible" to describe a relatively small vessel capable of operating beneath the ocean in contrast to "submarine" is not well documented. As is so often the case with other words it began appearing in the literature with increasing frequency, until now it is generally accepted as the preferred word. Dictionaries, as for example, the American Heritage Dictionary, 1976 Edition, also do not show any significant difference. A submersible is defined as: "a vessel capable of operating or remaining underwater"; and a submarine as: "a ship capable of operating submerged". However, the use of the word "ship", rather than "vessel" implies that a submarine is larger than a submersible, because a vessel is defined as: "a craft, especially one larger than a row boat, designed to navigate on water". This places the boundary limits of the size of a submersible as larger than a row boat, but not as large as a ship, because a ship is defined as: "any vessel of considerable size...". And a boat as: "a relatively small, usually open craft of a size that might be carried on a ship". Thus this semantic exercise may be summarized as follows: a submersible is to a submarine as a boat is to a ship. A distinction must also be made as the result of recent technological developments, between a manned and an unmanned submersible as discussed briefly in the Preface.

HISTORY

The history of the development and use of submersibles may be subdivided into several broad eras as noted in the Preface. The first era starts as far back as 4500 B.C., when mother of pearl was used to decorate vases in Mesopotamia; and other pottery from the Middle-East showed men swimming and diving. Similarly, vases dating from about 500 B.C. found in Greece showed a man diving from a ship for sponges. At the siege of Syracuse Greek divers were able to stay underwater long enough to saw through stakes which had been driven into the bottom to sink Greek ships; and in 196 A.D. divers also cut ship cables during the siege of Byzantum. The use of divers for salvage was also quite common at the time of Livy (59 B.C.), because he describes that the fees earned were a function of the depth of water in which these operations were conducted. For example, they obtained from 10% of the salvage value at depths of 12 feet to as much as 50% at only twice this depth.

All these references are to divers unaided by any type of breathing devices. However, as early as 375 A.D. in *De Re Militari* written by Vegetius divers are pictured with a hood and leather bag around their necks; and Aristotle mentioned two tools for divers to facilitate their breathing underwater. These consisted of a helmet to trap air and a hose to bring air from the surface. Similarly, about the time of Hamilcar in 480 B.C. divers off Assyria used an inflated animal-skin as an air-tank, as seen on a bas-relief. It was belted to the diver's chest and stomach with an air tube extended to the mouth. Perhaps the first monarch to become a diver in recorded history was Alexander of Macedon who in 300 B.C. was lowered in a glass barrel into shallow water to observe the marine life.

It was almost 2000 years later in the middle of the 17th century before the next major era in the history of man's descent into the sea was initiated. Van Drebbel is reported to have used a submersible in the Thames River in which King James I was a passenger. However Edward Halley, who discovered the comet named for him, is generally credited with building the first practical diving bell. In 1691 his bell which was about 2 m high and 1 m in diameter was used in salvage operations of a frigate at a depth of 20 m. Thirty years later his continued development of the diving bell resulted in one which enabled five men to spend two hours underwater at a depth of 20 m.

The first reported use of a submarine for military purposes is generally credited to Bushnell who in 1776 built one called TURTLE with which he attempted to sink a British frigate in New York Harbor. About 25 years later Fulton built a copper globe filled with compressed air which he called NAUTILUS. It had a crew of three and could stay underwater at a depth of 8 m for six hours. The first submarine using the semantic criteria presented in the discussion on definitions was built in 1856 by Bauer. It carried a crew of eleven, which was also its means of propulsion, and had sufficient air to sustain life for about seven hours.

Almost two thousand years after Alexander of Macedon was lowered in a glass chamber into shallow water Beebe developed the bathysphere. He was lowered to a depth of over 300 m in 1930 and within two years to more than twice this depth — it could also be recorded as the first tethered submersible, although it belongs in the manned rather than unmanned category. In 1960 Piccard and Walsh in the submersible TRIESTE dove to the bottom of the Challenger deep (ca. 12,000 m) in the Marianas Trench. Additional details of the history of the development of both manned and unmanned submersibles are presented in Chapters 1, 2 and 4. However, in conclusion it should be noted that in 1931 Sverdrup used Simon Lake's submarine NAUTILUS to obtain oceanographic information during cruises in Arctic waters. Its original mission envisioned by Sir Hubert Wilkens to sail directly under the Arctic ice to the North Pole was never accomplished.

MANNED SUBMERSIBLES

Unique capabilities

There has been considerable discussion and at times even controversy among proponents of manned and unmanned submersibles as to their respective roles and merits in studying and exploring the ocean as well as in the development of its resources. It is the purpose of the following sections of the introduction to summarize the opinions and reactions of the authors as they are developed in this book and to attempt to present some objective conclusions and generalizations. This has been done by quoting or paraphrasing some of their most pertinent observations and the reader may thus refer to the appropriate chapters for a more complete development of the following statements.

"If man's presence is necessary for a successful mission it is most likely because the mission requires real-time, high-resolution sight". (Talkington, Chapter 4)

Man, because he can make real-time observations, can also make real-time decisions that could turn a marginally successful dive into an outstanding discovery of some otherwise elusive but significant bit of knowledge; or for engineering operations successfully complete a critical task that otherwise might not be accomplished. The unmanned system might miss the opportunity or have difficulty in quickly returning to the area being studied or worked on. (Bline, Chapter 3)

Variations in shades of color, depth of features and other subtle but significant changes can be recognized most efficiently by man. To approach his response many man-made machines operating simultaneously and fully correlated would be required. (Palmer, Chapter 11)

Marine scientists have found that the manned submersible provides an unparalleled extension of surface capabilities because of the human factor. The scientist/observer can make decisions on the spot as to changes in scope and plan of observations especially in confined areas such as submarine canyons. The most economical use occurs when a trained observer is provided with a device to place him in direct visual contact with areas of the sea floor which would be unreachable through conventional surface sampling systems. (Palmer, Chapter 11)

"It provides a platform where a scientist can scrutinize small-scale features revealing the nature and magnitude of agents and processes at work on the sea floor. For example, in Hudson canyon tile fish weighing 45 kg were observed near excavations in soft rock of the canyon, but the largest tile fish ever caught in the area was less than one-half this size.

In the study of waste disposal on continental shelves it was assumed that canyons are "natural conduits" to the deep ocean basins and hence ideal depositories for refuse to be *carried away* from the coasts. However, current and sediment transport studies made from manned submersibles suggest the existence of significant transport at times *up* the canyons, as well as down". (Palmer, Chapter 11)

The human eye and brain can make observations on the spot which cannot be made by any instrument. (Heezen, Chapter 8)

Presence of biogenic fluff not noted with remote sensors such as cameras. This is a factor not previously considered in the study of sediment transport and suspension. (Keller, Chapter 9)

Keller attempted without success while in a submersible to cause slope failures and mass movements of surficial sediments downslope, on slopes as great as 70°. This is generally believed to be the mechanism for initiating turbidity currents. On the other hand manned submersible observations made by Keller and others have revealed what appears to be stable deposits on slopes exceeding the angle of repose. He also cites special insights gained from observations and discrete sampling on other processes such as sediment variability, bioturbation, and bathymetric mapping. This resulted in obtaining an entirely different concept and understanding of their modus operandi than previously believed. (Keller, Chapter 9)

Ability to collect samples from precise sites relative to one another is important. A trained observer can direct the sampling operation providing the only effective means to sample selectively both representative, as well as anomalous features which must be considered in their relative perspective, if a study is to have any significance. (Keller, Chapter 9)

In-situ measurements of acoustic properties of sediments from submersibles has advanced the field of acoustics markedly by providing the necessary ground truth to formulate concepts on sound attenuation and reflection characteristics of deep-sea sediments. (Keller, Chapter 9)

"Indeed as most of these results (Sedimentological and Faunal Investigations) could not be obtained in any other way, the advent of the manned submersible in the 1970's is as significant an advance as was the introduction of the naturalists's dredge in the 1830's (and just 100 years after the cruises of HMS PORCUPINE). For the first time the scientist can choose exactly for example where a sample should be collected to answer a particular question. Rare and delicate species can be located, observed, and photographed in situ collected and transported to the surface in virtually undamaged condition". (Wilson, Chapter 7)

The ability to remain on the bottom in studies of bottom sediment movement resulted in observations in the winter in the English Channel where the sand did not move when the currents were strongest, but occurred quite suddenly for periods of 20—30 seconds. At such times plumes of sand were seen to rise from the crests of the ripples and a "carpet" of sand moved with the current. (Wilson, Chapter 7)

The beam pattern of most echo sounders is quite broad and it is important to obtain an accurate profile of the canyon slopes and flow. Hence, the use of an observer, even with limited sampling facilities in a submersible becomes an obvious solution to the study of canyons. (Palmer, Chapter 11)

Fluctuations in the sediment-fill of the axis through discolorations and truncated growth of sessile forms at a depth of 250 m in Los Frailes canyon have been discovered from observations made in a submersible. This indicates a source of dynamic action in canyon sedimentation other than coastal run-off because southern Baja California is an arid area. (Palmer, Chapter 11)

There can be no more efficient method of detailed rapid and controlled data collection for ocean dumping site studies than by placing the trained observer on the sea floor. (Palmer, Chapter 13)

In table I on p. 318 are listed in summary form the advantages and disadvantages in using divers and/or submersibles for sea-floor surveys and inspections including dump sites for wastes and sewage outfalls. (Palmer, Chapter 13)

The primary advantage of a submersible lies in the ability to permit placing the human eye, hand and brain directly at the point of observation. This proximity permits an observer to discern details and thus frequently improves the accuracy and resolution, over results obtained by an integrating remote system. (Treadwell, Chapter 1)

Perhaps the most persuasive factor is that of the combination of the eye, mind and hand; the ability to do so selectively and to act on those factors which the brain perceives to be important. The ability to react, to pursue the unexpected, to alter plans quickly and continuously in response to a changing situation, may well be the most durable virtue of manned submersibles. (Treadwell, Chapter 1)

The presence of man in a submersible is both its greatest asset and its greatest liability. Humans provide the unique capacity for observation and interaction, but in turn they require life-support and safety systems which are complex, expensive and non-optional. (Treadwell, Chapter 1)

The impression left in the biological and geological scientists' minds during studies of reefs and banks have permitted a much more detailed and efficient synthesis of the data collected than would have resulted had these not actually been "seen" in the study areas. (Bright and Rezak, Chapter 6)

The manned submersible can be used to select meaningful locations at which to place a variety of specialized instruments needed to study dynamic processes such as sediment traps, time lapse cameras, and current meters. (Keller, Chapter 9)

In the study of naturally occurring hydrocarbons in the Gulf of Mexico which included the use of a submersible sitting on the bottom, it was determined that the gas seeps are intermittent. They characteristically emit repeated sharp bursts of several hundred bubbles at a time. Had this phenomenon not been observed accurate quantitative estimates of the total amount of gas produced in this manner could not be made, because from television observations made over the side of a ship it appeared that these streams of bubbles were continuous, rather than intermittent. (Bright and Rezak, Chapter 6; Geyer and Bright, Chapter 5)

A location of smooth glaciated solid rock beneath a pavement of glacial boulders was found and a sample was taken with a drill. This is a classic example of an important finding that could not have been achieved by any other method than a manned submersible. (Eden et al., Chapter 10)

Disadvantages

The primary disadvantages of submersibles are presented in the form of a concensus rather than as quotations. This has been done in the interest of avoiding a great deal of repetition, because most of the disadvantages that follow are mentioned by more than one author.

The necessity for life-support systems with adequate redundancy to overcome malfunction of components, increases its size, weight and maintenance requirements. Adequate safety provisions are also imperative during launch and recovery operations. This in turn requires a larger and more specialized support ship and decreases the number of ships of opportunity available to conduct specific missions. The life-support requirement also limits the time-on-bottom, because of the need for replenishment and to recharge power supplies. Safety requirements also restrict operating days because of sea state and weather constraints. Unmanned submersibles can be launched and retrieved in much higher sea states; and time-on-bottom need not be decreased to allow for the safe recovery time of a manned submersible, before the arrival of bad weather, nor to wait for the seas to subside before launching.

Although a major advantage is its independance of movement and operation from a support vessel this also results in severe disadvantages, because its range, speed, vertical control and maneuverability are in general more limited than an unmanned submersible.

Several authors have cited a great need to improve the navigational accu-

TABLE I

Uses of manned submersibles (major categories)

Oceanography	*Ocean Engineering*
Biological	Oil production facilities
Chemical	Pipeline facilities
Geological	Soil mechanics
Physical	Salvage operations
Geophysical	Civil
Deep scattering layer studies	Military
Radiation surveys	*Archaeology*
Ice studies	*Fisheries*
Acoustics	Research
Optics	Operations

racy in positioning submersibles as a major constraint to many of their present varied uses, as well as in developing new ones.

Uses

In the interest of completeness the diversified uses of manned submersibles that follow are not confined to those presented in this book. However,

TABLE II

Specific uses of manned submersibles in oceanography

Biological
*Lateral and vertical zonation of reefs can be studied successfully and unexpected details and events of the biotic community become evident (Chapter 6)
*Provide accurate description of faunal composition of specific taxonomic zones on reefs (Chapters 6 and 7).
*Selected specimens can be collected either because no preserved representative species is available, or for confirmation purposes in identification (Chapters 6 and 7)
*In-situ distribution studies for taxonomic purposes (Chapters 6 and 7)
*Obtain speed of movement of benthic organisms to be able to better design instruments and experiments to study them with remote equipment and also to verify results (Chapter 8)
*To study in detail the response of organisms to changes in currents and sediment patterns (Chapters 8 and 9)
*Fauna associated with iceberg plough marks (Chapters 7 and 10)
*Faunal and taxonomic studies of reefs and banks (Chapters 5, 6 and 7)
*Bioturbation studies (Chapters 6, 9 and 11)
Bioluminescence studies

Fisheries
Harvesting of coral (black and red) at 330 m depths
Underutilized species of fish at 700—1000 m depths
Underutilized species of crabs at depths of more than 200 m
Habitation and migration of deep-water lobsters and shrimp
Deployment and effectiveness of line arrays for lobster traps
Deep-ocean food-chain studies
Deep-ocean benthic fish and other organisms
Hover over school of fish and transmit its location as well as speed of movement and direction
Select optimum areas for trawl fishing

Geological
*General sea-floor characteristics (Chapters 7, 8, 9, 10, 11 and 13)
*Submarine canyons (Chapters 8, 9 and 11)
*Dynamic bottom processes
Biological fluff (Chapter 9)
Turbidity currents (Chapters 8, 9 and 13)
Suspended sediments (Chapters 6, 8, 9 and 11)
Sediment boundaries (Chapters 6, 7 and 10)
Slope variability (Chapters 8, 9, 10, 12 and 13)
Bioturbation (Chapter 6, 9 and 11)
Sand waves (Chapters 7, 8 and 11)

TABLE II (continued)

---

*Pore-water and heat-flow measurements (Chapter 9)
*Bathymetry (micro and macro) (Chapters 6, 7, 8, 9, 10 and 11)
*Sample collecting (cores, sediments) (Chapters 6, 7, 8, 9, 10 and 11)
*Subduction-zone studies (Chapter 8)
*Origin and distribution of manganese nodules (Chapters 8 and 9)
*Distribution and characteristics of lava flows including classification of various types (Chapter 8)
*Role of tidal currents in the erosion of submarine canyons (Chapters 8, 9 and 11)
*Iceberg plough marks (Chapters 7 and 10)
*Reefs and banks (Chapters 5, 6, 7, 10 and 12)
*Naturally occurring hydrocarbons (Chapters 5 and 6)

Geophysical
*Corroborate deep-ocean seismic studies of rift and fault areas (Chapters 8 and 9)
*Exploration for oil and gas (Chapters 5 and 6)
*Emplacement of heat-flow instruments (Chapter 8)

Physical
*Nepheloid layer (Chapter 6)
*Bottom tidal studies (Chapters 7, 8, 9, 10, 11, 12 and 13)
*Bottom-current studies (Chapters 7, 8, 9 10, 11, 12 and 13)
*Emplacement of current meters and disbursement of colored dyes (Chapter 11)

Chemical
Hypersaline salt ponds (ca. 250‰)
Sampling of bottom sediments in transition zone between base of water column and unconsolidated sediments
*Pollutants (Chapter 13)
*Manganese-nodule studies (Chapters 8 and 9)

Ocean engineering
*Monitoring sewer outfalls, dredge spoils, and solid waste sites (Chapter 13)
*Site surveys and inspection of pipelines and communication cables (Chapters 4 and 12)
*Repair of pipelines (Chapers 7 and 12)
*Ocean bottom mining exploration and production operations (Chapters 8 and 9)
*Geotectonic properties of marine sediments (Chapter 9)

Salvage
Civil:
Oil production operations
Aircraft
Ships

Military
Atomic-bombs (Chapter 4)
Aircraft
Ships

Archaeology (Chapter 14)

---

those uses preceded by an asterisk are discussed in the chapter or chapters noted. The purpose of Table I is to summarize the broad areas in which the use of manned submersibles has already made major contributions to our understanding and development of the oceans. Table II presents examples of more specific uses within the major categories listed in Table I.

UNMANNED SUBMERSIBLES

As stated in the Preface unmanned submersibles were not intended originally as a subject for this book, but it soon became evident that their increasing importance made it imperative that some consideration be given to them if a well-rounded concept of the role of submersibles in marine operations were to be presented. Therefore Chapters 2, 3 and 4 are directed primarily to this topic, although it is also touched on in other chapters, especially Chapter 1.

Advantages

The primary advantages of unmanned submersibles are listed as follows (although a comparison of the merits of tethered versus "free-swimming" unmanned submersibles is not discussed here):

Relative economy of development in time and equipment costs when compared with manned systems.

Unlimited operational endurance on the working site because of the cable-link to the surface.

Control and coordination at the surface of project efforts. This avoids clash of operational philosophies, "because he who is on the surface is in command."

Ability to perform in hazardous areas without endangering personnel.

Ability to change or modify all system components to meet individual tasks without affecting system safety or certification status.

Ease of changing crews without disrupting the mission. Men simply leave their places at the control consoles and immediately their replacements are available to continue the operation.

These systems are usually smaller and lighter, and because they are remotely manned, the handling problem is significantly reduced.

No operational time limit as long as support ship can maintain station and provide necessary power through cable. Therefore, vehicle operations and observers can be rotated in shifts and hence limits of effective human endurance are never reached. Because of these advantages a work vehicle to repair pipelines is being designed to remain on the bottom for up to 6 weeks. With several of these vehicles an unusually large continuity of effort could be provided.

Disadvantages

The disadvantages listed as follows again represent a concensus of those discussed in Chapters 2, 3 and 4;

Possible extensive loss to property if power should be lost and the vehicle would surface out-of-control, because it is operated with slightly positive buoyancy maintaining submerged positions using vertical thrusters.

Most vehicles have an open framework type of construction and hence are subject to entanglement if operating near cables, lines or protruding hardware.

The intense concentration required by the operator to guide and otherwise control the vehicle limits the optimum time to about 2 to 3 hours. This requires the availability of several trained operators per vehicle for continuous operation.

Many manipulative and observational tasks require three-dimensional viewing to be effective and existing television does not provide this need.

Missions are limited because of power-consumption constraints for untethered types.

The drag on the cable for tethered vehicles causes difficulty in maneuvering and requires added power requirements.

The use of unmanned submersibles involves the control of a vehicle using video and instrumentation and is analogous to flying an airplane in the clouds, while attempting to study the geology, biology, etc. of the earth.

Uses

The wide variety of unmanned submersibles now in operation can be classified into several categories including towed or self-propelled, and operating in the water column, as well as those crawling on the bottom. The latter at the moment do not have as many specialized applications, but do have additional disadvantages which limit their use. Busby in Chapter 2 lists over 20 different uses in those tasks and missions where the ability to view and maneuver is critical, as compared with those where in-situ manipulation and collection are involved. Other applications are discussed by Bline in Chapter 3 and by Talkington in Chapter 4. One of the most recent successes of CURV III described in Chapters 2 and 4 was in the recovery of a radioisotope thermoelectric generator (RTG) in 800 m of water on the San Juan Seamount located about 400 km west of San Diego.

CONCLUSIONS

Numerous examples of the uses of manned submersibles have been presented in detail in the case histories chapters and others discussed in the remaining ones. Similarly, examples of the uses of unmanned submersibles both tethered and free-swimming are emphasized in Chapters 2, 3 and 4. Comparisons have also been made throughout the book between the relative merits and the role of manned versus unmanned submersibles to develop the resources of the ocean and perform other tasks. However, it is difficult to conclusively demonstrate an overwhelming superiority for either type of submersible for uniquely conducting the myriad of tasks and missions to be performed in the oceans. The rationale for this generalization is evident in the following statements:

If comparisons between capabilities of manned versus unmanned are to be meaningful they must be qualified by so many ground rules the final conclusion becomes indecisive. It is like adding apples to oranges. The only unequivocal statement that can be made is that the unmanned submersible presents no hazard to human life. (Busby, Chapter 2)

Unmanned submersibles of the RCV-225 and CONSUB types are at about the same stage of development that the manned submersibles were in the mid-to-late 1960's. At that time they were going to be "all things to all men"; however, it took nearly a decade before the most practical designs were identified and used. The early manned submersibles were the "multi-purpose" vehicle, and they could survey, salvage, repair, support divers, and collect samples as well as do oceanographic research. This same philosophy is now being espoused for both types of unmanned submersibles, but it should be remembered "there is no all-purpose aircraft, ship, automobile, nor manned or unmanned submersible." (Busby, Chapter 2)

Only in manned submersibles can geological causes and effects be studied successfully. "The sight of sand carried in tidal flow undercutting the wall of a submarine canyon, the sight of ripple crests changing symmetry, or tidal currents changing from up canyon to down canyon, the view of the sharply defined narrow subduction-plane intersection with the trench floor, all give a clarity and sharpness of understanding unobtainable from remote observations." (Heezen, Chapter 8)

Man's desires in undersea exploitation exceed his ability to pay for them. Both manned and unmanned systems are necessary to achieve his goals. However, it is obligatory that unmanned systems be considered first and used whenever and wherever possible.

Man should be included in a system only if he is absolutely necessary for the success of a mission, because his presence in it markedly increases the cost, reflected not only in money but in more safety considerations. The latter in turn results in additional systems complexity, handling problems and time to accomplish a task. (Talkington, Chapter 4)

Much work in the sea at present is in relatively shallow depths within free-diving range, but the submersible has the advantages of extended residence time on the bottom, higher payload capacity and greater range and speed. (Treadwell, Chapter 1) This point has also been made quantitatively and in graphic form. (Palmer, Chapter 11, figs. 1, 2)

Manned submersibles can be made to look very good or very bad depending upon the mission, the location, the specific submersible to be used, and the other vehicles with which it is being compared. For example, selective sampling of rocks and detailed investigations of local geology in a deep trench can be done uniquely with a manned submersible. But it is probably not the most efficient way to make routine descriptions of water-column chemistry or physics. (Treadwell, Chapter 1)

The inescapable conclusion to be drawn from these varied comments regarding the role and merits for diversified uses in the oceans of manned and unmanned submersibles is that there are certain tasks and missions that can be performed only by a manned submersible; but there are many others that can be done by either. The two over-riding considerations as to which should be chosen for a specific task are safety of personnel and economics. Other major determining factors which must also be considered in reaching a final decision include: (1) operating depth, (2) duration of dives, (3) visibility and other environmental factors, (4) urgency of the specific task, (5) operating policies governing the user, and (6) possible loss of property if an unmanned submersible is used and it becomes unmanageable. It should be emphasized that the choice between selecting a manned versus an unmanned submersible to perform specific tasks should not always fall into an either/or category. There are certain tasks or missions such as the recovery of the bomb at Palomares described in Chapter 4 where the special capabilities of the two types of submersibles complemented one another, and working together as a team, this mission was accomplished successfully.

Ample evidence has been presented as to the major roles submersibles have played in developing the ocean's resources and performing other tasks of great significance to mankind. Several major improvements that could be made to further expand their capabilities include providing more sophisticated and efficient diver lock-out facilities, as well as navigation and propulsion methods. In addition, submersibles equipped with lock-out capabilities might eventually be developed to the point that they could act efficiently, when necessary, as mobile habitats. However, it would be meaningless to attempt to present a conjectural list of tasks and missions which the submersible of the future would be called upon to perform because these are limited only by man's needs and desires as determined by his fertile and creative imagination, and by the engineering and economic constraints of design criteria required for their construction.

Section I. History — Respective roles and merits of manned versus unmanned submersibles in solving oceanographic and ocean engineering problems

CHAPTER 1

THE RATIONALE FOR SUBMERSIBLES

T.K. Treadwell

INTRODUCTION

At the start of the 1960's — the decade of great oceanographic growth — there were only four deep research vehicles in operation in the world. At the end of the decade, the number had increased by an order of magnitude to 43. By the early 1970's, the bubble seemed to have burst; many submarines were sitting idle, being laid up, or scrapped. Now, in the mid 1970's the utilization of these undersea craft is again rising, and often for purposes somewhat different from those originally foreseen.

What is behind this curious cycle? Were the scientists, engineers, and businessmen who originally pushed the development of scientific submarines completely misguided? Are these boats merely novelties, of no real scientific merit? Or are they desirable tools which have unfortunately been almost priced out of the market? Let us examine the history of scientific submarines, the rationale behind their use, and speculate about their future.

Apart from diving bells and tethered chambers such as Otis Barton's of the 1930's, the first manned scientific submersible was the FNRS III. Built by the French Navy following the balloon-like principles of earlier models, it reached a depth of 4,050 m off Africa in 1954. Almost simultaneously the Piccards, long active in both ballooning and submarining, constructed the TRIESTE which reached the greatest known oceanic depth of 10,910 m in the Marianas Trench in 1960. The TRIESTE, the FNRS III and Barton's vehicle are described in Chapter 3.

Simultaneously with these achievements, the Committee on Oceanography of the National Academy of Sciences was writing a lengthy and detailed set of recommendations (Anonymous, 1959) designed to generate support for oceanographic research. In the key summary of recommendations, however, "Oceanography 1960 to 1970" had remarkably little to say about submersibles. The Committee noted that the Navy had brought TRIESTE to the United States, and suggested that funds be made available to provide a mother ship for it. They further recommended that a development program be initiated, aimed at building manned vehicles "as the need arises and as the state of the art progresses", but the writers gave far more attention to more conventional research platforms such as ships, aircraft, and buoys. It was speculated that submarine observation might be useful in biological studies, particularly in behavioral studies, and in investigations of mineral nodules.

In the separate chapter on future engineering needs, however, a group of specialists tried to come to grips not only with the many engineering and operational hurdles, but also with the problem of utilization. The engineering panel noted that there had been only limited experience on which to base projections of utilization, and observed that "Within five years our concept of the capabilities of manned deep submersibles will have broadened in scope." However, they felt safe in recommending the construction of a two-hull bathyscaph system, with one shell designed for a depth of about 6000 m, and the other for twice that, capable of going to the maximum oceanic depths. It was further suggested that design work be started on a small, simple, mid-depth submarine which "could be handled like a lifeboat" from a research ship.

The panel suggested a relatively modest level of funding for submarines rising to some $ 4,000,000 annually for construction and operations. Probably never in their wildest dreams did they visualize that in the next eight years not two, but forty, research submarines would be constructed, at costs totalling hundreds of millions of dollars, and practically all from investments by private industry and commercial concerns.

ROLE OF INDUSTRY IN DEVELOPMENT AND USE OF SUBMERSIBLES

The industrial commitment to support of oceanography is itself a striking anomaly. The vast bulk of funding for marine science in the United States came at that time, as it does now, from the Federal government, namely the Navy, National Science Foundation, Bureau of Commercial Fisheries, and others. Industrial activity was largely limited to proprietary work by oil companies, with minor additions from fishing interests. It was surely implicit in the recommendations of the Academy of Sciences committee that the burden of financial support for submarines would fall on the Federal Agencies.

Three factors joined to induce major investments by industry in scientific submarines. Perhaps most important was sheer competence. The aero-space program had spurred industrial development of a highly advanced technology related to orbital and outer-space activities, which was often directly transferable to submarine problems. The basic concept of small, self-contained, pressurized modules was as relevant to scientific submarines as it was to space capsules. A great deal of the detailed engineering on materials, structural design, and electronics were closely applicable. Many of the aero-space corporations felt, and probably quite correctly, that they could shift their talents and production capabilities from outer space to inner space with a minimum of dislocation.

Further, it was easily foreseeable by the mid-1960's that the government's aerospace program was going to peak out and decline. It was simply prudent business planning to seek to identify future areas of use for industrial talents, and it seemed entirely reasonable that government interest in the oceans might phase in to replace a fading space program.

Finally, quite apart from technical competence and a natural desire for continued business, there was a conviction that submarines would be of real use to scientists and engineers. This had been bolstered by the continued success of TRIESTE, as well as the arrival of TRIESTE II, Cousteau's DIVING SAUCER, and others during the early years of the decade. As a result of all of these factors, many of the most prestigious names in industry — Alcoa, General Dynamics, Grumman, Lockheed and Westinghouse — were quickly involved in the design and construction of research submarines.

Their activities were spurred on by other events. Following the loss of the combat submarine THRESHER, the Navy established a study to look into underwater problems. It made recommendations on both the naval and scientific aspects of small submersibles. The Interagency Committee on Oceanography produced a report (Anonymous, 1963) which was primarily concerned with research uses of deep-diving vehicles. The Committee on Oceanography of the National Academy of Sciences followed its earlier monograph on ocean research with a report (Anonymous, 1964) addressing possible expansion of oceanography as a part of the economy of the United States in the event of a thaw in the cold war. Terry (1966) contributed an excellent and extended treatment of deep submersibles, including historical and engineering information in addition to one of the first detailed analyses of potential utilization.

In 1968—1969 submersible activity reached a peak, with 46 of these boats either in operation or under construction. Unfortunately, it became quickly and painfully apparent that a good deal of money and effort had been expended, and was premised on the availability of technology and a certain amount of wishful thinking, but without the rigorous justification which should prudently precede such an investment. Funded research work for the submarines did not materialize to anything approaching the extent anticipated, and in addition to construction costs, corporate funds were being used to support their operations. Clearly, a stock-taking for scientific submarines was overdue.

RATIONALE FOR USE OF SUBMERSIBLES

The rationale for use of research submersibles was approached from entirely different directions by two studies in the late 1960's. Ballard and Emery (1970) examined 346 articles in the scientific literature through 1969, with a view to analyzing the results obtained from the use of research submarines. It was hoped that this might provide an indication of scientific utilization and productivity. They found, as would be expected, that the bulk of the articles based on work done in research submarines had been published in the last half of the 1960's, following closely on the availability of the platforms. Somewhat more surprisingly, they found that usage of submarines by individual investigators was highly concentrated. Of the articles examined, only 12% of the authors represented accounted for 50% of the number of papers; clearly, these vehicles were not widely-used research tools. Disciplinary focus was also not balanced; fisheries and biological studies accounted for 35% of the papers, and geology/geophysics for 32%. While part of this disparity can be accounted for by the numerical predominance of biologically oriented researchers, it also seemed to indicate that research submarines were not equally useful to all disciplines.

Taking a completely different approach, the Naval Oceanographic Office in the late 1960's carried out an evaluation of research submersibles involving actual operational usage. While their stated aim was to look at their effectiveness in "underwater surveying", the definition of surveying was so broadly constructed that the study was indeed an overall assessment of these vehicles for general scientific utilization.

A total of 95 working dives were made on boats ranging from the relatively small Perry submarine to giants such as ALUMINAUT and BEN FRANKLIN. Evaluations were made not only in the absolute sense of the submarine being able to do a given job, but more importantly, in comparison with surface ships and unmanned vehicles, and with reference to their cost-effectiveness. Experiments and performance checks were run in all the usual scientific disciplines, plus sea-floor engineering measurements and acoustics studies. In addition to their own evaluations, the Navy scientists drew on the experiences of other workers, and their conclusions, published in 1970, are among the most detailed available.

In recent years there has been increasing use of scientifically capable submersibles, particularly those of shallower-depth design, by industry. While no broad compilation seems to have been made, a review of the trade literature clearly indicates much heavier utilization of submarines in offshore oil activities, particularly in engineering studies and the conduct of actual work on well completions, platform installation and pipelining. Although much of this work is well within free diving range, the submarine's advantages of extended residence time on the bottom, more payload capacity, and greater speed appear to be attractive.

ADVANTAGES AND DISADVANTAGES OF MANNED SUBMERSIBLES

In spite of the major studies mentioned above, and the extensive range of views expressed by scientists in describing research and surveys done from submersibles, it is still remarkably difficult to arrive at valid generalizations about advantages and disadvantages. Some of this stems from what has been called the "gee-whiz" syndrome of the early days of scientific submarining.

There is no doubt that many early participants were, quite understandably, so impressed with seeing things that no human had ever viewed before that they tended initially to overlook the drawbacks which became increasingly irksome as the novelty wore off.

Part of the problem too lies with the necessity of viewing submersible research not in isolation, but in comparison with other platforms and sensors. Research submarines can be made to look either very good or very bad, depending upon (1) the type of research being done, (2) the location, (3) the particular submarine being used, and (4) the other vehicles it is being compared with. Selective sampling of rocks, for example, and detailed investigations of local geology in a deep trench area, can be done uniquely well from a submersible. On the other hand, for routine descriptions of water-column physics and chemistry, a submarine is probably not a very good choice.

Finally, quite apart from the merits of research submarines for any particular job, one must consider the questions of cost effectiveness and absolute cost. Getting the scientific job done is desirable, but getting it done within a realistic budget is something else again. The best tool in the world is useless, if the mechanic can't afford to buy it. In this context, it is worth noting that the scientist may well have a quite different opinion from those who fund his work. The researcher may be convinced that a submarine is the best platform for his study, but his sponsor may be driven by financial limitations toward an alternative method which is somewhat less productive, but far cheaper.

In examining the advantages and disadvantages of scientifically oriented submarines, it will therefore be more realistic to look at the submarine not in isolation, but in comparison to other vehicles usable for the same purpose. The key items for overall decision-making can be grouped into the interrelated categories of observational competence, operational capabilities, and cost effectiveness.

Conducting oceanography from a surface ship has been likened to exploring the United States from a balloon. If Lewis and Clark had carried out their expedition from a slowly moving blimp, riding five miles above a continuous cloud cover, buffeted by storms, and limited to sampling the air below and the surface of the continent by lowering instruments and nets on a wire, they would have been in about the same observational status as the usual oceanographic ship. Even with these constraints, the explorers would have identified the major mountain chains and rivers; they probably would have netted some of the predominant plants and the slower, more common animals; and they would have gotten a general idea of the geology and soil types from their random samplings. Certainly most of the rarer and fleeter animals would have easily eluded them. Their chances of discovering even major mineral deposits such as the mid-continent coal fields would have been low, and the likelihood of finding the localized metal ores of the mountains remote indeed.

In the category of observation the advantage of submersibles lies primarily in placing the human eye, hand and brain at the point of observation. A secondary consideration is placing the instruments themselves closer to the desired point of observation, so that things may be "seen" closely and directly, rather than through an observational telescope from the surface. This proximity often permits one to pick out details, and frequently improve accuracy, over the results obtained by an integrating remote system. Finally, getting away from the motion and turbulence associated with the air—sea interface is often desirable, and sometimes critically important.

These considerations may seem both limited in number and simple in concept. They can, however, in their ramifications, sometimes add up to a powerful case for putting a human directly on the spot, even given the costs and operational problems associated with doing so. Perhaps the most persuasive factor is that of the combination of the eye, mind, and hand; the ability not only to observe a multitude of things, but to do so selectively and to act on those factors which the brain perceives to be important. The ability to react, to pursue the unexpected, to alter plans quickly and continuously in response to a changing situation, may well be the most durable virtue of manned submarines.

In a purely observational sense, human participation in many of the instrumental observations has been most effective in the precise placement of sensors, and in monitoring and adjusting them for optimum results. Observing the performance of nets and trawls, or of the accurate siting of current meters, is typical of this sort of activity. Specifically checking the results of remote sensing — "ground truth" observations — is also vital when the reliability of the sensing system is not fully assured.

Submersibles can also, even with their limited mobility, fill a gap in the dimension of observations which frequently frustrates scientists. Surface-borne sensors will ordinarily give an adequate, generalized overview of the major features over a broad area; close-up or bottomed systems provide a detailed look at a very small space. In their ability to back away from the bottom, submarines can be helpful; they can observe a somewhat broader scope, but still pick up significant detail. Or, if one wants to transfer his detailed view from one area to another in a planned or opportunistic manner, the submarine permits this; remotely lowered systems can do it only very poorly.

OPERATIONAL CONSIDERATIONS

Turning to factors of an operational nature, it is perhaps paradoxical that advances in technology have worked both for and against the utilization of manned submersibles. On the one hand, technological and engineering advances have overcome some of the operational shortcomings which plagued these vehicles ten years ago. But the same sorts of advances have also improved the general quality of all kinds of instrumentation and accessory equipment, often greatly reducing the need for humans to actually accompany or monitor observations.

The following specific cases will help to illustrate this situation, which is particularly pertinent in the consideration of manned versus unmanned diving vehicles. The state of the art in television cameras, data transmission links, and command systems is now such that in many instances remote viewing and control of an unmanned vehicle is quite adequate. Low light level TV cameras now can see even better than the human eye; sophisticated command systems permit highly flexible and responsive operator control of a vehicle; and tools and manipulators on a vehicle can be adapted to a great many tasks. All these combine to lessen the need for a man-rated vehicle.

On the other hand, many of these advances, taken singly, also improve the capability of a manned submarine. Others, such as improvements in connectors, power plants, and electronics, are equally applicable to both manned and unmanned operations, and a good many of them to remote sensing or lowered sensor packages as well. What technological advances have achieved is to clarify the choices. To a very large extent, the basic engineering problems which caused serious operational concern in the 1960's are no longer an overriding consideration. One can now regard the vehicle options as being almost equally workable, in a technological sense, and can make a choice of methods based on operational utility and cost effectiveness without serious qualms as to safety or operability.

In spite of these advances in technology, and the clear advantages of manned submarines for many types of work, a considerable body of disadvantages still exists. Some of these can perhaps be overcome, but others unfortunately are inherent in the boats which will be available to scientists for the near to mid-range future.

Foremost among these are the basic operational limitations. Given the constraints that the craft is to be launched and recovered from a mother ship, and that it will use non-nuclear power, designers are driven toward relatively small hulls carrying limited payloads, and of low speed and endurance. This in turn leads to poor maneuverability, limited area coverage, and, in great depths of water, a high proportion of "elevator time" as compared to working time. The lack of space in small hulls not only may exclude desirable scientific capability such as precise navigation systems, but usually results also in cramped working conditions for the scientists and poor facilities for viewing and observation.

The small size and the basic submarine configuration, usually accompanied by a clutter of tender external appendages and instrumentation, combine to produce serious operational problems in the launching and retrieving operations and when the craft is at the air—sea interface. Submarines are notoriously unseaworthy and have very low freeboard, which make them hard to handle as an independent vessel. More importantly, the response of a submarine and its mother ship to waves and wind are totally different, making launch and retrieval a highly hazardous operation. As a result, the loss of operational time due to weather is usually very substantially higher than that encountered by a surface ship or an unmanned vehicle.

SUMMARY

Finally, the presence of people in the craft is both the greatest asset and the greatest liability. Humans provide the unique capacity for observation and intervention, but in return they require life-support and safety systems which are complex, expensive, and non-optional. One may perhaps proceed with a mission with some of the scientific gear inoperative, but no such alternative may be considered with regard to equipment needed for human safety. This inability to take risks is often the most serious operational constraint.

In spite of these problems, no one seriously questions the overall value of manned submersibles in marine science and engineering. What has developed in the last few years, however, is a clearer picture of both their utility and limitations, and from this is emerging a somewhat modified pattern of employment. For some types of scientific work, even in great depths of water, their value, and even their unique competence, is clear. Utilization on the continental shelves and in relatively shallow water has been increasing, due both to the shortened transit and elevator time and to the scientific and engineering tasks which have developed there. Operational combinations of submersibles and free divers, both independently and through diver lock-out configurations, have become increasingly common. There has been a distinct trend toward smaller and simpler craft. These are adequate for shallow depths, and do not require the large, expensive operational and maintenance staffs demanded by larger submarines.

Perhaps the most important change, however, has been attitudinal. The great majority of scientists now regard submarines as simply another tool available to them, rather than an end in themselves. The decision to do research from a submarine is made as unemotionally as the choice between nansen and niskin bottles, or piston corers and gravity corers. As a result, while the pattern of utilization for the future is far from clear, it seems certain that it will be based, as it should be, on rationality rather than on gimmicks or emotion.

REFERENCES

Anonymous, 1959. Oceanography 1960—1970. NAS/NRC Committee on Oceanography, U.S. Government Printing Office, Washington, D.C.

Anonymous, 1963. Oceanography, The Ten Years Ahead, Interagency Committee on Oceanography of the Federal Council for Science and Technology, Pamphlet No. 10. U.S. Government Printing Office, Washington, D.C.

Anonymous, 1964. Preliminary Plan for Expansion of Oceanographic Research as a Contribution to the Peacetime Economy of the United States Under Conditions of a Cold War Thaw. Special Report of the Committee on Oceanography of the NAS/NRC, U.S. Government Printing Office, Washington, D.C.

Ballard, R.D. and Emery, K.O., 1970. Research Submersibles in Oceanography. Marine Technology Society, Washington, D.C.

Terry, R.D., 1966. The Deep Submersible. Western Periodicals Co.

CHAPTER 2

UNMANNED SUBMERSIBLES

R. Frank Busby

INTRODUCTION

Ocean scientists of the 1960's witnessed an armada of vehicles and ideas to carry them safely and profitably to the ocean depths. The ocean scientist of the mid-1970's is witnessing just the opposite. For a variety of reasons, a host of commercial, government and academic activities offer alternatives to the in-situ human being. Instead of direct visual observation and on-the-scene participation, a means of remote viewing and manipulation is provided, namely, the unmanned submersible.

There is no rigorous definition of an unmanned submersible. The term generally refers to relatively small, self-propelled vehicles which are tethered by cable to a surface craft from which they derive their power and direction. Just as the manned submersible enjoys a number of aliases (as for example, mini-sub, undersea research vehicle, deep-submergence vehicle, submarine, etc.), the unmanned submersible has various names such as, remotely controlled vehicle, cable-controlled subsea vehicle or submersible, remotely manned undersea vehicle, deep-tethered vehicle, etc. The many different titles attempt to emphasize the fact that a human being is not inside the vehicle. Instead of being at the view port, he is on the surface observing a TV monitor. In essence, the unmanned submersible is a manned submersible with no capability for protecting or sustaining human life. Except for power, its operation closely parallels that of its manned counterpart. Indeed, in spite of its title, the entire "unmanned" system is no more productive than a lump of clay until the surface control and display console is "manned."

The unmanned submersible is the direct result of the ocean community's increasing need to selectively sample and study in more detail the sea floor, as well as to visually inspect, build and monitor undersea facilities. In the early 1950's various types of oceanographic instruments such as sediment samplers and underwater cameras were lowered individually from an anchored or drifting ship. Occasionally the two instruments were combined to work simultaneously to obtain both the sample and a photograph of the bottom. While these over-the-side procedures added substantially to our knowledge, they had serious drawbacks, with the major one being the lack of selectivity in sampling or photography. In short, the camera and the sampler performed their tasks at the whim of drifting vessels and ocean currents. There were a few isolated attempts to lower underwater television cameras, but the heave of the surface vessel caused the camera to alternately move up and down. The results were therefore confined to an occasional fleeting glance of the bottom as the camera went in and out of focus.

In the early nineteen-sixties instruments such as side-scan sonars and seismic sub-bottom profilers became commercially available. These were also used individually, but were towed by a surface craft underway. The individual use of data-gathering instruments is costly. Therefore, the next logical step was to combine two or more instruments into one package where they could effectively be used together.

DEEP TOW

The first reported effort to build and use a multi-instrumented package or "fish" occurred in 1960, at the Scripps Institution of Oceanography's Marine Physical Laboratory (MPL), San Diego, California, under the direction of Dr. F.N. Spiess. This initial attempt, combined an upward-looking and downward-looking echo sounder with an acoustic transponder system. The upward-looking echo sounder provided the depth of the fish, the transponder network gave its relative position, and the narrow-beamed, down-

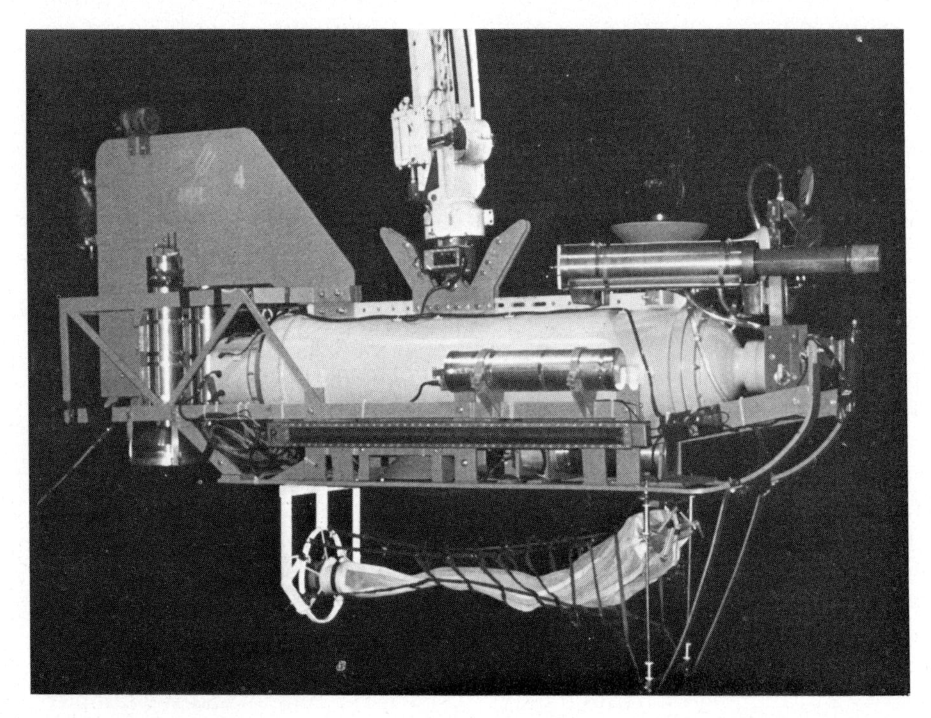

Fig. 1. DEEP TOW, a towed, unmanned vehicle operated by the Marine Physical Laboratory, Scripps Institution of Oceanography. (Courtesy, U.S. Office of Naval Research.)

ward-looking echo sounder provided a high-resolution recording of bottom slope topography. A few years later a proton magnetometer was added and in 1964 a side-scanning sonar. By 1972 this suite of instruments also included a seismic sub-bottom profiler, a stereo and wide-angle photo system, a temperature sensor and a snapshot television camera. The present DEEP TOW of MPL is shown in (Fig. 1). It is towed some 15—200 m above the bottom at a speed of about 1.5 knots. There are several other towed devices in operation today which are discussed subsequently. They, like DEEP TOW, are multi-instrumented and deployed in a similar fashion. Still, a requirement existed for a fish that could maneuver on its own and in the early 1960's the U.S. Navy's CURV was one of the first of such vehicles.

CURV I

CURV I (Cable-controlled Underwater Research Vehicle) was primarily designed to retrieve torpedos from depths of up to 305 m. It was developed by the U.S. Naval Underwater Warfare Center, Pasadena, California in 1964 and fulfilled its role in February 1965 by retrieving a torpedo from a depth of 168 m. The first CURV I (See fig. 3 in Chapter 4) was a rectangular-shaped, open-framework platform that transported and enclosed a TV camera, viewing lights, a passive/active sonar system, an altimeter, a deptho-meter and a hydraulically powered claw for grasping a torpedo. Torpedo-shaped cylinders on top of the framework provided a slight positive buoy-ancy when the vehicle was submerged. CURV I's reliance upon the support craft was not for propulsion, but for power and control which was supplied through a connecting cable. For mobility CURV I depended upon three, 10 hp electric motors, each powering a propeller. Two propellers were oriented to provide forward and reverse motion and one was oriented to provide ver-tical motion. Each propeller could be run independently of the other. There-fore, by running one of the horizontal propulsion motors in reverse, and the other forward, CURV I could be made to spin on its vertical axis. By applying differential forward thrust to the same propellers, it could change its head-ing. The vertical thruster provided an up and down degree-of-freedom to CURV I's mobility.

As an observer and grasper, CURV I fulfilled its role admirably. By 1968 successors to the original vehicle had recovered 50 torpedos from an average depth of 426 m and in an average in-water time of two and a half hours. Its most notable success, however, was the retrieval of an H-bomb off Palo-mares, Spain in 1966 from a depth of 869 m. In this task, CURV I was used to attach grapnels to the bomb's parachute. Twice CURV I attached grap-nels, but on the third dive it became entangled in the parachute. Nonethe-less, CURV I, grapnel, bomb and all were hauled up to the surface and recov-ered. The public debut of the unmanned submersible was a success. Both

DEEP TOW and CURV I are vehicles primarily designed for movement above the ocean floor.

RUM

A third type of unmanned submersible is designed to crawl along the bottom and several varieties are available. The U.S. Navy's RUM (Remote Underwater Manipulator) is a cable-tethered, tracked model and was completed in 1958. The system is controlled from the surface and now has a manipulator which is capable of lifting and positioning objects weighing up to 450 kg. Similar to its above-the-bottom counterparts, RUM (Fig. 2) has a closed-circuit television system, underwater lights and a sonar system for positioning. RUM is envisioned as an underwater construction unit. Several Japanese bottom crawlers provide a bulldozing capability as well and can obtain samples of the bottom and make geotechnical measurements. A recent addition to this fleet is the Irish bottom crawler, TRAMP, which uses six, independently suspended tires, instead of tracks. The tires allow it to

Fig. 2. RUM, a tracked, bottom crawler operated by the Marine Physical Laboratory, Scripps Institution of Oceanography. (Courtesy, U.S. Office of Naval Research.)

crawl over large boulders as well as over soft mud or sand bottoms. All of these unmanned vehicles depend upon a cable for power, control and data telemetry.

UARS

There is another unmanned submersible that operates without a tether and offers a potential for under-ice studies and other forms of research as well. The Unmanned Arctic Research Submersible System (UARS) is a modification of the University of Washington's, Applied Physics Laboratory's SPURV. The development of this vehicle began in 1959 and has continued to the present. The advantages of such vehicles are their lack of a restraining tether. This precludes the need of sophisticated handling gear and cable storage devices and permits almost 37 km to be surveyed in only a few hours. In the course of its travels, the UARS vehicle can collect bathymetric data, under-ice echo-sounding profiles and can measure a variety of sea-water properties.

The preceding general overview delineates the variety of unmanned submersibles now in operation. There are other types of vehicles which are also unmanned and are used for specific data gathering and work tasks. According to Baxter and Mueser (1971) underwater cable burial, for example, is often preceded by towing an instrumented sled over a proposed cable route to determine its "plowability." Subsequently, an unmanned, towed plow weighing up to 15,000 kg and equipped with television, hydrophones, tensionmeters and plowshare is employed to bury the cable 60 cm below the sediment—water interface.

Bline and McDonald (1976) describe a vertical transport vehicle having a length of 22 m, a width of 6.7 m and weighing 63,600 kg. A work vehicle is also used simultaneously to locate, assess, repair, transport and position pipeline sections under the direction of a surface-based operator. It is 13.7 m long, 6.7 m wide and weighs 48,989 kg. Banzoli et al. (1976) describe a 190,000-kg unmanned, tracked vehicle designed to dig a trench 4 m deep in bottom sediments having a compression strength of up to 50 kg/cm^2.

These and other devices have been developed in an effort to either provide more detailed and accurate data, or to offer a more effective, safer alternative to the ambient diver and the manned submersible. Whether all of these will be successful remains to be seen because the entire field is in its infancy. Additional experience is required before any of the unmanned systems can demonstrate their superiority. Unmanned systems are constantly being modified as they find greater utilization and this trend can be expected to continue for some time to come. Until all their shortcomings are revealed and corrected, any description of a particular unmanned system will be subject to change based on operational experiences. Therefore the following descriptions are subject to change.

TABLE I

Towed, unmanned vehicles

Name	Owner	Operating depth (ft/m)	Dry weight (lbs/kg)	Maximum speed (kts/ km per hr)	Power requirement	Work equipment
BATFISH	Bedford Inst. of Oceanology Halifax, N.S.	650 198	156 71	5.0 9.3	110 VAC, 6 Hz	TV, still camera, side-scan sonar
CRAB	Institute of Oceanology, USSR, Moscow	13,123 4,000	1,548 702	NA	lead—acid batteries	TV
DEEP TOW	Marine Physics Lab., San Diego, Calif.	20,000 6,096	2,000 907	1.5 2.8		TV, still camera, magnetometer, side-scan sonar, sub-bottom profiler, echo sounder, diff. pressure gauge
DSS-125	Hydro Products, San Diego, Calif.	20,000 6,096	5,100 2,313	1.5 2.8	115 VAC, 60 Hz, 20 A, or 220 VAC, 50 Hz, 10 A	TV, still camera

NRL System	Naval Research Lab., Washington, D.C.	20,000 6,096	2,200 998	3 5.6	20 A hr, Ni—Cd batteries, 28 VDC	TV, still cameras, magnetometer, side-scan sonar, water sampler, sub-bottom profiler
RUFAS II	National Marine Fisheries Services	2,400 731	1,000 453	6 11.1	115 VAC, 12 VDC, 4 kW total	TV, still camera, scanning sonar
S^3	University of Georgia, Athens, Ga.	6,000 1,829	700 317	6 11.1	NA	TV, side-scan sonar, magnetometer, sub-bottom profiler, dredge
SEA PROBE *	Alcoa Marine Corp., Washington, D.C.	10,000 3,048	—	0.5 0.9	NA	TV (2 ea), still camera (35 mm), side-scan sonar scanning
STOVE	Submarine Development Group One, San Diego, Calif.	20,000 6,096	NA	NA	lead—acid batteries	still cameras, side-scan sonar
TELEPROBE	NAVOCEANO Suitland, Md.	20,000 6,096	3,500 1,588	3 5.6	50 A, 120 V regulated power	TV, side-scan sonar, magnetometer, stereophotography

* SEA PROBE consists of a surface ship, drill string and equipment pods; the latter can only operate from its specially designed ship.

TOWED VEHICLES

To differentiate a towed, unmanned submersible from a towed, multi-instrumented package, a television camera with a real-time surface monitor must be on the unmanned system. This distinction, although arbitrary, must be made, otherwise every side-scan sonar, seismic sub-bottom profiler, water or bottom sampling device can be classed as an unmanned submersible. By this definition, there are ten unmanned, towed submersibles reportedly in existence; they are described in Table I.

The majority of towed vehicles operate to depths of between 3,048 and 6,096 m, and receive their power from the surface. The construction and at-sea deployment of these vehicles differs in detail, but in many ways they follow the same general system configuration and method of operation. The U.S. Naval Oceanographic Office's TELEPROBE will be used as a representative of the towed submersibles to portray their general characteristics.

TELEPROBE

TELEPROBE is an unmanned deep-towed search and identification system designed to find and identify objects lost at sea. It is also capable of performing detailed bottom surveys because of its various sensing devices.

The basic structure of TELEPROBE is an open framework of aluminum which encloses and supports its various pieces of equipment. This "fish" or the underwater portion of the system is 3.0 m long, 1.7 m high and 1.2 m wide. With the magnetometer added the length of the fish increases to 6.1 m. Fully equipped the fish weighs about 1,633 kg.

To conduct its search tasks, TELEPROBE is equipped with a magnetometer, a side-scan sonar and a forward-looking sonar. These instruments are initially used to locate the target or object. Once contact has been made, they are then used to close-in on the target. As the vehicle approaches the target, its identification instruments are then used: a closed-circuit television camera and a 35-mm stereo camera system. The TV provides real-time visual information and the stereo-camera system provides higher-resolution, hard-copy documentation. If further photographic runs are necessary, acoustic pingers can be dropped from the fish to mark the target and subsequently aid in its relocation.

In its search mode of operation, TELEPROBE is towed at an altitude of about 60 m and a speed of 1.5 knots. For photographic and television use, the vehicle is held about 7.6 m above the bottom and towed at the same speed. The distance off the bottom of the fish is determined acoustically by a depth sonar. Because the television system requires the entire bandwidth of the cable, a switching arrangement turns off all other systems when the television is on. In this mode a pinger on the vehicle is activated, and both the direct and bottom-reflected pulse is received by the support craft, and the difference in arrival time is displayed on a CRT as a measurement of distance off the bottom.

The position of the fish relative to the surface craft is obtained by (1) interrogating a transponder on the fish to gain slant-range and bearing, and by (2) interrogating a bottom-mounted transponder network from the fish to gain its position relative to the transponders.

Thus far only the submerged components of the vehicle have been discussed. There is more to this system than the "fish." It is connected to the surface by 9,144 m of 1.7-cm cable. This provides the required scope if TELEPROBE is operating at a depth of 6,096 m. A frequency-multiplexing, telemetry system allows all systems to use the same cable.

The support ship also must be quite specialized if TELEPROBE is to realize its ultimate potential. First, it must have a station-keeping capability that calls for a bow thruster and an active rudder, because it must be able to maintain headway at 1.5 knots. Second, it must have a winch capable of handling 9.3 km of cable or whatever shorter length is required to perform the task. And finally it must have a deck crane or stern A-frame which can handle the fish and fair-lead the cable. These capabilities are not found on many ships, and at present only the USNS DESTEIGUER and USNS KANE can accommodate the TELEPROBE system without major modifications.

A further support ship requirement is that of enclosed space for housing the display and control equipment and accommodations for personnel. The TELEPROBE operating and support crew requires fifteen highly-trained and experienced personnel. Some idea of the scope of equipment can be gained by the fact that two tractor-trailer trucks are normally used for long distance land shipment, and that 24 hours is required for mobilization, i.e., to load the entire system aboard a truck.

TELEPROBE and DEEP TOW are the most sophisticated examples of towed, unmanned vehicles operating today. For the commercial customer there is only one available system, the DSS-125 or Deep Sea Survey System.

Hydro Products' DSS-125 is a deep ocean television/photographic system which was used for several months in 1974 aboard the German research vessel VALDIVA for manganese-nodule exploration (Vigil et al., 1975). The basic system is shown in Figs. 3 and 4 and consists of the instrument platform (or fish), a depressor, a winch and a control/display console.

The instrument platform consists of a welded aluminum frame which is made about 23 kg positively buoyant by attaching syntactic foam. Encompassed and supported by the frame is a 102 light level TV, a 70-mm still camera and strobe light, two thallium-iodide lights, two high-intensity spot lights, an electronic compass and a telemetry cannister.

The depressor frame is constructed of galvanized, welded steel and two trays are included to carry individual blocks of lead which serve as the main weight for the depressor action of the platform. The 1,089 kg depressor weight is designed to produce a large angle between the ship-to-depressor

Fig. 3. The DSS-125 towed system showing from left to right the control console, depressor platform and instrument platform. (Courtesy, Hydro Products a Tetra Tech. Co.)

portion of the cable and the depressor-to-instrument platform portion. The ship's vertical motions are largely damped by the time they are translated into relatively small angular changes in the instrument cable.

The cable serves as the towing element and also carries power and signals from the surface ship to the fish. It consists of two layers of armored wire with breaking strength 23,134 kg and a coaxial assembly enclosed by a waterproof jacket under the strength members.

The distance of the instrument platform above the bottom is adjusted by the winch operator based on input from the TV monitor and the collimated spot lights. When the TV camera is at the desired position, the spot lights can be used to determine the size of objects on the monitor and the distance off the bottom. Ideally, the platform is towed a distance of 10 m above the bottom at speeds of from 0.5 to 1.5 knots.

The fully equipped platform and depressor can weigh up to 2,313 kg in air, and at 6,100 m, the weight of the cable in water is 7,258 kg. It is obvious that special winching and handling systems are required aboard the support craft to accommodate the DSS-125. A further refinement to this system envisions the attachment of side-scan sonar transducers to the depressor.

RUFAS II

RUFAS II (Remote Underwater Fishery Assessment System) is a deeper version of an earlier prototype designed for the single objective of assessing

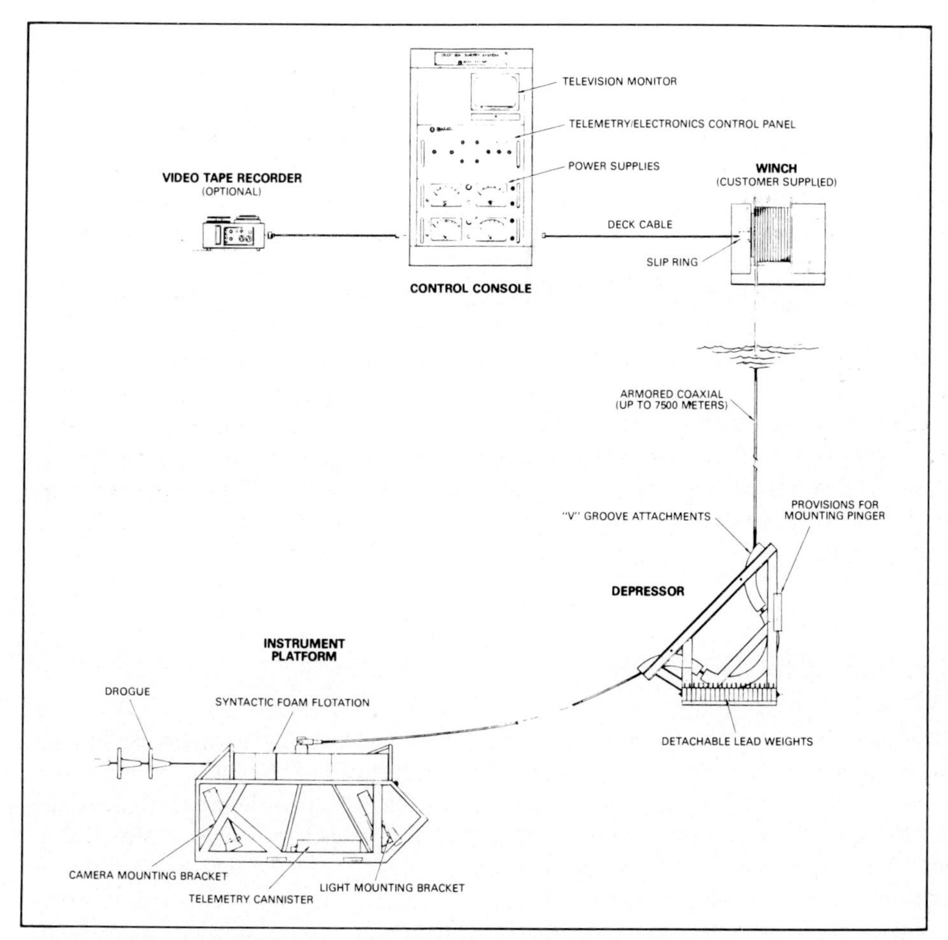

Fig. 4. The total DSS-125 system showing shipboard and underwater components. (Courtesy, Hydro Products a Tetra Tech. Co.)

calico scallops in situ. The RUFAS II system is similar in many respects to the vehicles discussed above. It consists of the underwater vehicle (with television camera on a pan and tilt mechanism, 35-mm cine camera, lights and forward-looking obstacle-avoidance sonar), a control/display console, and an interconnecting cable providing power, telemetry commands and data transfer. Whereas the other systems rely upon the winch operator to control the vehicle's distance above the bottom, RUFAS II's depth is controlled by stern-mounted dive planes, one port and one starboard. The operator obtains roll, pitch and depth information from the control/display console and uses this information to position the sled at a desired depth. A further refinement includes an automatic depth control system, in which an analog signal (produced by a downward-looking echo sounder) is directly proportional to the vehicle's distance above the bottom. Preliminary sea trials of RUFAS II's bottom topography-following control system showed that the vehicle automatically followed the sea-floor contours to within the ±0.3-m resolution of the height-above-the-bottom telemetry readout (Jue, 1974). Additionally, a roll-stabilizer control system, derived from the bottom topography-following control system, maintained the roll between two and five degrees, when monitors indicated external roll forces of orders of magnitude greater acting upon the vehicle.

The design of the University of Georgia's S^3 (Sea Floor Surveillance System) is based on that of the RUFAS II vehicle. However, it is used for Outer Continental Shelf Oil and gas lease surveys and mineral exploration (Noakes et al., 1975). A major difference between RUFAS II and S^3 is that the latter is designed to be towed both on and above the bottom. During its bottom excursions a wire mesh dredge can be lowered and raised to collect bottom samples. The remaining instrument suite on S^3 consists of standard bottom and sub-bottom data-collection devices packaged to operate individually or concurrently.

SELF-PROPELLED, TETHERED VEHICLES

Within the past two years the number of CURV-like vehicles produced and under construction has increased remarkably. In 1974 eight of these vehicles were available, but now, there are at least 31 (Table II). The impetus for this growth was the development of oil and gas discoveries in the North Sea. Prior to the late 1960's, most offshore diving support was conducted in relatively warm, shallow, waters favorable for diving. The North Sea, however is deeper, colder and more hostile than the Gulf of Mexico, or the offshore areas of the Middle East. The manned submersible, which was fast becoming a historical curiosity during the 1960's, found in the North Sea an area where it could compete with the ambient diver, both financially and operationally. Virtually overnight the demand for manned vehicles burgeoned. The nature of the work was mainly observation and inspection of facilities on the bottom. It occasionally included relatively simple manipulative tasks involving recovery of lost gear or detachment or attachment of components. Such tasks were the reason for CURV's existence, and commercial operators saw a potential market in which a CURV-like vehicle could successfully compete. The consequence of such reasoning is reflected in Table II, where 17 of the 31 vehicles listed are, or will be, operated commercially.

The contents of this section demonstrates that there is no "average" unmanned submersible. Therefore, the following discussion presents boundary conditions and ranges to gain an appreciation for the diversity of their

TABLE II

Self-propelled, tethered vehicles

Name	Owner	Operating depth (ft/m)	Dry weight (lbs/kg)	Maximum speed (kts/km per hr)	Power requirements	Work equipment
ANGUS	Heriot-Watt Univ., Edinburgh, Scotland	984 300	850 386	2 3.7	440 VAC, 50 Hz, 3-phase	TV, 16 mm cine camera, directional hydrophone, wide-band hydrophone, transponder interrogator
CONSUB 1	British Aircraft Corp., Bristol, England	2,000 610	3,000 1,360	2.5 4.6	240 V, 50 Hz, 1-phase, 3 KVA; 415 V, 50 Hz, 3-phase, 35 KVA	TV (2 ea) color and black and white, rock drill, stereo cameras
CONSUB 2	British Aircraft Corp., Bristol, England	2,000 610	4,400 1,996	2.5 4.6	380/415/440 V 50/60 Hz, 3-phase, 150 kW	TV (2 ea) color and black and white, stereo cameras
CORD	Harbor Branch Foundation, Ft. Pierce, Fla.	1,500 457	720 327	5 9	500 VAC, 60 Hz, 3-phase, 5 kW	TV, current meter, scanning sonar, temperature sensor, echo sounder
CURV II[1]	Naval Undersea Ctr., San Diego, Calif.	2,500 762	3,450 1,565	4 7.4	440 VAC	TV (2 ea), still camera, scanning sonar
CURV III	Naval Undersea Ctr., San Diego, Calif.	10,000 3,048	4,000 1,814	4 7.4	120–140 VAC, 50 kW	TV (2 ea), still camera, scanning sonar
CUTLET	A.U.W.E., Portland, U.K.	1,000 305	NA	NA	NA	TV, directional hydrophone
DEEP DRONE	U.S. Navy SUPSAL	2,000 610	1,200 544	3.5 6.5	115 VAC, 1-phase, 2 KVA, 440 VAC, 3-phase	TV with 320° pan and 190° tilt, CTFM sonar, altimeter

(cont.)

TABLE II (continued)

Name	Owner	Operating depth (ft/m)	Dry weight (lbs/kg)	Maximum speed (kts/ km per hr)	Power requirements	Work equipment
ERIC	French Navy, Toulon, France	3,281 1,000	4,409 2,000	2 3.7	50 kW, 60 Hz, 3-phase	TV, still camera, echo sounder
MANTA 1.5	Institute of Oceanology, USSR, Moscow	4,921 1,500	2,200 998	NA	380 VAC, 50 Hz, 3-phase	TV, various manipulator claws
RCV-225	Various [2]	6,600 2,012	180 82	1.7 3.1	220 VAC, 3-phase 50—60 Hz or 440 VAC, 3-phase, 50—60 Hz	TV, still camera, automatic constant depth control
RCV-150	Various [3]	6,000 1,829	450 204	2 3.7	220/440 VAC 50—60 Hz, 3-phase	TV on pan/tilt device, automatic constant depth control
RECON II	Perry Ocean Group, Riviera Beach, Fla.	1,500 457	620 281	2.5 4.6	440 VAC, 60 Hz, 3-phase, or 220 VAC	TV on pan/tilt device, current meter
RUWS	Naval Undersea Ctr., Hawaii	20,000 6,096	5,000 2,268	1 1.8		TV, scanning sonar
SCARAB	AT and T, Inc.,	6,000	5,000	0.5	480 VAC, 3-phase,	TV, camera, magnetometer,

I and II	NYC, N.Y.	1,829	2,268	0.9	100 kW	CTFM sonar, altimeter
SEA SURVEYOR	Rębikoff Underwater Products, Ft. Lauderdale, Fla.	660 200	385 175	5 9.3	230 VAC, 60 Hz, 4 kW	TV, still camera (35 mm), side-scan sonar, magnetometer, depth indicator
SNOOPY	Naval Undersea Ctr., San Diego, Calif.	1,500 457	150 68	1 1.8	120 VAC, 3 kW	TV, 8 mm cine camera
SNOOPY	NAVFAC, Washington, D.C.	1,500 457	300 136	1 1.8	120 VAC, 3 kW	TV, 70 mm still camera
SNURRE	Royal Norwegian Council for Scientific Research	1,969 600	2,645 1,200	1.5 2.8	420 VAC, 3-phase, 40 KVA	TV, echo sounder, compass, depth gauge, still cameras, 8 mm cine camera, directional hydrophones
TELE-NAUTE 1000	I.F.P., Paris, France	3,300 1,006	2,420 1,097	3 5.6	380 VAC, 50 cycles 3-phase, 50 kW	TV, constant depth control, magnetometer
TROV	Canadian Ctr. for Inland Waters, Burlington, Ont.	1,200 366	1,130 513	1 1.8	lead—acid batteries	TV, magnetic compass, echo sounder, transponder, transponder interrogator
TROV-01	Underground Location Services, Ltd., Stonehouse, Glasgow, Scotland	1,200 366	2,000 907	1.5 2.8	30 KVA generator	TV, magnetic compass, echo sounder, transponder, transponder interrogator

[1] A second CURV II is operated by the Naval Torpedo Station, Keyport, Wash.
[2] Seven vehicles total: Seaway Diving, Bergen, Norway (2 vehicles); Martech International, Houston, Texas (2 vehicles); SESAM, Paris, France (1 vehicle); Taylor Diving and Salvage, Belle Chase, La. (1 vehicle); Esso Australia Ltd., Sale, Australia (1 vehicle).
[3] Two vehicles total: Martech International, Houston, Texas; and Scandive, Stavanger, Norway.

group. A more detailed description of CONSUB 2 and RCV-225 is provided subsequently.

Components

The basic tethered, self-propelled vehicle system consists of the vehicle and sometimes an underwater clump or launcher, and a cable in addition to a shipboard control-display console. Supporting equipment includes a launch—retrieval device, a cable winch, sheltered areas for the vehicle operators and shipboard components and, if shipboard power is not available or suitable, a power-supply unit. A unit such as SNOOPY (Chapter 4) can be operated with much less equipment and facilities than those listed. However SCARAB vehicles and CURV III require far more.

Operating depth

Vehicles owned by industry range in depth capabilities from 200 m to 2,012 m with an average depth of 1,306 m. Military vehicles range from 457 m to 6,096 m with an average depth of 1,350 m.

Structure

Most vehicles consist of an open, metal framework that supports and encloses, for protection, its various components. Aluminum is the most widely used material because of its light weight. Buoyancy is generally positive by a few pounds when the vehicle is submerged providing a fail-safe assurance that it will surface in the event of a power failure. Generally, but not always, syntactic foam blocks mounted on top of the framework provide the required buoyancy.

Launch/retrieval

The underwater component(s) or "vehicle" of these systems vary in weight from 68 kg to as much as 2,268 kg. One manufacturer states that a man can easily launch the vehicle without the aid of a mechanical advantage. It would be safe to assume that most men would find even the lightest vehicle a challenge to pick up and lower into the water, regardless of the amount of freeboard. For the others a handling frame is required on the support craft or carried as part of the system.

The sea-state limitations on launch/retrieval directly reflects the nature and sophistication of the shipboard handling equipment. The unmanned system, in spite of the fact that the retrieval line is always attached, is as vulnerable as a manned system when contact with the sides of the support platform is a possibility. Because most of these systems are designed for deployment from a ship of opportunity, it is not possible to place a sea-state limit on the system until the handling gear is specified.

Some indication of sea-state limits can be gained from the following operator statements: CONSUB 1 can be launched/retrieved through sea state 4. DEEP DRONE is designed to be handled up through sea state 5 if "normal" handling equipment is available which is generally employed to handle manned submersibles. These two are not the heaviest vehicles in operation weighing 1,360 kg and 544 kg, respectively, but they do fall around the average vehicle weight of 961 kg. Operational data under heavy-sea conditions is sparse, but it would appear that up to and including sea state 5 is a limit for normal operations from any ship of opportunity. In an emergency situation this limit may be exceeded, but not without the possibility of severely damaging the components by slamming the vehicle into the support craft. A further complicating factor accompanying high-sea states is the ability of the support craft to maintain station. Launch/retrieval in sea state 6 or 7 may be possible, but the ship may not be able to maintain station for controlling the vehicle.

Speed

The speed of unmanned vehicles is quite similar to those of manned vehicles, and ranges, at the surface, from one to five knots. An unmanned vehicle's surface speed, assuming little or no cable drag decreases considerably with depth. The decrease in speed with depth or with increase in currents ranges from 20% to 84% of the surface speed, and averages about 42%. The SCARAB vehicles, for example, can make three knots on the surface, but only 0.5 knots at 1,829 m, an 84% decrease. The RCV-225 makes 1.7 knots on the surface and 1.0 knots at 2,012 m a 42% decrease. This difference in reduction of speed is caused by the different modes of deploying the vehicles. The SCARAB vehicles are designed to cruise along the bottom while it tows the entire length of cable, in conjunction with the surface ship. The RCV-225 is deployed from a launching cage, and works around the launcher on 120 m of tether cable, thereby substantially reducing the amount of cable drag. For this reason, many of the unmanned vehicles employ a launcher or clump.

The reduction in speed with depth does not, in most open ocean areas, detract from the unmanned vehicle's capability to perform because 0.5—1 knot currents are not generally encountered except in areas such as the Gulf Stream or Kuroshio Current. This does not apply to areas such as the North Sea. In the Frigg field, for example, tidal-induced current speeds of 0.6 knots and greater are common, whereas farther south in the Deborah, Dottie and Lemon Bank fields currents as much as 3 knots may occur. In such areas the operating capabilities of the unmanned vehicles, and manned as well, are reduced considerably, if not completely.

Maneuverability

All but a few vehicles are capable of two translation motions and one rotational motion; namely, thrust (forward/reverse) and heave (up/down); and yaw (left/right heading changes), respectively. These motions are provided by the arrangement of two horizontal or forward thrusters and one vertical thruster. By adding a forth lateral or side thruster, a third translational motion is obtained, namely, sway or sidle. If the lateral thruster is mounted forward, it is used to augment yawing, rather than providing a sideward translational motion. Pitch motion is not common and is only found on the CONSUB vehicles where two forward/aft-mounted vertical thrusters can impart pitch to the vehicle by operating in opposing directions. By designing its television camera mounting such that it can be trained ±90° from the horizontal plane, the manufacturers of RCV-225 have provided a substitute for pitching the entire vehicle. No unmanned vehicle is known to be capable of roll motion.

The umbilical cable offers an obvious potential for fouling, but this is a constraint under which the operator must work and he has no choice but to be selective in his job applications. Less obvious is the possibility for imparting too many twists in the cable. Hydro Products has included a counter/display unit for measuring the number of twists and their direction.

Crew

For routine operations the support crew ranges from one to seven, with three to four being average. Commercial operators of unmanned vehicles do not specifically define a routine operation but it may be assumed that eight to ten hours is routine. In an emergency situation a 24-hour capability is required. If the CURV vehicles, which require ten personnel, can be used as a guide, then between four to six personnel would be necessary to operate and support most vehicles over a 24-hour period.

Work instrumentation

The instruments listed in Table III are those which are standard onboard equipment. There are numerous equipment options on virtually all unmanned vehicles. However, these options are not listed because they reflect what the vehicle could be, not what it is. The British Aircraft Corporation's CONSUB 1 has a manipulator-held rock drill which has successfully operated in the field, but is not listed in Table III.

A quick glance at Table III reveals that the primary task of the tethered, self-propelled vehicles is observation, and in many cases, photographic documentation. The secondmost frequent capability is manipulation or, more specifically, grasping. The majority of manipulators are simple devices

TABLE III

Work instruments of self-propelled, tethered vehicles

	Viewing/Photography					Sonar			
	TV	still	stereo	cine	manipulator	search	homing	current meter	thermistor
ANGUS	X			X			X		
CONSUB 1	X		X						
CONSUB 2	X		X						
CORD	X				X	X		X	X
CURV II (2 ea)	X	X			X	X	X		
CURV III	X	X			X	X	X		
DEEP DRONE	X	X				X	X		
ERIC	X	X			X				
MANTA 1.5	X				X				
RCV-225 (5 ea)	X								
RECON-II	X				X			X	
RUWS	X					X	X		
SCARAB I and II	X	X			X (2)				
SEA SURVEYOR	X								
SNOOPY (2 ea)	X			X					
TELENAUTE	X			X	X				
TROV (2 ea)	X				X (2)				

which can do no more than extend and open/rotate its claw. The limited orientor and locator motions are, in reality, not a liability, because the vehicles themselves can provide several more degrees-of-freedom to the manipulator because of their excellent maneuvering capability. The active sonar equipment can serve two purposes: the first is to locate acoustically reflective targets, e.g., hardware or outcrops, and the second to interrogate bottom-mounted transponders or markers for positioning. The homing-sonar provides the capability for acquiring active acoustic targets such as marker pingers, which may then be used to guide the vehicle to the source of the pulse.

These are the general characteristics and capabilities of this type of unmanned vehicle. Space considerations do not permit an individual description of all the vehicles listed in Table II. Instead, the two vehicles RCV-225 and CONSUB 1, are described below and will be taken as representatives of the tethered, self-propelled unmanned submersibles.

CONSUB 1

British Aircraft Corporation's CONSUB 1 was built in 1974. It was funded jointly by the Department of Trade and Industry and the Institute of Geological Sciences to encourage development of unmanned, self-propelled vehicles in the United Kingdom. CONSUB 1's original purpose was to minimize downtime for bottom sampling due to rough weather, and to extend the range of operations beyond the conventional diver's limits (Skidmore and Bircham, 1975). A description of CONSUB 1 is presented in Table IV and a photograph of the vehicle and its shipboard control-display panel is included in Figs. 5 and 6, respectively.

The CONSUB 1 vehicle has conducted a number of different tasks in the offshore areas of the U.K., including geological surveys and pipeline and hardware inspections. The experiences with this vehicle provided the design and performance criteria for CONSUB 2, which is now under construction.

The CONSUB vehicles operate from the support craft by towing the cable as the support ship maintains station overhead or follows the vehicle as it transits along the bottom. Because of the high drag area of a positively buoyant cable, a negatively buoyant cable is used. But the section near the vehicle is positively buoyant to prohibit entanglement (Bircham and Skidmore, 1976). A typical range of "footprints" as a function of depth is shown in fig. 19 of Chapter 10 for a uniform 2-knot current, with a neutrally buoyed cable with a scope of three and negligible vehicle drag compared with cable drag.

RCV-225

This system is manufactured by Hydro Products, San Diego, California and was the first commercially available self-propelled, tethered submersible.

TABLE IV

CONSUB 1 specifications

Operating depth: 610 m.
Dimensions (l × w × l): 271 cm × 182 cm × 145 cm.
Weight (dry): 1,360 kg.
Speed: (Max. surface 2.5 kts
(Max. operating current) 2 kts at 2,000 ft.
Structure: Tubular aluminum alloy HE 130.
Buoyancy: Two cylindrical, pressure-resistant, fiberglass cylinders provide a positive buoyancy of 18 kg when vehicle is submerged
Power requirements: 240 V, 50 Hz, single phase, 3 KVA; 415 V, 50 Hz, 3-phase, 50 KVA (to the control cabin transformer). If the latter is not available a diesel generator can be used. Surface transformer converts supply voltage to 415 V/1000 V 3-phase for transmission to vehicle.
Propulsion: Two lateral and two vertical fixed, reversible thrusters. All are electro-hydraulically powered, 5 hp each, and capable of independent operation.
Instrumentation: Compass, inclinometer, depth gauge, two TV cameras (1 color; 1 black and white), stereo camera system, rock drill capable of taking 1.3 cm diameter, 13 cm long core. Stereo and TV cameras are mounted on a pan and tilt unit which trains ±150° in azimuth and tilts +30° to –90° from the horizontal.
Navigation: A system called SCANTIE (Submersible Craft Acoustic Navigation and Tract Indication Equipment) can be fitted to the vehicle. In its basic form SCANTIE measures horizontal range, compass bearing, depth and heading of the vehicle relative to the support craft. Slant range accuracy is ±0.75 m, horizontal range accuracy is ±0.3% out to 1.5 km (0.8 nm) and bearing is accurate to ±1° within 360°. The SCANTIE system can be expanded to operate in a range—range mode relative to a bottom-mounted transponder array. Basic display is a color CRT on which graphic and alpha-numeric data are presented.
Shipboard components: Control console (2 TV monitors; vehicle/instrument controls), Transformer, System Distribution Box (connects transformer, ship junction box and consoles) Ship Junction Box (terminates umbilical cable of support craft) and Faking Frame (for storage and deployment of umbilical).
Support-craft requirements: Launch/retrieval system capable of supporting vehicle dry weight. Freeboard not to exceed 2.7 m. Deck space: 3 m × 3 m clear space with tie-downs for vehicle; area of 6 m × 6 m required for umbilical cable faking frame. Cabin space: 1) must be large enough to contain a 2.1 m long bench for control console at which two operators sit, and must also provide a view of operational deck area: 2) bulkhead area 0.9 m × 1.3 m for distribution box; 3) deck space in cabin of 0.6 m × 0.6 m for transformer and 4) an access port of 12.7 cm diameter is required for electrical service cables.
Operating/Maintenance Crew: Three to four.

At present there are seven such systems in operation (Table II). Initially the system was called RCV-125, but operational experience resulted in various modifications leading to the redesignation RCV-225. A description of the basic system is presented in Table V and in Fig. 7.

The vehicle itself is quite different from other vehicles in this category. It is not an open framework, but is composed of syntactic foam which encloses

Fig. 5. The 2,000-ft. (610-m) CONSUB 1. (Courtesy, British Aircraft Corp. Ltd.)

and supports its components. The vehicle's propulsion arrangement provides an exceptionally high degree of maneuverability and quick response. Its depth and heading can also be automatically maintained by servo-controls (Fig. 8).

The RCV system and others in this class use an underwater deployment unit or launcher from which the vehicle operates. A launcher is not used with the CONSUB system. The RCV launcher is used to protect the vehicle from damage during launching and retrieval, and to transport it to working depth. At this depth the vehicle "flies" out of the launcher to perform its task. The vehicle-to-launcher tether cable is winched in or out of the vehicle by remote control. If the vehicle becomes entangled, the tether cable can be separated on command. The positive buoyancy provided by the syntactic foam then will bring it to the surface where a strobe light and pinger aid in its location. In practice, the vehicle operates within the length of the tether radius from the launcher. The launcher unit reduces drag on the surface cable and absorbs surface heave which would be imparted to the vehicle if it operated independent of the launcher. However, this arrangement also reduces the range of the vehicle's operation. The U.S. Navy's RUWS system also employs a launcher, but propulsion motors have been incorporated into the launcher to increase the range of the vehicle's operation.

TABLE V

RCV-225 specifications

Operating depth: 2,012 m.
Dimensions (l × w × d): 51 cm × 66 cm × 51 cm).
Launcher (*l* × *w* × *d*): 1.1 m × 1.0 m × 1.0 m).
Weight (dry): 82 kg; launcher (dry): 80 kg.
Speed: (Max. surface) 1.7 kts.
(Max. operating current) 1 kt at 2000 m.
Structure: A syntactic foam hull shaped into a prolate spheroid encloses the motors and the camera/electronics pressure housing.
Bouyancy: Vehicle is positively buoyant by 2 kg.
Power Requirements: 220 VAC, 3-phase, 50—60 Hz; or 440 VAC, 3-phase, 50—60 Hz(5 kW maximum).
Propulsion: Four oil-filled electric motors, two provide thrust and yaw and two provide heave, a forth motion (sway) is provided by the vertical thruster configuration. A desired depth and heading can be automatically maintained.
Instrumentation: TV camera (low light level) which is capable of being tilted ±90° in the vertical, two 45 W tungsten—halogen lights, compass and depth sensor.
Navigation: By compass heading and visual sighting.
Shipboard Components: Control/display console, power supply, hand controller, deployment unit (winch/skid/A-frame).
Support Ship Requirements: Enclosed area for control console.
Operation/Maintenance Crew: Two to three, depending upon length of task.

Fig. 6. The control console of CONSUB 1. (Courtesy, British Aircraft Corp. Ltd.)

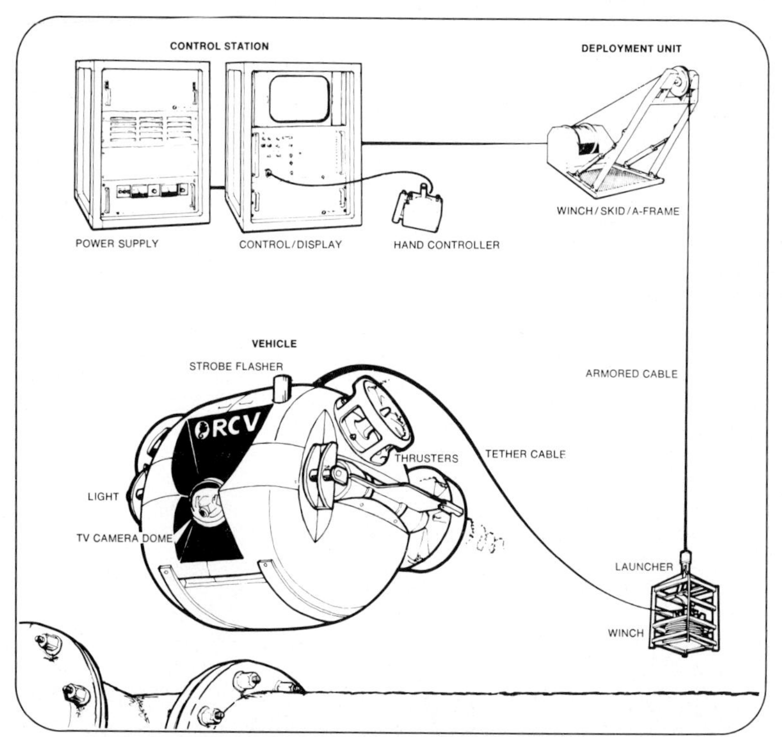

Fig. 7. The RCV-225 System. (Courtesy, Hydro Products a Tetra Tech. Co.)

Fig. 8. RCV-225 Maneuvering and camera controls. (Courtesy Hydro Products a Tetra Tech. Co.)

BOTTOM CRAWLERS

The unmanned bottom-crawling vehicles have received a mixed reception from the ocean community. Technical literature of the late 1960's and early 1970's contains descriptions of a variety of bottom-crawling bulldozers, surveyors and work vehicles; but in the past three to four years there has been a noticeable absence of such papers.

RUM

The Scripps Institution of Oceanography's RUM (Fig. 2) has not been in operation since November of 1973. H. Kawasaki et al. (1972) described two bulldozer systems built by Hitachi Ltd., Tokyo. The first system was designed for a maximum working depth of 5 m and the second from 5 to 60 m. Both systems had digging, grading and compacting capabilities. Two tracked vehicles were constructed by the Japanese Marine Science and Technology Center (JAMSTEC), Yokosuka in the early 1970's, but were not in operation in September, 1975. In 1971 the West German firm of Hagenuk vormals Neufeldt and Kuhnke GmbH, Kiel constructed a shallow, test model of a tracked vehicle. It is designed to conduct sampling and survey operations (von Reibnitz, 1972), but nothing further has been reported of its activities.

TRAMP

The most recent vehicle is TRAMP, designed and built by Winn Technology, in Kilbrittain (Ireland). TRAMP's six independently suspended wheels allow it to traverse ocean bottoms ranging from unconsolidated muds to large boulders. It is equipped with television and a force-feedback manipulator. The foot controls (forward, reverse, turn) and the operator's seat are equipped to sense pressures experienced by TRAMP. The force-feedback is accomplished by monitoring the pressures on the vehicle and translating them, first by a multiplexed signal and then hydraulically to the operator's console. A three-dimensional image of the vehicle appears on the control console as a line drawing, and video tapes record the actions of the vehicle and its manipulator functions. If a task which has been recorded must be repeated, the tapes can be replayed to automatically control the vehicle in reiterating the task.

Assets and liabilities of tracked vehicles

Construction and operation of the earlier tracked vehicles is similar to tracked, land vehicles, with the obvious exception that everything is either water-resistant or pressure-compensated. A closed-circuit television, sometimes more than one, replaces the operator's eyes. A unique feature of RUM is that it can be made lighter while on the bottom to inhibit sinking. Motions of the vehicle are forward, reverse and turning. The turning motion is obtained by independent operation of the tracks.

The reasons are not immediately apparent why bottom-crawling vehicles have not been used more. A capability for heavy lifting and moving should find application in a variety of undersea construction and emplantment tasks. Indeed, Anderson et al. (1970) show that with reasonable amortization, the RUM system is financially competitive with small manned submersibles. But, perhaps a strictly financial comparison is not the ultimate yardstick, because the bottom crawlers have deployment and operational constraints which seriously affect their applicability and performance.

Most of the bottom crawlers are quite heavy in the water: RUM weighs 6,120 kg and the shallow Hatachi bulldozer weighs 2,970 kg. Such mass requires very specialized equipment and surface support for handling, but this is a minor consideration compared to its effect on the bottom.

In most deep-sea areas the bottom is composed of unconsolidated muds. They are easily disturbed and create a cloud of sediment which obscures visibility. From personal experience with manned submersibles, it can take from five to ten minutes for this cloud to drift off or settle after gently bumping the bottom and then, remaining stationary. The bottom crawler, however, does not remain stationary and the sediment disturbance is therefore constant. The result is a continuous potential for obscuring or completely obliterating visibility. Deep-sea currents cannot be depended upon to carry off this visibility-obscuring cloud, because they are generally not of sufficient speed to act with the desired rapidity. Of course, on sand or cobble bottoms the bottom crawler may not create this problem.

Another area of difficulty is produced by the vehicle's weight. Anderson et al. (1970) described a situation wherein RUM, although set to yield a track pressure of 0.1 kg/cm^2 sunk an estimated 15 cm into the sediment, and during further movement increased the track pressure to 0.2 kg/cm^2. The additional weight was believed to be caused by accumulation of sediment in the track suspension system.

Many of the present bottom-crawler problems can be overcome by redesign and technological advances. But the problem of sediment disturbance, with subsequent obliteration of visibility, appears almost insurmountable.

SELF-PROPELLED, UNTETHERED VEHICLES

The only vehicles known to have actually been built and operated belonging in this classification are the University of Washington's SPURV (Self Propelled, Underwater Research Vehicle) and UARS (Unmanned Arctic

Research Vehicle). In 1970 Mickelsen of Oceanic Industries, Seattle, Washington described a design concept for an untethered, self-propelled research or survey vehicle designated as SEA DRONE I, but construction has not begun. Likewise, Tiemon (1972) of Dornier System, Friedrichshafen, West Germany described a similar system concept for the SF 3. Nothing further has been heard of its development.

UARS

The UARS was financed by the Advanced Research Projects Agency and the Office of Naval Research. It is 3 m long, 48 cm in diameter. Its hull is fashioned from aluminum forgings and filament wound fiberglass resulting in a 457-m maximum operating depth (Francois, 1973). UARS is powered with silver—zinc batteries and its operating speed in the Arctic tests was 3.7 knots. It can operate at this speed for ten hours. The vehicle carries acoustic systems for communications, tracking, homing, obstacle avoidance and emergency recovery. The data collection instrumentation is also acoustic and consists of an under-ice profiler system which measures and digitally records the ice profile five times a second.

Launching of UARS is through a hole in the ice. It is lowered in a horizontal position and the motor is started before its release. The vehicle controls its own depth to within 7.6 cm and pitch and roll to within 0.5°. All control functions and measurements are recorded on a 9-track digital tape.

Recovery of UARS is by ensnaring it in a net containing an acoustic beacon on which the vehicle homes when commanded. If command function is lost, the vehicle will automatically begin homing after a preset period of time. In the event that all power is lost, the vehicle's positive buoyancy allows it to ascend to the undersurface of the ice. It then automatically lowers an acoustic beacon to aid an on-top-of-the-ice search party.

UARS underwent final development testing in March 1973 off Fletcher's Ice Island (T-3), about 275 km north of Ellesmere Island. The vehicle operated approximately 0.9 km from the ice hole and followed a rosette pattern which totaled 31.5 km. During this time over 200,000 position-correlated under-ice profiles were made.

The UARS vehicle was funded only as a developmental project, and no further field work has been conducted since its successful sea trials in 1972. Subsequently, only two U.S. developments in unmanned, self-propelled, untethered vehicles have been conducted; one at the Massachusetts Institute of Technology where a small untethered vehicle is presently being developed and the second at the Naval Research Laboratory.

UFSS

According to Johnson et al. (1976), the Naval Research Laboratory's prototype self-propelled, untethered submersible is intended to replace the "slow and costly" towed systems. The vehicle is called the UFSS (Unmanned Free-Swimming Submersible), and is characterized as a "smart" submersible. The source ofthe UFFS's "smartness" is a miniaturized computer made to react somewhat like the human brain, by using stored information and adapting it to information provided by the vehicle's sensors. Power will be provided by the isotope of Plutonium, ^{238}P. Its radioactive decay generates thermal energy which can be converted to electricity and excess thermal energy can be dissipated through the flooded hull.

The prototype UFSS will be ready by 1978 and will initially be used for ocean-science data collection. Preliminary design envisions a flooded, low-drag fairing shaped somewhat as a tapered oar blade. It will be 5.5 m long, 1.2 m in diameter with an enclosed volume of 3.5 m^3 and a payload volume of 1.4 m^3. The vehicle will be capable of autonomous operations, but can be over-ridden by the surface controller to obtain data and to transmit commands. To determine its position, it will periodically "pop up" from its 457-m cruising depth and obtain an OMEGA fix which would be fed into its OMEGA/DR (Dead Reckoning) system. The improved UFSS envisions 6,096 m depth, a mechanical hand, inertial navigation, and pattern recognition and artificial intelligence capability.

APPLICATIONS OF UNMANNED, TOWED, SELF-PROPELLED VEHICLES

Rather than theorize what could be done with unmanned, towed and self-propelled vehicles, this section is devoted to what has been done. Few work accomplishments of the tracked and untethered vehicles have been reported. Anderson et al. (1970) described a few equipment handling tasks performed by RUM during its developmental dives. The UARS vehicle's under-ice profiling was discussed and this is all that is reported of its activities. Consequently, the following deals only with the towed, and self-propelled tethered vehicles.

Towed vehicles

The applications of towed submersibles can be divided into three general categories: research, search and identification, and survey. The first, research, entails studies of the sea floor primarily for academic purposes. Search and identification tasks involve the search for, and location of, objects of high national interest, or the monitoring of wastes deposited in the deep ocean. Survey tasks involve the mapping or charting of ocean-floor areas on, or in which, cables and pipelines will be installed, or for the assaying of a particular chemical constituent or organism.

A few incidents wherein specific vehicles were used for these tasks will be discussed. All these vehicles may, and sometimes have, been used to conduct tasks in all three categories, although they may have been constructed to satisfy only one.

Spiess and Mudie (1970) define the DEEP TOW's role as being a measurer of meso-scale aspects of the deep-sea floor. DEEP TOW was designed to fill an observational gap identified by Laughton (1963) as "... a gap in the size spectrum of observable features where they are too small to be detected by conventional echo-sounding, but too large to observe visually." To fill this gap DEEP TOW is an adaptation of surface-ship techniques operated close to the sea floor to observe bottom topography and roughness as well as shallow sub-bottom structure and magnetics. A wealth of detailed bottom and sub-bottom information has been collected by DEEP TOW, and is described by Spiess and Mudie (1970). These studies include:

(1) block faulting,
(2) trends of magnetic anomalies,
(3) bottom-slope measurements,
(4) growth patterns of deep-sea fans,
(5) geophysical and geological studies of deep oceanic arches,
(6) deep-sea erosion and sedimentation studies,
(7) investigations of submerged wreckage.

In 1973 DEEP TOW was used in a joint French-American study (Project FAMOUS) of the mid-Atlantic Ridge. Its role was to study and confirm the micro-topographic relief and localized variations in magnetic anomalies which had been revealed earlier by surface ships.

TELEPROBE and NRL systems

The U.S. Navy's TELEPROBE and NRL (Naval Research Laboratory) systems are primarily designed as search and identification systems. The first application of the NRL system was in the search for the sunken submarine THRESHER (SSN-593) in 1963—1964. Using a slow-scan television camera and a stereo-camera system mounted and towed on an open framework of "Unistrut" racks, the NRL system took more than 100,000 bottom photographs of the search area (Brundage et al., 1967). In June 1969 this system played a significant role in the salvage operation for the sunken submersible ALVIN. "Flying" over ALVIN at a depth of 1,555 m the NRL system took photographs of the submersible which demonstrated it had sustained little damage and was in an upright position. This photographic evidence was critical in reaching the final decision to salvage.

In 1970 an ex-Liberty ship was scuttled in a depth of 4,877 m about 370 km east of Daytona Beach, Florida. Inside the ship were concrete lined vaults of a toxic, anti-personnel nerve gas. In 1970—1974, the NRL system surveyed the hulk and obtained photographs and water samples which were later analyzed for indications of leakage. A scanning sonar was used to find the hulk by locating the target at a range of about 610 m and at a distance of approximately 152 m a magnetometer survey was made. No indications of leakage were observed over this four year monitoring program.

Toward the later stages of this program, a photographic system termed LIBEC (Light Behind Camera) was developed by NRL to reduce backscatter and increase the area of ocean floor that could be covered. The principles and techniques of LIBEC are explained briefly by Patterson et al. (1975). The camera is lowered about 10 m below the light source which consists of six pulsed xenon lights having a storage capacity of 52,000 J and the bottom 25 m below the camera is photographed.

The LIBEC system was also used in Mid-Atlantic Rift preliminary investigations in 1973. Heirtzler and Bryan (1975) report that the resultant photographs covered areas over 100 times as large as those from conventional techniques. The resultant photographs were formed into a mosaic on the floor of a gymnasium, where scientists scheduled to study the same area from manned submersibles were able to preview the terrain before diving.

The activities of TELEPROBE have concentrated in the military sector and fewer publications are available than for those of DEEP TOW and the NRL system. The system, however, is quite active in deep-ocean surveying and mapping to provide data to be used to construct limited area, large-scale charts, on which side-scan sonar imagery is superimposed on extremely detailed bathymetric charts.

DSS-125 was employed off the Hawaiian Islands in 1974 to assay the bottom for manganese nodule concentrations at a depth of 5,182 m. Several hundred miles were surveyed and documented by television and still-camera photography.

In many, if not most applications of the towed submersibles, their work is of a supportive nature. Much of their contributions are contained in tasks where the data, and not the means of its collection, is the subject. Therefore, the accomplishments of towed vehicles are not always evident.

Self-propelled, tethered submersibles

The uses to which this category of submersibles have been applied is readily imagined by reviewing the onboard instrumentation listed in Table III. With few exceptions, the work of self-propelled tethered submersibles has stressed their capability to view and maneuver. They have been used to a lesser extent for manipulation and collection tasks, but these have been few, and thus far mainly for purposes of demonstration.

The majority of reports relating the accomplishments of self-propelled vehicles are found in trade journals as half-column news articles or in manufacturer's brochures. Hence, there is a dearth of information regarding the details of these tasks. From a variety of such sources the following inspection/documentation tasks have been performed:

— Inspection of flooded mine shafts

— Identification of prominent geological features
— Inspection of sacrificial anodes on oil and gas facilities
— Pipeline inspection
— Site inspection of sea bottom
— Reconnaissance survey of sea bottom
— Inspection of oil storage and load buoy
— Inspection for degree and nature of marine fouling organisms
— Observation of platform installations both during and after emplantment
— Subsea completion system inspection
— Observation of bottom to assess anchor dragging
— Identification and confirmation of historic wrecks

Manipulative/collection tasks include the following:

— Torpedo retrieval
— Rock fragment collection
— Benthic organism (sea urchin) collection
— Hard rock drill sampling
— Drill bit recovery
— Connection of surface line to hardware for recovery

In this last category, probably the most publicized feat was the attachment of a lift line in 1973 to the manned submersible PISCES III by CURV III at a depth of 419 m. The ways and means of CURV III's involvement in the PISCES rescue is discussed by Talkington (1973) and also in Chapter 10. A summary of CURV II's work accomplishments are related by Smith (1968), and its role as a biological sampling platform is explained briefly by Hoyt (1970).

The potential and actual tasks which these unmanned vehicles can and have performed closely parallel those of the ambient diver, with whom they are intended to compete. Indeed, in deep and hostile waters they can offer formidable competition in viewing and photodocumentation tasks. Unfortunately, all but a few accounts of their activities repeat a litany of successes with no failures (one exception is that of Holmes and Dunbar, 1975). However, anyone who has worked with undersea electronics knows inherently that some problems must have occurred, especially with vehicles which have had so little time at sea. But as their use expands and their idiosyncrasies become known, a multitude of additional work tasks other than viewing and mere grasping will be successfully accomplished.

ASSETS AND LIABILITIES

It is, perhaps, unavoidable that the present upsurge in unmanned submersible activities is attended by papers (see, for example, Busby, 1972; Jalbert and Jacobsen, 1976) and brochures outlining their advantages over manned submersibles and the ambient diver. Such comparisons, if they are to be meaningful, must be qualified by so many ground rules that the final result becomes indecisive. To apply a well-worn adage "it is comparing apples to oranges," and the only unequivocal statement that can be made is that the unmanned vehicle presents no hazard to human life. From that point on, every comparison must be made on a task-by-task basis which includes definitive criteria of (1) depth, (2) visibility, (3) length of dive, (4) environment, (5) the operating policies governing the user, and (6) the urgency or importance of the task, to mention but a few. Consequently, this section has been entitled "assets and liabilities" rather than "advantages and disadvantages."

Johnson et al. (1974) refer to the slowness of the towed system and the same criticism has been made of the manned submersible and the diver. In any system where the prime asset is either direct or remote viewing by the human and where the visibility is limited to tens of feet at best, slowness is not a liability, it is a tremendous asset. If one has ever observed the ocean bottom from a distance of 1.5—9 m, either directly or remotely, at a speed of 2—3 knots, features or objects as large as soccer balls fairly zoom by the view port or camera lens. Therefore speeds as low as 0.5 to 1.0 knot in deep-ocean, real-time observational systems cannot be accepted as a liability. Indeed, there are many instances when one does not wish to move at all, and the ability to hover in one spot can be the solution to the problem.

One last aspect that should be mentioned is the adversary relationship between sea water and electronics. The undersea user of anything electronic is always at a disadvantage, because, for a variety of reasons, connectors, penetrators, cables or housings seem destined to leak at one time or another. The cause may be on the part of the manufacturer or of the user. Failure mode analyses of these components conducted by the U.S. Navy showed where each is equally at fault. Perhaps, with time, both will learn to cope with this problem.

Towed vehicle assets

(1) Real-time data: Provided by the television camera to the surface operator. Can be recorded and replayed for on-site analyses.

(2) Selective photography: Objects selected by the television camera. Reduces photographic film usage.

(3) Concurrent multi-measurements: Reduces ship time, precludes retracing tracks with subsequent navigation and control difficulties inherent in attempting to precisely retrace a previous track.

(4) Increases measurement accuracy: Sensors are closer to the subject; resolution is greater.

(5) Extended operating endurance: Operators are only limited by surface ship's at-sea endurance, the weather, and the number of operating personnel.

(6) No depth limitations: State-of-the-art instrumentation can be available to operate at any known ocean depth.

Towed vehicle liabilities

(1) Support craft facilities: Few ships are available that carry suitable cable winches and handling gear as standard equipment. Ships with slow speed and capable of maintaining course at slow speed are also limited.

(2) Cable breaks: Non-destructive methods for indicating severe deterioration of a cable's inner strands are not available. Breaks, with consequent loss of the fish, have occurred which could have been avoided were such methods available, see Spiess (1974).

(3) Irregular surge: Vehicles with depressors (e.g., DSS-125), impart an alternately fast—slow surge motion to the fish. Televised viewing is optimum at slow and marginal at fast. For measurements requiring near-constant forward speed, such as forward-lapping stereophotography, side-scan sonar mapping, the alternating motion makes both data collection and interpretation difficult.

(4) Heave: Vehicles with no depressor are alternately raised or lowered in accordance with the heave experienced by towing ships. The resulting data are subject to errors similar to that described for surge.

(5) Inspection: Limited capability for detailed inspection of a small area target; essentially no capability for object recovery.

The foregoing generalizes the assets and liabilities accompanying the use of towed systems. For comparative purposes, perhaps some indication of the cost of these systems should be included. In 1972 the cost of TELEPROBE's fish and onboard processing electronics was $500,000 U.S. dollars, which does not include the winch or deck-handling gear. The DEEP TOW in 1972, which incorporated more instruments and sophistication at that time, approached $1 million. For a 24-hour operation, a support operating crew of 15 and 13 for TELEPROBE and DEEP TOW, respectively, is required. When the cost of the handling gear and support ship is included, towed systems of this magnitude are expensive to use.

Self-propelled, tethered vehicle assets

Because they are both connected and powered by a cable to a surface ship, several assets and liabilities of the towed systems also apply to the self-propelled, tethered vehicles. But the fact that this class can maneuver and is generally much smaller and lighter provides unique assets and liabilities. It must be realized that the factors listed below are generalizations, for if the 2,268-kg RUWS and the 68-kg SNOOPY are compared equally, there will naturally be many exceptions.

Assets provided by the self-propelled, tethered vehicle system are listed as follows:

(1) Mobility: Can be crated and shipped aboard most commercial aircraft

or forms of land transportation. Usually can be installed aboard ship within a few hours.

(2) Support craft: Can be accommodated by virtually any type of surface platform.

(3) Safety: Does not jeopardize human life. No classification requirements or liability insurance required; results in lower operating costs.

(4) Operating endurance: Limited only by support craft at-sea endurance and weather.

(5) Manipulator design: Requirements for sophisticated, multi-degrees-of-freedom manipulator greatly reduced owing to maneuverability of vehicle.

(6) Selective photography: Objects selected by television, reduces film usage, maneuverability of vehicle provides a variety of viewing angles for TV and still or motion picture camera.

(7) Depth: No hardware limitations.

(8) Real-time data: Provided by television camera to the surface operator. Can be recorded or replayed for on-site analysis.

(9) Viewing range and resolution: Some of the television cameras employed provide greater resolution of objects than does the human eye and can do so with lower levels of illumination.

(10) Power failure: In the event of total power or buoyancy failure, the vehicle can be hauled to the surface by its cable.

Self-propelled, tethered vehicle liabilities

The liabilities of these systems are also generalizations. Some vehicles overcome several of the factors listed below because of their propulsive arrangements, control or structure. All, however, have a common problem, namely, the surface-to-vehicle cable. Paradoxically, the very feature that makes these vehicles unique and competitive, also limits their performance. Eight self-propelled, tethered vehicle liabilities are listed:

(1) Power and control cable: Creates high drag; limits speed through water and maximum current in which it can maneuver. Subject to fouling in external appendages. Can be severed by support craft's propellers. An early model of CURV terminated operations while working off the coast of Washington when the support ship's screws severed its cable; it was never found.

(2) Fouling: Most vehicles are of an open-framework construction. This arrangement is susceptible to entanglement if operating in the vicinity of cables, lines or protruding hardware.

(3) Operator endurance: The operators of ANGUS and TROV state that two and three hours, respectively, is the limit of operator endurance owing to the high degree of concentration required. While this in itself is not a liability, the effectiveness of the operator to search and observe can be questioned well before he reaches the limit of endurance.

(4) Sediment disturbance: The orientation of the vertical thruster(s) is such that visibility-obscuring sediment clouds may be generated by the propwash of the thruster as it operates close to the bottom.

(5) Hovering: The intense and total concentration required of the operator to hold the vehicle motionless in mid-water prompts the need for automatic "flight" control during long-term inspection missions.

(6) Manipulation/lifting: The only means of gaining torque or lifting an object with the manipulator is to dynamically counterbalance the force applied by use of the thrusters. This procedure calls for an exceptionally skilled operator and force-feedback and the latter is not a feature of present vehicles.

(7) Viewing: Many manipulative and observational tasks require 3-dimensional viewing for effective performance; this is not a feature of present television arrangements.

(8) Support platform: The required station keeping capabilities cannot always be found on ships of opportunity. Likewise, suitable handling gear for the larger vehicles is not always available.

Some idea of the cost of these vehicles is pertinent. The total cost of both SCARAB systems, including deck handling equipment, is $5.8 million. The cost of RCV-225, including deck handling equipment, is approximately $380,000. The support ship cost and crew salaries must be added to these costs.

CONCLUSIONS

The ways and means of viewing and working beneath the oceans are many and varied. For the present, it appears that the towed and self-propelled, tethered vehicles are seeing greatest utilization and, of these two devices, the latter by far has the most advantages.

Unmanned submersibles of the RCV-225 and CONSUB variety are at about the same stage of development as the manned submersibles of the mid- and late 1960's. At that time the manned submersible was going to be all things to all men, but it took almost a decade before the most practical designs were identified and put to work. A hallmark of the early manned submersibles was the "multi-purpose" vehicle. It could survey, salvage, repair, support divers, collect samples and make movies, and, when it wasn't doing these, it could conduct oceanographic research. A student of the nineteen-sixties manned submersible scene experiences a sense of *déjà vu* when reading or hearing of the multi-purpose unmanned submersible. But there is much to be done before this vehicle becomes a reality, if ever. There is no all-purpose aircraft, or ship, or automobile, or manned submersible. Each variety is designed to do a particular task and to do it well. It seems unavoidable that each new undersea vehicle or capability must experience growing pains, until finally it finds the task for which it is best suited and concentrates thereon. This realization will come to the operators of the unmanned submersible as it did to those of the manned submersible. Many of the present unmanned problems will be solved as their utilization increases and in the process each will find the role in which it excells.

Predicting the future of unmanned submersibles is dangerous. But the exploitation of offshore oil and gas in areas of great depth and hostility to the ambient diver will continue to increase as will the legal requirements for periodic inspection of undersea pipelines and storage tanks. Therefore, the potential work for both remote and direct manned employment seems quite promising. In a parallel vein, the need to obtain greater detail and knowledge of the ocean bottom over and through which these vehicles must pass is greater now than ever before. At a cost of the order of $1 million per pipeline mile in some areas one cannot depend upon inferences based on data obtained from over-the-side methods of the 1950's. For these reasons the human eye, either direct or remote, will be called upon increasingly to pass final judgement.

ACKNOWLEDGEMENTS

The following individuals and organizations graciously provided advice, information and illustrations: Mr. Art Sepin and Ms. Kathy Murphy, Hydro Products, a Tetra Tech Co., San Diego, Calif.; Messrs. Edward Sutter and Clint Mangum, Sutter & Co., Vienna, Va.; Mr. Robin Holmes, Heriot-Watt Univ., Edinburgh; Mr. F.M. Daughtery, Naval Oceanographic Office, Wash., D.C.; Dr. F.N. Spiess, Marine Physics Laboratory, Scripps Institution of Oceanography; Mr. William Greenert, Naval Material Command, Wash., D.C.; Mr. S.A. Tomlinson, Canadian Dept. of Environment, Ottawa; Mr. R.W. Cook, Harbor Branch Foundation, Ft. Pierce, Fla. and Messrs. T.C. Bickerton and Leo Schafer, British Aircraft Corp., Bristol.

REFERENCES

Anderson, V.C., Gibson, D.K. and Kirsten, O.H., 1970. RUM II — Remote underwater manipulator (A program report). Proc. 6th Annu. Conf. Exposition, Mar. Technol. Soc., Washington, D.C., 1: 717—731 (preprints).

Banzoli, V. et al., 1976. New concept of underwater remote controlled tracked vehicles for deep water trenching operations. Proc. Offshore Technol. Conf., Houston, Texas, 1: 647—664.

Baxter, H.A. and Mueser, R.E., 1971. The development of ocean-cable plows. IEEE Trans. Commun. Technol., COM-19 (No. 6): 1233—1241.

Bircham, J. and Skidmore, G., 1976. Operational use and techniques with unmanned cable controlled submersible in the North Sea environment. Proc. Offshore Technol. Conf., Houston, Texas, 3: 1049—1058.

Bline, D.B. and McDonald, W.M., 1976. Development of an unmanned deepwater pipeline repair system. Proc. Offshore Technol. Conf., Houston, Texas, 3: 732—739.

Brundage, W.L., Buchanan, C.L. and Patterson, R.B., 1967. Search and serendipity. In: J.B. Hersey (Editor), Deep-Sea Photography. Johns Hopkins Press, Baltimore, Md., pp. 75—87.

Busby, R.F., 1972. Diver, submersible or instrument package. Underwater J., 4: 115—123.

Francois, R.E., 1973. The unmanned Arctic research submersible system. Mar. Technol. Soc. J., 7: 46—48.

Heirtzler, J.R. and Bryan, W.B., 1975. The floor of the Mid-Atlantic Rift. Sci. Am., 232: 79—90.

Holmes, R.T. and Dunbar, R.M., 1975. Sea-bed surveying by remote control. In: Conf. Pap., Oceanology International 75, Brighton, pp. 75—80.

Hoyt, J.W., 1970. Biological collecting with a remote-controlled underwater vehicle. Mar. Technol. Soc. J., 4: 65—66.

Jalbert, P.E. and Jacobsen, L.R., 1976. Unmanned cable controlled subsea vehicles — status and potential in the offshore oil industry. Proc. 8th Annu. Offshore Technol. Conf., Houston, Texas, 3: 1037—1047.

Johnson, H.A., Verderese, A.J. and Hansen, R.J., 1976. A "smart" multimission unmanned free swimming submersible. Nav. Eng. J., pp. 84—95.

Jue, M.F. et al., 1974. An automatic terrain-following control system for a deep-towed submersible. Proc. Offshore Technol. Conf., Houston, Texas, 1: 747—754 (preprints).

Kawasaki, H. et al., 1972. Development of underwater bulldozer systems. Proc. 8th Annu. Conf. Exposition, Mar. Technol. Soc., Washington, D.C., pp. 669—680 (preprints).

Laughton, A.S., 1963. Microtopography. In: M.N. Hill (Editor), The Sea, 3. Wiley, New York, N.Y., pp. 437—472.

Mickelsen, P.W., 1970. SEA DRONE I and the unmanned system. Proc. 6th Annu. Conf. Exposition, Mar. Technol. Soc., Washington, D.C., 2: 837—845 (preprints).

Noakes, J.E. et al., 1975. Surveillance system for subsea survey and mineral exploration. Proc. Offshore Technol. Conf., Houston, Texas, 1: 909—914.

Patterson, R.B., Buchanan, C.L. and Gennari, J.J., 1975. LIBEC system engineering. Mar. Technol. Soc. J., 9: 3—13.

Skidmore, G. and Bircham, J., 1975. Development of an unmanned cable controlled submersible vehicle for surveying and sampling exposed subsea rock shelves. Proc. Offshore Technol. Conf., Houston, Texas, 3: 127—138.

Smith, H.D., 1968. Useful underwater work with CURV. Mar. Technol. Soc. J., pp. 19—23 (October issue).

Spiess, F.N., 1974. Recovery of equipment from the ocean floor. Ocean Eng., 2: 243—249.

Spiess, F.N. and Mudie, J.D., 1970. Small scale topographic and magnetic features. In: A.E. Maxwell (Editor), The Sea, 4. Wiley, New York, N.Y., pp. 205—250.

Talkington, H.R., 1974. The U.S. Navy participation in the rescue of PISCES III. Mar. Technol. Soc. J., 8: 63—67.

Tiemon, A., 1972. Exploration by unmanned submersibles. Conf. Pap. Oceanology International 72, Brighton, pp. 252—257.

Vigil, A.E., Frisbie, H.L. and Hatchett., 1975. Deep sea survey system. Proc. 7th Annu. Offshore Technol. Conf., Houston, Texas, 1: 915—922 (preprints).

von Reibnitz, G., 1972. Cable controlled sea bottom measuring vehicle. In: Conf. Pap. Oceanology International 72, Brighton, pp. 274—277.

CHAPTER 3

THE UNIQUE ROLES AND HISTORY OF MANNED AND UNMANNED SUBMERSIBLES IN OCEANOGRAPHY

Don Bline

INTRODUCTION

The necessary technology has existed for the past hundred years to enable manned craft to operate submerged in the ocean at relatively shallow depths. Innovations of the past several decades have resulted in rapidly improving designs of military submarines and the appearance of deep-operating manned vehicles for civilian use, generally referred to as submersibles.

MANNED TETHERED VEHICLES

The simplest deep-diving vehicle is a bathysphere. First appearing in 1929, this vehicle consisted of a cast-steel pressure-resistant sphere, fitted with viewing ports and equipped with life-support, basic-instrumentation and communication systems. To operate this non-buoyant device it was suspended in the water column from a surface support ship by a steel cable. Positioning of the vehicle was achieved by adjusting the length of the cable and the speed of the support ship.

The hazards of operating this vehicle became somewhat frightening with the realization of the total dependence upon the umbilical cable for return to the surface. If the steel cable were severed the vehicle would simply sink due to a lack of emergency buoyancy capability. It becomes even more frightening to contemplate being a passenger in the vehicle being towed, at even very slow speeds, near any obstacles. Upon visually detecting the obstacle the observer calls the support ship and directs evasive action to be taken. "A large coral head is dead ahead" would be the observer's calm remark. Hopefully, a winch would be quickly engaged to haul the bathysphere up to avoid the pending collision. This was not a very satisfactory arrangement.

Compared with the technology of the 1970's, which provided obstacle-avoidance sonars, acoustic-navigation systems, satellite-navigation systems, constant-tension heave-compensated winches, dynamic-positioning systems, etc., the bathysphere is an unattractive vehicle for any purposes other than simple viewing in a mid-water situation. It was the first means of man's direct viewing of the ocean beyond diver depths and is regarded by many as the genesis of today's deep-diving non-combatant submersible vehicles.

The next tethered vehicle known to have been built was completed in 1950. KUROSHIO I was basically a diving bell configured for support from the surface. This vehicle was operated from 1951 thru 1960 when KUROSHIO II was launched. KUROSHIO II, while supplied with power from the surface, was equipped with propulsion and ballast systems. A limited suite of scientific instruments was provided in addition to a compass, and vertical and horizontal sonar. It was a significant advance beyond the bathysphere, but still dependent upon the umbilical, both for power and as a means of being hauled to the surface in case of an emergency situation. The drag of the cable created a problem in maneuvering.

In 1970 another tethered vehicle made its appearance. GUPPY built by Sun Dry Dock and Ship Building Co. depended totally upon the umbilical cable to deliver electrical energy to the vehicle for propulsion and maneuvering. Again difficulties were experienced in maneuvering the vehicle due to the drag of the umbilical cable. GUPPY has not been operated for several years.

The objectives for building the bathysphere, such as KUROSHIO I and KUROSHIO II, are quite clear. The bathysphere was used for many years by William Beebe in his studies of deep-sea organisms. KUROSHIO I and KUROSHIO II were built for fishery studies off the coast of Japan. The KUROSHIO vehicles have made hundreds of successful dives performing the functions for which they were designed. GUPPY was designed to be a low-cost, manned submersible for long-endurance missions. Demand is a measure of success and hence the inference can be made that the GUPPY system was not successful in meeting the objectives due either to market forces or system deficiencies.

MANNED SUBMERSIBLES

An accepted definition of a manned submersible requires a non-combatant craft be capable of independent operation on and beneath the water surface and to have a direct means of viewing for the occupants who are in a dry atmosphere. The phrases independent operation and dry atmosphere clearly differentiate the manned submersible from tethered vehicles, wet vehicles and subsea habitats.

FNRS II

The first true submersible may have been built in 1929 by T. Nishimura, but little is known of his design. The vehicle had glass view ports and was designed for fisheries research. FNRS II followed, conceived by the noted physicist and balloonist Auguste Piccard in 1939. Its completion was interrupted by World War II and FNRS II made its first test dive in 1948. FNRS II was designed for a working depth of 4,500 m. A pair of cast-steel hemispheres joined at the equator and fitted with two plastic view ports provided a dry environment for the two occupants. The personnel capsule was carried below on a thin-walled steel float in a fashion similar to a gondola beneath a balloon. The float was divided into several tanks filled with gasoline, which was used as positive-buoyancy material. Other tanks filled with steel shot, gravel and scrap iron were available to provide negative buoyancy. All of the tanks were controllable thus providing vertical maneuverability. Power was provided using pressure-compensated lead—acid batteries mounted externally and exposed to the sea. Propulsion was provided using electronic motors permitting a cruising speed of 0.2 knots.

During testing, the frailty of the float design and the general unsuitability of the vehicle for towing became apparent when the system first experienced heavy weather. Unable to safely retrieve the vehicle the support ship replaced the gasoline with CO_2 and began towing operations. The vehicle experienced extensive damage during towing operations. Shortly after this experience the vehicle was donated to the French Navy. After redesigning the float for surface towing, the vehicle, now known as FNRS III was launched in 1953.

FNRS II was approximately 6.9 m long and 5.7 m high and was slow and not very maneuverable, whereas FNRS III was approximately 15.6 m long. The float closely resembled a small conventional submarine complete with conning tower. FNRS III was significantly faster and more maneuverable than FNRS II. FNRS III was also equipped with (1) sonar, (2) underwater phone (UQC) radiotelephone, (3) water sampler, (4) depth gauges, (5) temperature sensor, (6) still camera and, (7) light. It dove successfully to 4,500 m and remained in service for seven years. Little is known of its mission or degree of success.

TRIESTE I AND II

Piccard, benefiting from the experiences with FNRS II launched TRIESTE I in 1953. The principles of FNRS II were designed into TRIESTE I. Substantial improvements were made including a redesign of the float, making it larger and more suitable for towing. The vehicle dove successfully to 3,333 m in 1954 and four years later it was sold to the U.S. Navy.

After extensive refitting, including replacement of the personnel sphere, TRIESTE II made the record dive to 11,933 m in 1960 and successfully demonstrated that any depth in the ocean could be reached by man.

TRIESTE II is approximately 26.3 m in length, has a beam of 5 m and a height of 9 m and a dry weight of more than 79,000 kg. A suite of instruments including sonars, sound velocitymeter, cameras, Doppler navigator, x—y plotter, gyrocompass, etc. has been provided. Because of the size of the vehicle, operations while submerged are conducted in areas relatively free of obstructions. The surface-support requirements of TRIESTE II are met using a modified floating dry dock towed by an ocean-going tugboat.

DIVING SAUCER

Cousteau as early as 1953 began the design of a small submersible capable of operating to a depth of 333 m. It would enable scientists to observe and photograph the oceanographic phenomena of the undersea valleys and canyons where larger vehicles could not maneuver. Cousteau had dived several times in FNRS III and noted its weak and strong points. The pressure-compensated batteries and conical plastic view ports of FNRS III were features he included in his DIVING SAUCER. However, he insisted upon a high degree of transportability, better viewing and greater maneuverability than were possible with FNRS III.

The resulting vehicle is 3 m long, has a 3-m beam, and is 1.6 m high. It weighs only 3,800 kg, making it easily transportable, and it can be supported by any support ship having a lifting device of a capacity of 4,545 kg. The vehicle can provide underwater observation in relative safety and comfort. A limited payload capacity restricts the extra equipment that might be required for experiments. In the performance of design objectives the vehicle was a success and is still in use after many refits and redesignations.

The DIVING SAUCER is believed to be the first vehicle to use the positively buoyant hull as the means of surfacing. During a dive the vehicle was trimmed to neutral buoyancy while carrying an iron weight. To surface, the weight was manually jettisoned causing the vehicle, now positively buoyant, to surface. This mode of operation while positive, does have some disadvantages. Prior to jettisoning the weight, the area above the vehicle must be determined to be free of ships or other hazards. If for any reason the pilot wishes to remain submerged he is prevented from doing so after jettisoning his weight because he cannot alter his buoyancy condition.

In the early 1960's a number of small relatively shallow diving submersibles appeared. The incentive for private industry to construct these submersibles is not clear, but for larger companies perhaps the lure of new markets proved enticing. The reason for the construction of vehicles by some of the smaller companies seems difficult to assess. In the years that followed from 1963 to the present (1976) many submersibles were launched both in the United States and abroad. They were designed for depths of hundreds to several thousands of feet, for missions ranging from poorly defined general observation types to special-purpose subsea work platforms. Multiple generations of vehicles were built occasionally by the same builder and payloads and performance were increased. Builders were requested by potential users to provide solutions to special requirements. New instruments and devices were built to capitalize on the availability of these new vehicles as subsea laboratories or work platforms. The rush to build submersibles resulted in the construction of more than 120 vehicles by 1976.

DEEPSTAR 2000 (Fig. 1) and SHELFDIVER (Fig. 2) are examples of typical small manned submersibles. Modern technology has allowed man to overcome the deep-ocean environment permitting prolonged excursions by scientists and engineers to learn at first hand and in real time about this frontier. For the first time man can observe the biota of their natural habitat; he can cruise or loiter as dictated by the situation, with only seconds of reaction time being required for decision making. Physical measurements can be made and samples for laboratory analysis can be taken with confidence and assurance that the location is optimum and the sample is valid for the inherent purpose.

As a tool for subsea engineering purposes, the manned submersible allows controlled close observation, measurement and evaluation. The present heavy usage of manned submersibles in the North Sea has clearly indicated the worth of these vehicles in missions involving pipeline-route surveys,

Fig. 1. DEEPSTAR 2000.

Fig. 2. SHELFDIVER, which will be used during habitat site surveys on the Continental Shelf, to be conducted by the Virgin Islands Government. The 265-m depth submarine, which can lock out two divers for work on the seabed, was constructed by Perry Submarine Builders. (Photo by Submarine Equipment.)

immediate post-pipe-lay surveys and in routine surveys at regular intervals after the pipeline has been in service. The manned submersible is an important and valuable tool for use in oceanography. The shortcomings and limitations of the present generation of vehicles are certain to be overcome, with increased usage and profitability spurring development in the military and private sectors.

DEEP-SUBMERGENCE RESCUE VESSELS (DSRV)

At least two special modifications of the "standard" manned submersible deserve separate consideration. The U.S. Navy's deep-submergence rescue vehicles are unique, special-function submersibles. They are potentially useful to accomplish a wide variety of tasks because of the availability of a relatively large payload and on-board power. DSRV's are fitted with transfer tanks and air locks and can mate with a military submarine to effect a deep-water, dry rescue of the occupants. These vehicles can be transported and operated from the deck of a military submarine while submerged. The submerged speed of the submarine is restricted to 15 knots with a DSRV on board. The problems associated with surface operations that have plagued the operators of other submersible systems have been eliminated. The requirements of an emergency rescue operation could not be jeopardized because of the inability of the system to be employed because of inclement weather. This is clearly a case of need and priorities resulting in the development of a solution to a previously unsolved problem.

A number of diver-lockout submersibles have been successfully constructed and are in use. These vehicles provide a mobile deep-water platform from which divers can make excursions to perform required tasks. By using mixed breathing-gas techniques together with support-ship deck chambers, saturation diving well below the depth limits of ambient-pressure air systems is being performed routinely. The submersible in this application performs as a diver-transport and submerged support station. The construction of these vehicles varies from the normal submersible in that two or more isolatable pressure-resistant capsules are provided within the envelope. The vehicle operator and non-diving passengers remain in a one-atmosphere environment throughout the mission, whereas the divers in the other capsule(s), fitted with pressure controls, gas supplies and access hatch, experience up to ambient pressures at working depths. Other modifications to the "normal" submersible system are required to compensate for the changes in buoyancy trim moments that occur when operating the vehicles in a diver-lockout mode. The vehicle in Fig. 2 is representative of a typical lockout submersible.

HABITATS

Habitats per se should only be mentioned in any discussion of submersibles. As diver hotels or way stations, habitats have proven their usefulness particularly to those interested in relatively shallow waters. The U.S. Navy's Sea Lab III program is the deepest habitat venture for which information is available. It was useful in that knowledge was gained and the experiences used in the planning of future diving activities using mixed breathing gases. The program was stopped prior to planned completion and has not been reactivated.

A habitat in its simplest form consists of a bubble of breathing gas captured at the desired working depth by a non-permeable membrane. Refinements have added lights, beds, and cooking facilities, as well as sanitary and heating provisions. Total support to the bottom-anchored habitat is supplied from a surface element of the system. For this purpose, ships or barges fitted with the necessary machinery and equipment have been used.

UNMANNED SUBMERSIBLE VEHICLES

For purposes of this discussion unmanned submersible vehicles will be considered to be remotely controlled, umbilically supplied and tethered vehicles having propulsion and maneuvering capabilities. The most widely known vehicles of this general type are the U.S. Navy's CURV series. Members of this family of vehicles have received wide public exposure due to their involvement in various salvage and rescue missions. They do not require the large, heavy pressure-resistant capsule needed to protect human passengers as in a manned vehicle. In addition, by supplying power by electrical energy through the umbilical cable, the size and weight of these vehicles can be made much smaller than a manned submersible having a similar capability.

The CURV vehicles, along with others, have been designed to be slightly positive in buoyancy. A vertical thruster provides the forces required to control the position of the vehicles within the water column. The advantage of this arrangement in an emergency condition is in a self-surfacing vehicle, should power be lost or the cable severed. A further benefit is realized by the elimination of variable-ballast systems. An example of this type of vehicle is shown in Fig. 3.

The operation of these vehicles depends upon the on-board instrumentation and video signals being transmitted and displayed to the operator on the support ship. Extensive use is made of sonar devices to determine depth and distance above the bottom, as well as to detect objects. Video equipment, lighting, pingers and other sensing devices, are also provided. The control of the vehicle using video and instrumentation can be compared with flying an aircraft in clouds. It is markedly different than direct visual observations, but considerable skill can be achieved through practice and effort.

In addition to providing observation and image-recording capabilities, vehicles of this type have been equipped with a wide variety of devices. These include manipulators, deburial devices, cutters, magnetometers and other special-function apparatus. Builders of these vehicles maintain that in addition to salvage and rescue operations, site surveys, exploration of the sea floor, mining, cable and pipeline repair, as well as other underwater construction functions can be accomplished.

While the majority of the vehicles in this category are relatively small, at least one system presently under design will have an air weight in excess of 45,000 kg. This system is being designed as a work vehicle to be used in the repair of deep-water oil and gas pipelines. It will be provided with a very large variable-ballast system, and will have a payload of more than 13,636 kg.

A second submersible vehicle is required to complete the submerged portion of the system. It is designed to deliver and install very large pipeline-repair devices in a pipeline at depths ranging up to 1,333 m. The payload will

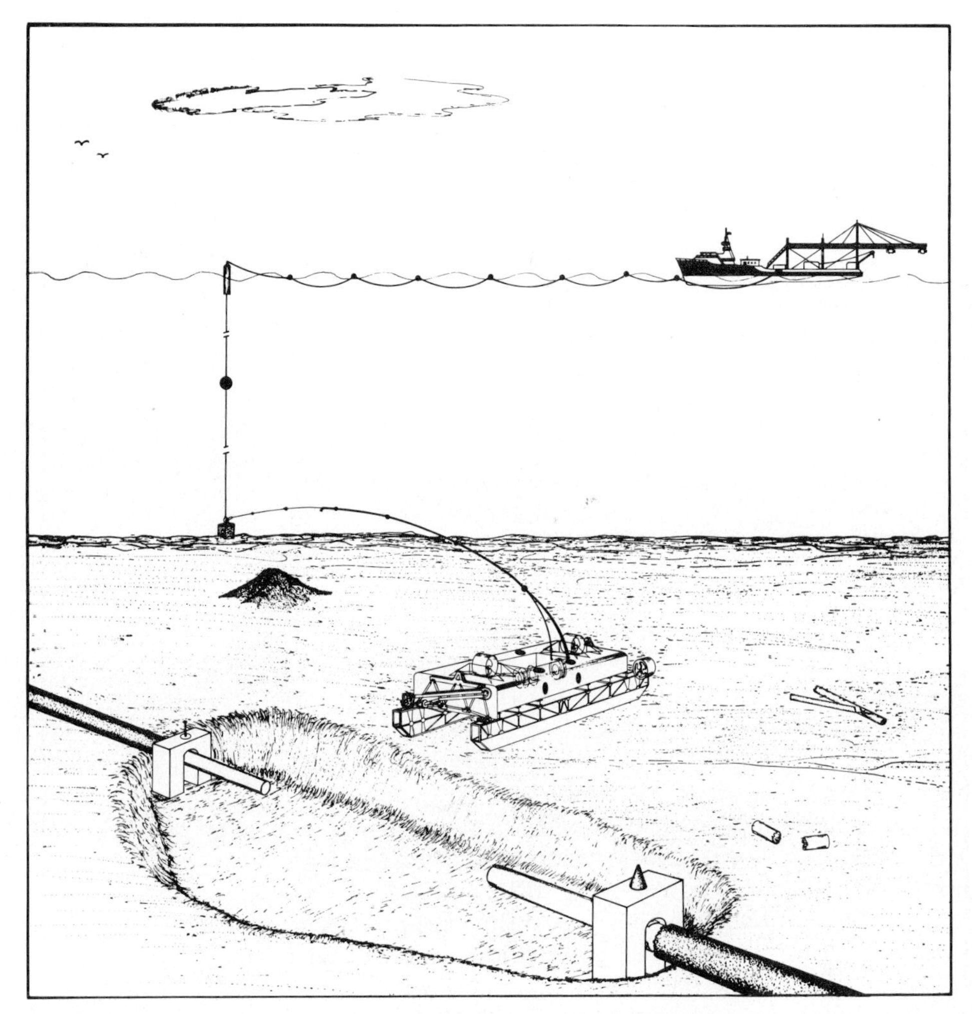

Fig. 3. Artist's impression of unmanned submersible vehicles engaged in pipeline repair.

be greater than 34,000 kg. Figs. 3 and 4 represent an artist's impression of these vehicles in operation.

Any manner of special-purpose robot vehicles is possible. A number of unmanned submersible vehicles fitting this category has been built. Wheels and tracks have been used for running on the sea floor. Excavation and reburial devices have been mounted to be used in submarine cable and pipeline-repair efforts. These are described in Chapter 12.

In general, the support requirements of unmanned vehicles are not as demanding as might be expected for a manned system. The smaller dimensions and lighter weights of these vehicles require less capacity in the

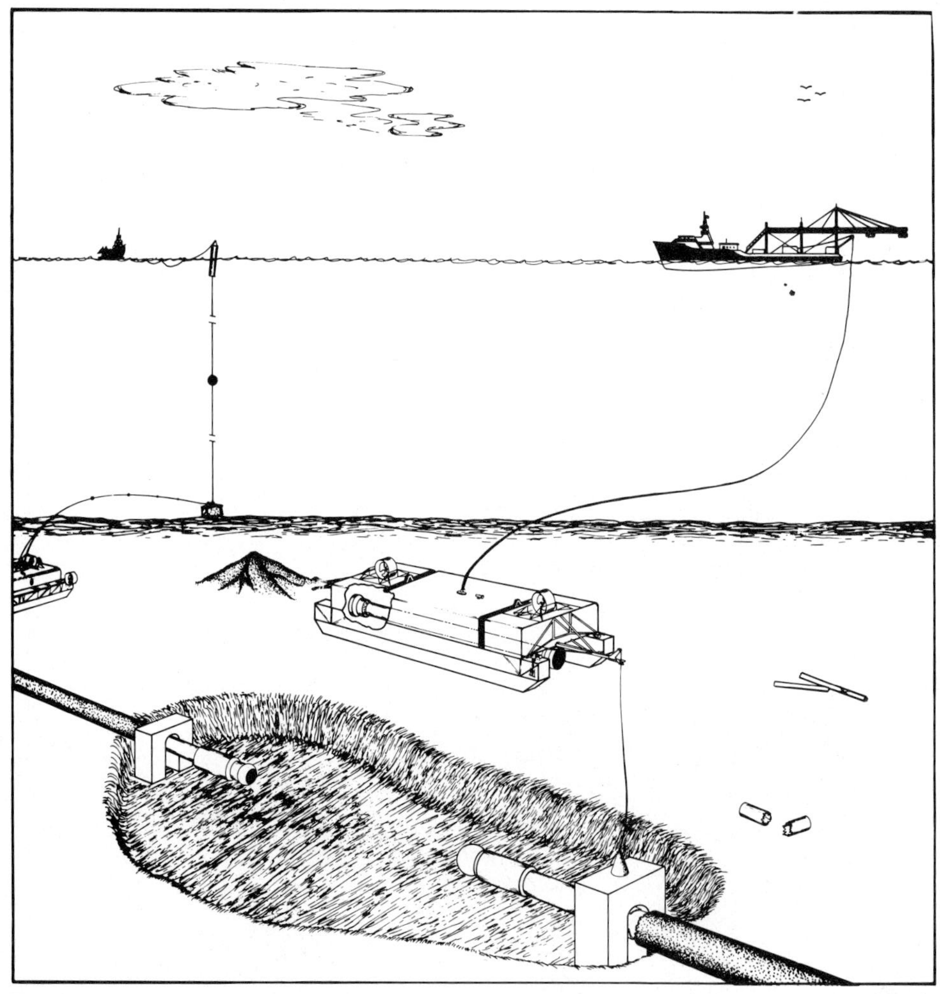

Fig. 4. Same as Fig. 3. See also text.

handling systems. A smaller and hopefully less expensive ship can be used without any sacrifice of safety.

HYBRID SYSTEMS

Recently, a new one-atmosphere diving-suit category has been introduced. Suits for depths of 333, 417 and 500 m are now operational, and a suit for 1000 m is under development. These resemble an oversized suit of armour fitted with port holes, and can be used to perform most diver tasks, but at a slower rate. Diving times are limited by the fatigue experience of the diver/operator.

The use of this apparatus for the moment at least seems limited to well-trained diving personnel. Principal use of the system is almost certain to be in the offshore oil fields of the world.

COMBINATION OF MANNED AND UNMANNED SUBMERSIBLES

Another recent development combines the principles of an unmanned device with a manned submersible. The umbilically powered unmanned device can be controlled by the occupants of a manned submersible. The submersible is designed to couple with the unmanned device while both are on the sea floor using magnetic coupling and inductive controls. This arrangement allows the maximum utilization of the best features of each system. The sea-floor device is a high-power consumer with the power supplied by the umbilical from a surface generator. The manned vehicle allows the operator the benefit of direct visual observation to be used in controlling the unmanned device. A novel feature provided allows the manned submersible to recharge batteries while on the bottom drawing energy from the unmanned device.

SUBMERSIBLE MISSION

It would be meaningless to attempt a complete listing of submersible missions. Submersibles will continue to be used to accomplish tasks limited only by the user's needs, his creative imagination and the basic limitations imposed by the vehicle design, and the basic laws of economics.

The earlier submersibles were basically observation platforms that enabled users to see the deep ocean. Descriptions mumbled into microphones soon gave way to photographic techniques used to document what was seen. In turn, and in a similar manner, engineers and scientists collaborated developing new, or modifying existing, instrumentation for submersible use. This was an effort to provide supporting data to help toward obtaining a complete understanding of visual observations.

Biologists desired information on the physical and chemical characteristics of the water where specimens were seen or taken. Geologists in addition to wanting samples, desired information pertaining to in-situ conditions to verify laboratory results. Various bottom-bearing strength devices were built, and shear-measurement instruments were developed.

Subsea engineers seeing submersibles as possible underwater work platforms were interested in manipulators and special-purpose tools for cutting, and joining, as pertained to their particular need.

Much of the effort directed toward improving the usefulness of the submersible was done in small shops on limited budgets. The submersible user must dedicate a major portion of the project's budget to the submersible and its operation. The remainder is available for salaries, instrumentation, laboratory time and all of the other functions that are a part of such an endeavor. The vendors, developers, manufacturers and sellers of instrumentation and devices are normally motivated by economic pressures. In short, if there is a market for his wares he will provide whatever is required, if he, the vendor, can make a profit from the venture.

One-of-a-kind production is extremely expensive and is frequently unprofitable. The submersible industry is not very strong, and there has been a lack of clearly defined purpose or mission for most submersibles. The application of specially designed equipment is generally very limited. In addition, new developments in materials and techniques, particularly in the field of electronics, often served to make a prototype obsolete before production could begin. Private industry is therefore understandably hesitant to invest money to develop devices for which no market could be assured.

Design of instrumentation or work devices for manned or unmanned vehicles is governed by the common constraints of working depth, entanglement avoidance, wave slap, attachment, corrosion, weight and volume. Unmanned vehicles with power provided by the cable from a support ship are less restrictive in terms of power consumption than the battery-powered, manned, free-swimming submersible. The unmanned vehicle must be provided with telemetry equipment to provide instrumentation data to the operator on the support ship for control purposes. The manned-vehicle instrumentation outputs are available within the personnel compartment for direct viewing. Instrumentation for the manned vehicle must be so arranged as to place at least the sensors outside of the personnel space. Cabling, power supplies, interference, interior space, and personnel hatch-size, all become meaningful considerations. The foregoing and other unmentioned constraints tend to cause a continuation of one-of-a-kind designs.

Submersibles have been used in sedimentology, biology, optics, soil mechanics, geophysics, geology, physics, physical chemistry, acoustics, soil mechanics, deep scattering layer studies, currents, radiation, and others tasks. Deep-ocean salvage and recovery, extensive surveys, ice investigations, archaeology, mining, engineering tasks in the world's offshore oil fields, and diver-support stations, all have been missions assigned at one time or another to submersible vehicles. At this time most of the operational manned submersibles are gainfully occupied at relatively shallow depths in oil-field-related activities in the North Sea. There are exceptions of course, to every rule. Where budgets permit and priorities dictate, scientific investigations are continuing in deep waters, but at a comparatively very low level of effort. Hopefully, a more even distribution in the level of expenditure by goals may not be too far distant. Should present trends continue unchecked it is not inconceivable that builders will be exclusively meeting the needs of their prime users with specially designed vehicles. The result could be that the individuals or small institutions then would be faced with the continuing problems of adaptations to meet the needs of their particular mission.

MANNED VEHICLE OR DRONE

Were a poll taken today of those in favor of one type of system or the other, proponents of both systems would be found. Both factions can provide strong arguments to support their case to the detriment of the other. For the most part, these arguments are subjective, based on the limited use of one system or another.

Without a doubt there are times when a particular mission, as it was conceived and planned, could not be executed without a manned submersible vehicle. Could the same goals be reached using a different approach and an unmanned system? If it were possible to totally predict the events and happenings experienced during the dive the answer would be in the affirmative. A very real benefit of the manned system has to be the presence of the man. The man making observations, can make real-time decisions that could turn a marginally successful dive into outstanding discovery of some previously elusive bit of knowledge. The man in this case eliminates the telemetry, remote viewing and control systems. Upon perceiving something the man can immediately direct the vehicle to allow better viewing or sampling. The unmanned system might miss the initial opportunity or have difficulty in quickly returning to the sighting position.

For missions where data and impressions are important there can be no doubt of man's importance on the scene. Television viewing in many cases is technically better in some respects than the unaided human eye. Video images can be generated using less available light than man requires. In addition, viewing ranges can be markedly extended using various image-processing techniques. Man's impressions are lacking, however, to complement the camera's perfection. Man's senses functioning together, perceiving the dimension, the relationships and other inputs create an impression that simply is not possible with video alone. Variations in shades of colors, depth of feature, and other subtle changes can be sensed by the human. To approach man's response, many man-made machines operating simultaneously and fully correlated would be required.

For missions involving diving operations from a lockout submersible, substitute means were required prior to the development of vehicles with this capability. The advantages realized using the submersible include, greater flexibility, increased mobility, and supervisor viewing of work in process. These advantages were motivating forces leading to the construction of this type of vehicle. No substitute vehicle system known can provide greater advantages.

FREE-SWIMMING MANNED VEHICLES

Free-swimming manned vehicles are inherently more maneuverable than the umbilically teathered, unmanned vehicle. The drag forces on the umbilical along with required consideration of entanglement and fouling are all limiting factors. The ability to divorce vehicle motion from surface-wave effects with a coupling umbilical varies from system to system.

The free-swimming vehicle has serious operational limitations because it is totally free of the support ship thus requiring it to be completely self-sufficient. Technology has not yet advanced to the point where these small vehicles can be fitted with the on-board power-generation capability required. Battery-stored electrical energy is used therefore as a power supply. The particular construction of the battery serves to modify the power per unit density and discharge curves, but it can do nothing toward supplying more energy than was initially stored. Missions are therefore limited in terms of power consumption. By carefully controlling the use of the high power-consuming propulsion systems and photographic and viewing lighting systems, some mission extension is possible. This sort of conservation can be followed, until man becomes the limiting factor ultimately governing mission endurance.

TETHERED UNMANNED VEHICLE

The tethered, unmanned vehicle has no operational time limits. As long as the support ship is able to maintain station and provide the necessary support, the mission can continue without interruption. Vehicle operators and observers can be rotated in shifts, and the limits of effective human endurance are never approached. Unmanned vehicles are ideally suited to missions requiring extended bottom times. Capitalizing on this characteristic, at least two efforts are in the preliminary design stages to provide very large unmanned vehicles that would be used in pipeline-repair efforts. One of these proposed systems is shown in Figs. 3 and 4, consisting of two special-function vehicles. The first, a work vehicle will locate, excavate and prepare the pipeline to receive repair components. This vehicle has an air weight of approximately 50,000 kg. The umbilical is provided with a taut mid-section to divorce the vehicle from surface-wave effects. The function of the second vehicle is to transport loads of up to 45,455 kg from the support ship to precise alignment in the pipeline. This system is expected to remain on the bottom for maximum periods of six weeks. The use of a manned vehicle would be impractical, if assigned this kind of mission. Even if a series of vehicles, alternate crews and quick-change battery packs to provide continuity of effort on the sea floor were to be used, the magnitude of the tasks is beyond the capacity of any existing manned system.

SELECTION CRITERIA FOR VARIOUS TYPES OF VEHICLES

A task trade-off is necessary prior to making any decision on which type of vehicle is best suited to the particular mission requirements. Occasionally modification of the details of the mission may be possible that could in turn affect vehicle requirements. The selection of a submersible system should be made using suitability to task accomplishment as one criterion.

A second selection criterion to be considered is total cost. Owners can easily give day-rate costs comparable to a car rental agency. Based on his initial costs, operational costs and profit margins, a day-rate cost for a manned system will probably be significantly different from those applicable to unmanned vehicles. This is to be expected, but the initial costs of a manned vehicle must include certain features not required in an unmanned system. The personnel capsule, variable-buoyancy system and certification costs are the beginning of a potentially long list. The surface-support requirements of a manned system will in general tend to be greater than those of an unmanned system, because it is larger and heavier.

Day-rate costs alone are meaningless. The potential user should have a means of assessing the amount of meaningful work he can expect for his money, but unfortunately, there is no realistic guarantee available. Owners will talk in terms of operational days, weather days, system down-times, and offer prices for each. The amount of work the user can expect, the owners can argue, is under the user's control. A possible aid in determining the actual cost per unit of work might be to consult with past users of the system(s) under consideration. Under limited-budget operations cost may well be the deciding factor.

A third consideration is safety consistent with mission requirements. Human safety must, of course, be considered in any operation. Manned submersibles must be designed, constructed and tested in accordance with guidelines and requirements established by various groups and government agencies. Unmanned vehicles have no such requirements. A submersible pilot once said in response to a question regarding the hazardous nature of his work, that the most dangerous aspect of his job was driving the freeway to work. This statement may not be capable of being supported by statistics, but submersibles at that time had been free of any accidental deaths.

Losses to property must also be considered. Many unmanned vehicles are operated slightly positive in buoyancy, and maintain a submerged position using vertical thrusters. Should power interruptions be experienced, the vehicle, obeying Archimedes principle, will surface. At least one vehicle of this type has been lost having experienced a severed umbilical cable. The vehicle surfaced, but could not be located. Manned submersibles are fitted with various means at the pilot's command to cause positive buoyancy in an emergency situation. Entanglement with either type of vehicle on firmly fixed bottom obstructions presents a major problem, generally requiring the assistance of other vehicles to gain freedom.

SUMMARY

The potential user of submersible vehicles has a wide variety of types, sizes, capabilities, etc. from which to select the one best suited to his particular requirement. The selection criteria should include mission performance capabilities, as well as cost and safety considerations. Manned and unmanned submersibles each have unique roles in oceanographic pursuits based on their capabilities and limitations. The future of either system depends on the development of these capabilities through imaginative and creative use.

CHAPTER 4

MANNED AND REMOTELY OPERATED SUBMERSIBLE SYSTEMS: A COMPARISON

Howard Talkington

INTRODUCTION

The age of exploration is not over. Even as outer space beckons, the majority of our own planet remains unexplored: hydrospace has yet to be fully developed. The United States is now conducting a program which has a declared goal of developing, promoting, and supporting a national operational capability for man to work under the sea. The goals are to achieve a better understanding, assessment, and use of the marine environment and its resources. Whenever undersea work and exploration are discussed, manned systems engender the most attention and interest in the participants. Here it is that we must first ask, why man? Although manned systems are useful, exciting, and, many times, necessary, the majority of undersea tasks facing man can be accomplished more safely and economically, and as thoroughly, with remotely manned systems. Guidelines for making the decision to use a manned or unmanned system for the execution of a specific undersea task are proposed and explained.

UNDERSEA TASKS

Three examples of undersea tasks are presented to provide a context for the following paragraphs. These fall into the general categories of exploration, search and recovery, and work, but the vehicles described in each case can be used for other tasks as well. The projects cited are intended only to suggest the kinds of tasks that must be performed beneath the sea and the types of vehicles that might be available to accomplish them. The first example involves TRIESTE, which was utilized in the Navy's pioneering efforts in the field of deep ocean engineering. TRIESTE was the first successful manned, deep-diving, free-swimming submersible. It was an innovation because it enabled man to dive into the depths of the sea in the relative safety and comfort of a one-atmosphere pressure hull. Because the hull was heavy steel, it required a large gasoline-filled float to give the submersible an overall neutral buoyancy. For looking at the undersea world outside the TRIESTE there was one view port 10 cm in diameter in the steel pressure hull. This is the vehicle that carried man into the deepest part of the world's oceans — to the bottom of the Marianas Trench.

A later development of the design is TRIESTE II, presently the Navy's deepest-diving submersible. Improvements in the electronic, acoustic, photographic, and high-pressure systems have extended TRIESTE II's ability to operate in the deep ocean (Fig. 1).

For the second example we must return to early 1966 and to the Mediterranean Sea where it touches Spain near the village of Palomares. Two aircraft of the U.S. Strategic Air Command had collided in midair and scattered wreckage and four H-bombs around Palomares. Three of the bombs were quickly found on land, but the fourth was apparently lost in the sea. A fisherman had reported seeing a bomb-like object fall into the waves, and for almost three months search and recovery efforts were diligently pursued. The efforts embraced every way man can extend himself under the sea including divers as well as manned and remotely manned systems. While divers worked the relatively shallow water, the manned Perry submarines, ALVIN (Fig. 2) and ALUMINAUT, searched the deeper, more rugged areas. The U.S.N.S. MIZAR provided an instrumented, unmanned sled which enabled the searchers to examine a large area (about 65 km^2) to depths, if necessary, of 6,100 meters. MIZAR has a center well through which the sled is lowered and then towed at the selected depth.

The manned ALVIN twice found the lost bomb, and the remotely

Fig. 1. Design improvements have extended TRIESTE II's ability to operate in the deep ocean.

Fig. 2. The manned submersible ALVIN participated in the recovery of the H-bomb lost at sea off Palomares, Spain.

manned CURV I ("Cable-Controlled Underwater Recovery Vehicle") was used to recover it. CURV I (Fig. 3) had been developed for recovering test ordnance at the Naval Undersea Center's Long Beach and San Clemente Island test ranges to depths of 610 m. To meet the need at Palomares, CURV I was modified so it could work at greater depths.

The bomb was resting tenuously on a craggy slope at the brink of an undersea canyon, and the parachute still attached was drifting back and forth in the current. There were two dangers here for those attempting a recovery; (1) getting entangled in the parachute shrouds and (2) dislodging the bomb and possibly losing it deeper in the sea. When the bomb was first discovered, the ALVIN attached a marking pinger, but it became entangled and there were some anxious moments before ALVIN was able to disengage. After that incident ALVIN preferred not to become involved and the remotely manned CURV I made the necessary attachments raising the lost bomb to the surface from a depth of 869 m. This was an intricate, tense, and diagnostic example of different types of systems working together to conduct a successful operation.

The third example involves a complicated task for which a remotely manned system, CURV III was used. A major overhaul was scheduled for the Azores Fixed Acoustic Range (AFAR), and CURV III was selected as the underwater work platform. CURV III (Fig. 4), the most versatile in the

Fig. 3. CURV I raised the lost H-bomb to the surface.

Fig. 4. CURV III has proven to be a versatile, reliable system able to operate to depths of 2,300 m.

CURV series of remotely manned vehicles, has all the necessary equipment for searching for, locating, recovering, and documenting the recovery of a lost item, or the completion of a particular support task at depths to 2,300 m. This equipment comprised both active and passive sonar, two closed-circuit television systems, a 35-mm documentary camera and strobe, and an underwater lighting system. The standard work tool is an electro-hydraulically operated manipulator, but special work tools and equipment can be readily attached to the vehicle.

Before CURV III performed the tasks it was assigned to do at AFAR, engineers reviewed the requirements and supervised the special modifications which equipped CURV III to accomplish its mission. The tasks accomplished by CURV III at AFAR included: (1) rigging one of the 38-m acoustic towers so that it could be lifted from the sea floor; (2) cutting various underwater electric cables that were from 38 to 89 mm in diameter; (3) retrieving underwater electric cables from the ocean floor; (4) sonar mapping of the acoustic tower sites; and (5) inspecting the underwater range, once all the other tasks had been successfully completed.

WHY MAN?

While keeping the above examples of undersea tasks in mind, let us return to the question — why man? Man's attempt to learn about the world he lives in has most often been conditioned by the clash between desire and economics. What he wants to do usually far exceeds what he can afford to do. Columbus spent years in search of funding before he was able to finally set sail for the New World. The Apollo Program has become history and the absence of funds truncated the list of desired goals. In considering our goal of fully using the marine environment and resources, we must investigate the effect of putting man into a submersible system. How does he interact in the relationship between desire and economics? This question should be answered before any system is made the focus of time, effort, and money.

First, we must be honest with ourselves about ourselves. Man has the desire to see, to know, to be there. He has an ego and he wishes to leave his personal mark, he wants others to acknowledge that achievement, and then he pushes on. A flag could be planted on top of Mount Everest by dropping it from an aircraft, and that would indeed bring one level of satisfaction. However, to set the flag at the summit — after having scaled the heights of the icy mountain — that is the supreme satisfaction, the supreme accomplishment. This is the glory of a goal personally attained. That man is a searching, conquering, proud being must be taken into account. This conviction affects the thinking of everyone who establishes goals for an undersea project, especially those who always insist that man must be present at the work site. It is not being said here that this conviction is good or bad, but only that it exists and must be recognized.

Beyond the desire for personal accomplishment there are other reasons man should or could be included in an undersea work or exploration system. The poet, Dylan Thomas, has a line which reads "when all my five and country senses see." Man is a sensing creature possessing an integrated, coordinated, active intellect. When a man's trained intellect is part of a system, he is able to repair, reset, adjust, and adapt, and in short, to respond intelligently to the unusual situation. He can perform a variety of tasks because of his general orientation and versatility. The free-swimming diver comes closest to exercising directly his senses in the ocean (primarily seeing, touching, and hearing). The man in the manned submersible, however, is sensing his environment remotely, except for one sense — that of sight. In the unmanned system all sense data is perceived remotely. Thus, this system is "remotely manned," for man's intellect and senses are still a part of the overall system, but they are applied remotely to the work site. Therefore, the primary reason for placing man at the scene is to make use of his active, interpretive ability to see.

COST OF MANNED SYSTEMS

This seeing-man is the one that is placed in a manned system; but there should be irrefutable reasons for putting him here, because the cost is high for risking a human life in a hostile environment. There is the safety factor, which makes it necessary that the system sustain and support human life. Therefore, those funds which must be allocated to support man are not available for accomplishing the basic goal. In addition, an adequate life-support system substantially increases the weight and complexity of the whole system, and, therefore, the cost. Because manned systems are not powered from the surface, they require a self-contained power supply comprising special high-energy storage and charging systems. The power supply increases the weight and volume of the system, and it generates power for only a relatively short time. These factors thus severely limit mission endurance and together significantly affect the effectiveness of the system.

When man is in the system he must be protected from the hostile environment by a pressure hull. Since the pressure hull is usually made of steel, it becomes the largest, heaviest, and most costly part of a manned submersible. Once the manned submersible is constructed it must undergo man-rating certification. This procedure of tests and documentation is not only costly in itself, but it imposes necessary and costly design constraints that all support components and subsystems must meet. Along with the safety factor is the anxiety factor. When a manned vehicle becomes entangled there is a great deal of concern for the safety of those on board. However, if a remotely manned system had been entangled that parameter of anxiety would not have existed.

A man in a system also complicates the already difficult problem of handling. Because manned systems, in addition to being larger and heavier, require a special fail-safe handling capability and any accidental rough handling could result in injury or death. This handling capability also increases the cost of the system. Therefore, the following questions must be considered when designing a system for undersea tasks. Where do we need man in the system? Do we really require his presence at the work site? Could he be used more effectively at the surface by taking advantage of the longer mission-duration potential for instance?

Experience with the DEEPSTAR-4000 illustrates what has been said. Many dives made use of the man inside, by his ability to be an active observer, but that was not always so. To meet some specific test objectives, DEEPSTAR carried a full complement of scientific instrumentation (Fig. 5), including sound velocitymeters, salinometers, water sampling devices, and a coring device. It was noted that during many of the test dives the scientist inside the submersible was so busy that he never looked out the view port. Of course, the question must be asked: Did the "observer" need to be there on a site? He used none of his senses to learn about the environment. Could

Fig. 5. Manned submersible DEEPSTAR-4000 has carried a variety of instruments for taking oceanographic data; these data could be taken by remotely operated systems.

these particular tasks have been accomplished equally as well with a remotely controlled system as well or more safely and economically?

REMOTELY MANNED SYSTEMS AND USE OF THE OCEANS

Table I presents a list of ocean exploration and survey parameters compiled by the Panel on Platforms for Ocean Exploration and Surveying of the National Academy of Engineering's Marine Board. The list shows which parameters are pertinent at each of five separate levels: (1) the air—sea interface

TABLE I

Ocean exploration and survey parameters

Parameter	Air—sea interface (+10 to —10 m)	Upper water column (—10 to —500 m)	Lower water column (—500 m and deeper	Bottom	Sub-bottom
1. Ice	X				
2. Sea-swell-surf	X				
3. Surface meteorology	X				
4. Surge	X				
5. Tides	X				
6. Currents	X	X	X		
7. Hydrodynamic forces	X	X	X		
8. Noise	X	X	X		
9. Salinity	X	X	X		
10. Temperature	X	X	X		
11. Turbidity	X	X	X		
12. Biomass	X	X	X	X	
13. Nutrients	X	X	X	X	
14. Oxygen	X	X	X	X	
15. Pollutants	X	X	X	X	
16. Electrical		X	X	X	
17. Bathymetry				X	
18. Geomorphology				X	
19. Rheology				X	
20. Engineering properties				X	X
21. Geochemistry				X	X
22. Geology				X	X
23. Geothermal				X	X
24. Physical properties				X	X
25. Radiometric				X	X
26. Gravity					X
27. Magnetics					X
28. Seismic					X

(+10 m to —10 m); (2) the upper water column (—10 m to —500 m); (3) the lower water column (—500 m to bottom); (4) the ocean floor; and (5) the subbottom. This illustrates what scientists feel is necessary to better understand, assess, and use the marine environment and its resources. Not only are there so many parameters to be measured, but these must be measured in many areas of the world before the oceans can be fully utilized. Many measurements in many areas is the desired goal, but once again economics affects the degree of accomplishment. The panel concluded that buoy systems and unmanned systems should be used whenever possible, because they would enable scientists to get the maximum amount of information for the available dollars. This method would avoid the expense of using a manned system such as DEEPSTAR when the only responsibility of those on board is to ferry the instrumentation to the appropriate level to gather data. In the final analysis, when man is put into a system there must be a specific, necessary purpose for having him there, and he must achieve that purpose.

SONODIVER AND SPARBUOY

Buoy and unmanned systems are available now for the data gathering that will yield the information most useful to man. Two such systems are SONO-

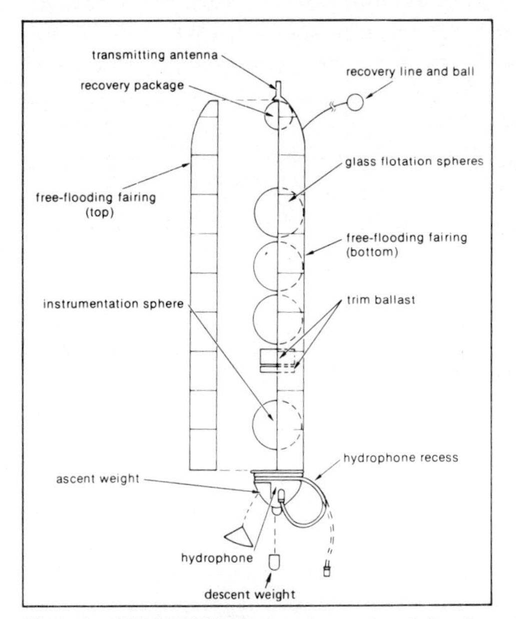

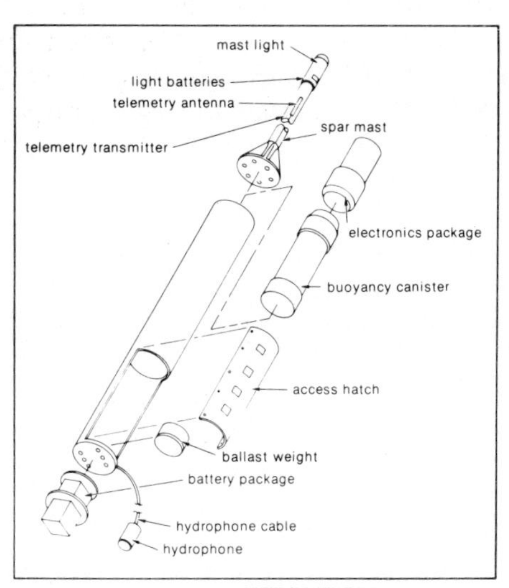

Fig. 6. SONODIVER is approximately 3 m long and 0.46 m in diameter. It releases weights to hover, then returns to the surface.

Fig. 7. SPARBUOY's main unit is the same size as SONODIVER, but it carries a mast 6 m long.

DIVER and SPARBUOY. SONODIVER is a buoyancy-actuated system designed to gather acoustic and other environmental data at predetermined levels to a depth of 6,100 m. It is approximately 3 m long and 0.46 m in diameter (Fig. 6). In operation, SONODIVER is first launched. Then it descends, releases its descent weight, hovers, takes data, releases its ascent weight, and returns to the surface. Its data are recorded on magnetic tape that can be played back aboard the support ship after recovery.

SPARBUOY is a surface unit that deploys a hydrophone to depths up to 100 m. The hydrophone is decoupled from wave action by the catenary configuration of its cable. SPARBUOY is the same size as SONODIVER but carries a mast 6 m long, and transmits data continuously to shipboard recorders (Fig. 7). When the two units are used together, SPARBUOY's data help to determine whether changes in ambient noise measured by SONODIVER are caused by changes in depth or are the result of a general variation in the ambient noise level.

SEAPROBE

Another example of the present capabilities of unmanned systems is that provided by SEAPROBE. The SEAPROBE ship has a drillstring with an instrument pod attached which has a built-in large manipulating capability. This system demonstrated that man can work at extreme ocean depths and that he can extend his senses of hearing and seeing and his manipulative abilities, from the safety of the surface to the location requiring his attention. SEAPROBE has operated effectively and proved to be a very good remotely operated system. For example, it has successfully completed a task which required its capabilities for the handling of array systems in the Bahamas.

CURV AND RUWS

Other remotely manned systems exist in a variety of shapes and sizes depending upon their intended applications. The larger systems include CURV III, discussed on pp. 79—81, and RUWS ("Remote Unmanned Work System"). They were constructed under the Deep Ocean Technology Program for experimental tasks in the deep ocean. The dimensions of CURV III are 2.1 by 2.1 by 4.9 m. It weighs 2,040 kg in air and can operate to depths of 2,300 m. The vehicle is designed so that all its major operational components can be disassembled and installed on any surface craft with adequate deck space. This feature has enabled the vehicle to perform successfully under emergency conditions. When the manned submersible PISCES III sank off Cork, Ireland, in 1973, CURV III was flown from North Island

Naval Air Station to Cork together with its support equipment and crew by two U.S. Air Force C-141 transports (Talkington, 1974). The men and equipment embarked on the Canadian Coast Guard Ship JOHN CABOT. It reached the location of the sinking less than 48 hours after the Naval Undersea Center was asked to assist in the rescue effort. In very rough water, estimated as sea state six, CURV III found the downed submersible at a depth of 460 m and attached a line by which it was raised. The two men aboard were recovered in good condition. This operation, performed from a ship of opportunity under harsh time constraints and in bad weather, demonstrated CURV III's versatility in a gratifying fashion.

RUWS, unlike CURV III, is not tethered directly to its support ship. This experimental system includes a primary cable termination (PCT) frame that serves as a launch and recovery platform for the work vehicle (Fig. 8), which then operates on a buoyancy tether to distances up to 300 m from the PCT. The dimensions of PCT are 1.5 by 1.8 by 3.8 m, while those of the work-vehicle are 1.2 by 1.8 by 3.8 m. The total weight of the system is approximately 1,600 kg. The goal of this program is to provide a vehicle to operate at depths of as much as 6,100 m, thereby providing access to more than 98% of the ocean floor.

Fig. 8. RUWS ("Remote Unmanned Work System") consists of two major units, the primary cable termination, shown here at the right, and the work vehicle. The design goal is a depth capability of 6,100 m.

Fig. 9. The RUWS control console aboard its support ship is representative of those used with large, remotely operated systems.

RUWS is designed in modules so that components can be interchanged for specialized experiments. The work-vehicle carries two manipulators, one a heavy grabber and the other a highly articulated manipulator. Additional major components include television cameras, incorporating a head-coupled system giving the remote operator a sense of being present at the work site, and other instrumentation required for the successful completion of its tasks (Fig. 9).

SNOOPY

Small, lightweight submersibles are typified by the SNOOPY vehicles. ELECTRIC SNOOPY is intended primarily to provide a remotely controlled underwater observation vehicle. Although it is only 1.07 m long and 0.76 m wide and weighs approximately 90.7 kg in air, it can operate to depths of 460 m. A similar vehicle, NAVFAC SNOOPY, has been designed for use by the Naval Facilities Engineering Command during ocean construction work.

Fig. 10. One of the smallest remotely operated systems is the Hydroproducts RCV-125. This vehicle weighs only 82 kg.

SNOOPY carries a neutrally buoyant reel and strong, lightweight, Kevlar line for implanting and recovering items from the sea floor.

The basketball-sized RCV-125 ("Remotely Controlled Vehicle") shown in (Fig. 10) was developed by Hydroproducts, Inc. The vehicle weighs 82 kg in air and can operate to a maximum depth of 2,000 m. It carries a low-light level television camera and two tungsten halogen lamps. Four electric motors give the vehicle mobility in all directions.

CONCLUSIONS

It is recognized that to meet the challenge of making a thorough and effective use of the marine environment and its resources, a full complement of manned and remotely manned systems will be required. However, it is imperative that remotely manned systems be used as much as possible. These systems are better suited to most undersea work and exploration tasks for at least six reasons: (1) relative economy of development in time and equipment costs when compared with manned systems; (2) unlimited operational endurance on the working site because of the cable link to the surface; (3) control and coordination at the surface of project efforts (this avoids a clash of operational philosophies because he who is on the surface is in command);

(4) ability to perform in hazardous areas without endangering personnel; (5) ability to change or modify all system components to meet individual tasks without affecting system safety or certification status; and (6) ease of changing crews without disrupting the mission. Men simply leave their places at the control consoles and immediately their replacements are available to continue the operation. In addition, because these systems are usually smaller and lighter, as well as remotely manned, the handling problem is significantly reduced.

Man should be included in a system only if he is absolutely necessary for the success of the mission, because his presence in a system drastically increases its cost. This cost is reflected not only in dollars, but also in more safety considerations, system complexity, handling problems, and time. Also, if man's presence is necessary for a successful mission, it is most likely because the mission requires real-time, high-resolution sight. A corollary to this observation is that, if a man is needed for seeing, then provide him with a system which offers maximum visibility (Forman, 1971).

The Navy's TURTLE and SEA CLIFF (Fig. 11) are versatile research submersibles capable of performing search, recovery, photographic, and scientific tasks to depths of 1,980 m. However, they have only relatively small view ports through which the observer can exercise his ability to see. Some-

Fig. 11. Manned research submersible SEA CLIFF is a versatile vehicle.

thing different from this type of submersible configuration is often required. At the present time there is a group of fully instrumented submersibles (Figs. 12—15), which also provide maximum or panoramic visibility. Among this group are the totally transparent-hulled NEMO (Fig. 12), SEA-LINK (Fig. 13) and MAKAKAI (Fig. 14). NEMO is the first fully operating and certified submersible having an acrylic hull. It is a self-contained system with a one-atmosphere environment, and carries its crew of two on missions to depths of 180 m. Its acrylic sphere affords the crew the all-round visibility that makes NEMO a superb observation platform. SEA-LINK also has an acrylic sphere providing the required visibility, in addition to a welded aluminum hull permitting diver transport and lock-out. SEA-LINK is designed to operate at depths more than 900 m and can enable a team of three divers to work

Fig. 12. Manned submersible offering panoramic visibility: NEMO.

Fig. 13. Manned submersible offering panoramic visibility: SEA-LINK.

at 500-m depths. MAKAKAI, "eye of the sea," lives up to its name and also uses a transparent acrylic sphere as its pressure hull, which permits all-round visibility. The MAKAKAI (Murphy et al., 1971a,b) is a two-man, free-swimming submersible with an operating depth of 180 m. Its two pi-pitch cycloidal thrusters give the submersible a cruising speed of 0.3—0.4 m/sec with a maximum speed of 1.5 m/sec. At cruising speed MAKAKAI can operate for 6 hours.

Additionally, several manned submersibles have been constructed with very large ports of transparent materials such as acrylic or glass (Stachiw, 1971, 1974). The Perry submersibles PC-8, 14, and 15, several of the later Hyco (International Hydrodynamics Co.) submersibles, and the U.S. Navy's DEEPVIEW (Fig. 15) all fall into this category. The Perry PC-8 is typical of the commercial boats. Equipped with navigation and control instrumentation, a communication system, and a manipulator arm, PC-8 can operate to depths of 230 m for 2 hours of continuous running at a maximum speed of 2 m/s or for 8 to 10 hours at 0.5 m/s. DEEPVIEW, a two-man submersible with a

Fig. 14. Manned submersible offering panoramic visibility: MAKAKAI.

transparent bow, is the first submersible to make use of massive glass as a significant portion of the pressure hull. Its nose is a large glass hemisphere 38 mm thick. DEEPVIEW currently operates to a depth of 33 m at speeds

Fig. 15. Manned submersible offering panoramic visibility: DEEPVIEW.

from 0.50 to 1.5 m/s for 6 hours. As viewports cast from glass ceramic or chemically surface strengthened glass become available submersibles such as DEEPVIEW may dive deeper than the 610-m limit of acrylic plastic hulls. Ceramic windows 200 mm in diameter have already been fabricated for use in unmanned systems. (Stachiw, 1974)

When man is required in a system for his active seeing ability, these are the types of systems that make him most effective.

SUMMARY

In summary, this paper has acknowledged the overall goal of developing, promoting, and supporting a national operational capability for man to work under the sea to achieve a better understanding, assessment, and use of the marine environment and its resources. At the same time, some of the particular data requirements are noted in Table I that have to be met if the overall goal is to be attained. It gives examples of tasks confronting various systems as the marine environment becomes increasingly available to man. Then the questions were asked, Why man? Why do we need man in a system? Or, more specifically, *where* in the system should the man be? Should he operate at the work site, or remotely, from a surface craft? Why does he want to be at the scene? Is he necessary? Is he superfluous? The answers to these questions reveal that all exploration, research, and work represent a compromise between desire and economics.

Man's desires in undersea exploitation exceed his ability to pay for them. To put man under the sea entails high costs in terms of money, time, and system complexity. Thus, the following conclusions are reached. Both manned and unmanned systems are necessary to attain the overall goals. However, it is obligatory that unmanned systems be considered first and used whenever and wherever possible. Man should be considered for systems only if it is essential to the mission's success. Furthermore, what makes man essential in a system is his ability to provide active, real-time, high-resolution sight. Therefore that system should enable him to exercise this ability to the greatest degree. As Aristotle wrote in his *Metaphysics*, Book I:

> "All men by nature desire to know. An indication of this is the delight we take in our senses; for even apart from their usefulness they are loved for themselves; and above all others the sense of sight. For not only with a view to action, but even when we are not going to do anything, we prefer seeing (one might say) to everything else. The reason is that this [seeing], most of all the senses, makes us know and brings to light many differences between things."

Therefore, if man must be in the system, give him visibility — panoramic visibility! But regardless of how useful, exciting, and necessary manned systems may be, the majority of undersea tasks facing man can be performed more safely, economically, and as thoroughly, with remotely operated, unmanned systems.

REFERENCES

Forman, W.R., 1971. The DEEP VIEW Submersible. In: Engineering in the Ocean Environment, Record of the 1971 IEEE Conference Held at San Diego, Calif., 21—24 September 1971. Institute of Electrical and Electronic Engineers, New York, N.Y.

Murphy, D.W. et al., 1971a. A systems description of the transparent hulled submersible. In: Eighth U.S. Navy Symposium on Military Oceanography, 1: 200—224.

Murphy, D.W. et al., 1971b. The transparent hull submersible MAKAKAI. In: Engineering in the Ocean Environment, Record of the 1971 IEEE Conference Held at San Diego, Calif., 21—24 September 1971. Institute of Electrical and Electronic Engineers, New York, N.Y.

Stachiw, J.D., 1971a. Spherical acrylic pressure hulls for undersea exploration. J. Eng. Ind., 93.

Stachiw, J.D., 1971b. Acrylic pressure hull of submersible NEMO. In: Am. Soc. Mech. Eng., Fourth Underwater Technol. Conf., Houston, Texas, Pap. No. 71-UnT-2.

Stachiw, J.D., 1971c. Development of acrylic plastic submersible hulls with panoramic visibility. In: Engineering in the Ocean Environment, Record of the 1971 IEEE Conference Held at San Diego, Calif., 21—24 September 1971. Institute of Electrical and Electronic Engineers, New York, N.Y.

Stachiw, J.D. and Gray, K., 1971. Procurement of safe viewports for hyperbaric chambers. American Society of Mechanical Engineers. J. Eng. Ind.

Stachiw, J.D. and Gray, K., 1971. Acrylic pressure hull for JOHNSON—SEA-LINK submersible. Am. Soc. Mech. Eng., Annu. Winter Meet. Washington, D.C. Pap. No. 71-WA/UnT-6.

Stachiw, J.D. and Gray, K., 1974. Transparent structural materials for underwater research and exploration. Ind. Atom. Spat., 3.

Talkington, H.R., 1974. The U.S. Navy participation in the rescue of PISCES III. Mar. Technol. J. 8 (No. 1).

Talkington, H.R., 1974. The rescue of PISCES III. U.S. Nav. Inst. Proc., 100 (No. 4).

Section II. Case histories of diversified uses

CHAPTER 5

THE ROLE OF SUBMERSIBLES IN A UNIVERSITY

Richard A. Geyer and Thomas Bright

INTRODUCTION

A submersible can, by providing unique capabilities, play a significant role in supporting a wide variety of teaching and research objectives of a university. However, it should be considered primarily as still another highly specialized instrument to be used to help to achieve them. It may be considered as analogous, for example, with a special-purpose computer used in teaching and research in the basic branches of classical oceanography. Similarly in ocean engineering, it would be comparable to a facility such as a small wave tank or flume. A submersible can be used also to provide information for graduate students to apply to their theses and dissertations.

One unique capability is to provide a means to observe directly oceanographic phenomena and processes which otherwise must be studied remotely and less effectively, using instruments or underwater television. A submersible can also implant specialized instrumentation in and on ocean sediments and within the water column, as well as monitoring and replacing them.

The term submersible covers a wide variety of underwater vehicles with respect to size, depth capabilities and number of personnel who can be transported for varying durations and at different speeds. Attempts to define the term more specifically are made in chapters 2 and 3. In general, a support ship is also required to operate this type of vehicle; and the significant cost of the submersible and of the support vehicle must be considered in the ultimate justification of its use for a given task. Therefore, the first factor to consider is the optimum size, and depth capabilities needed to permit it to efficiently make a significant contribution toward achieving some of the primary objectives and missions of a university.

The problems for which submersibles can best provide solutions are becoming more sophisticated. No longer is the objective merely to see how far down one can go into the depths of the ocean and observe strange-looking creatures or phenomena. A wide variety of problems need to be solved in the major branches of oceanography and ocean engineering which do not require submersibles with great depth capability, size and endurance. Hence the initial as well as operating costs and the size of the support vessel need not be inordinately large. The era of large submersibles with great depth and duration capabilities which started in the late 1960's and continued to the early 1970's is drawing to a close. Some of the early submersibles of this period could operate at depths of as much as 5,000 m, carry a crew of 6 men and drift submerged for as long as 30 days. Daily operating costs including support vessels of some of the larger submersibles were of the order of $15,000 per day.

BRIEF DESCRIPTION OF OTHER TYPES OF SUBMERSIBLES

Specific examples for comparative purposes include ALUMINAUT, built by the Reynolds Corp., having a maximum operating depth of 5,000 m, BEN FRANKLIN built by the Grumman Corp., weighing 150 tons and carrying six men submerged for a period of 30 days, and the Lockheed Corporation's DEEP QUEST, with an operating depth of 2,600 m. Only ALVIN, with a maximum depth capability of 4000 m, operated by Woods Hole Oceanographic Institution remains in limited operation at a cost, including the support vessel of approximately $1,000,000 a year, with funds provided by ONR, NSF and NOAA.

Submersibles built and operating during the past few years generally have more realistic operating depths and are much smaller in size. This markedly reduces both the original cost, depreciation, normal operating expenses and the cost of the support vessel. They are generally capable of operating at depths of from 350 to 600 m, and vary in length between 7 and 10 m and weigh between 5,000 and 20,000 kg. Two, and at most, three people can stay submerged up to eight hours cruising at speeds of 1—3 knots. Many are equipped with hydraulic arms or manipulators that can implant instruments on or into the ocean bottom, retrieve lost objects, and can take rock as well as biological samples, as well as bottom cores.

DIAPHUS (Fig. 1) costing $160,000 in 1973 operated by the Department of Oceanography at Texas A & M University belongs in this category. Its specifications and capabilities are as follows:

Manufacturer: Perry Submarine Builders.
Name: DIAPHUS; length: 6 m; weight: 4540 kg.
Payload: 400 kg; depth capability: 365 m.
Life-support capability: 96 hours for two people.
Speed: up to 3 knots.
Power: 36 V DC system.
Passengers: carries one pilot and one observer.
View ports: seven 20-cm view ports in conning tower and one large 90-cm hemispherical view port in nose.
Photography: one externally mounted strobe-light which will couple with standard camera systems located inside the submersible.
Scientific sampling: hydraulic manipulator arm with four functions: forward—aft, left—right, in—out, claw open—claw close, (Fig. 2).
Communications: underwater telephone, ship to submersible, citizens band radio, ship to submersible.
Dive duration: 4 to 8 hours depending on electrical power used.

Fig. 1. DIAPHUS aboard support ship R/V GYRE. It is hoisted out of the water and "A" frame starts to retract bringing the submersible directly over the cradle. Then the submersible hoist operator lowers it into its place on deck. The "V" shaped anti-swing yoke attached to the "A" frame beneath its override block prevents the submersible from swinging unduly during launch and retrieval.

Although the space for personnel is small, the vessel is comfortable enough for two grown men to occupy, even on extended dives. The controls and instruments are compactly and conveniently arranged (Fig. 3) and there is ample space aft of the pilot to carry several hundred kilograms of additional gear and equipment. Therefore, a number of blank through-hull connectors were provided during construction. The 36-V power plant, composed of banks of 12-V electric vehicle batteries is capable of supporting dives of up to 8-hours duration with a considerable reserve of power left. The four-day life-support system for two people is a significant safety factor.

If the scientists had to designate the one most outstanding feature of the submersible, it would be the acrylic hemispherical forward window seen in Fig. 2. This large window provides excellent visibility forward and to a large extent laterally. Visibility aft, however, is only through the small conning tower windows seen in Fig. 1.

When not at sea, DIAPHUS and its support van are kept in a warehouse-like building at the Texas A & M Research Annex, located several kilometers

Fig. 2. Hydraulic manipulator arm (with a sponge in its claw), sample basket with "doggie door", subsea photographic strobe attached to sample basket, and floodlamp attached to bow plane.

from the main campus at College Station, Texas (240 km inland). It is transported under contract to whatever shore base is to be used on a standard lo-boy flatbed truck. In the past, the research-oriented submersible operations have been conducted using the 58-m long Texas A & M Oceanography Department research vessel R/V GYRE.

Mobilization aboard R/V GYRE involves installation on deck of the submersible support van, hoist, and two air tugger winches. The submersible and its cradle and an anti-swing yoke are located on the after "A" frame of the ship. A period of one or two days total mobilization time is required. A portable handling system is currently being developed to launch and retrieve DIAPHUS from vessels other than R/V GYRE.

Underway, the submersible is chained to the deck at all times, except

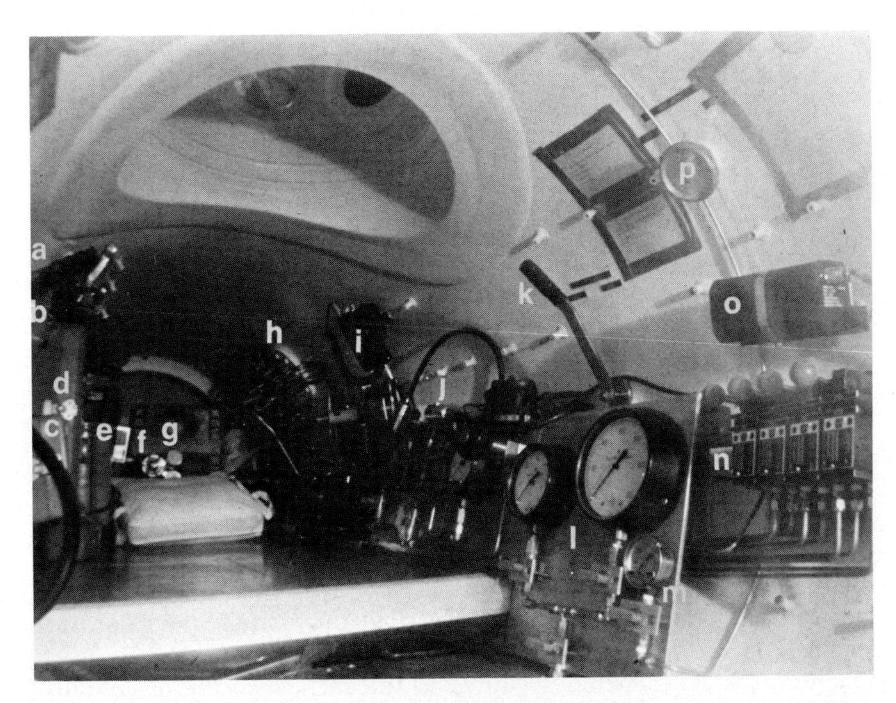

Fig. 3. Inside DIAPHUS: *a* = rudder control handle; *b* = propulsion control box; *c* = leak detector box; *d* = power panel; *e* = CO_2 scrubber (upright, cylindrical); *f* = O_2 flow meter; *g* = O_2 regulator; *h* = through-hull penetrator plate; *i* = citizens-band radio (underwater telephone not shown); *j* = compressed-air system; pressure guage and controls with emergency breathing regulators attached; *k* = bow plane control handle; *l* = two depth guages; *m* = hydraulic-pressure gauages; *n* = hydraulic manipulator arm controls; *o* = gyrocompass; *p* = port main ballast tank vent control (between registration papers and observer instructions taped to hull). Pilot usually sits on a box directly beneath conning tower which has windows allowing visibility in all directions and straight up. Recorders and other scientific equipment are generally kept aft of the pilot (where the cushion is in this picture). The observer sits next to the large forward viewport.

during launch, dive and retrieval. Undoubtedly the most critical points in the entire operation of the submersible are launches and retrievals.

LAUNCHING PROCEDURES

A procedure aboard R/V GYRE was developed permitting launch and retrieval of DIAPHUS from the ship in up to approximately 1.5-m seas. This is a fair-weather operation at best in the Gulf of Mexico. Personnel required includes one deck-chief with a walkie-talkie radio, two on-deck tag-line handlers at the air tuggers, a swimmer (or two men in an outboard-powered "ZODIAC" rubber boat with a walkie-talkie radio), a submersible hoist operator, an "A" frame operator, a ships' officer on the bridge with a walkie-talkie, and the submersible pilot with a citizens-band radio inside the submersible. Prior to launch the submersible receives a thorough pre-dive check by the pilot. At the deck-chief's direction the submersible is unchained, and hoisted out of its cradle with the submersible hoist. It is then extended over stern with the "A" frame, as the tag-line handlers hold it parallel to the stern of the ship. DIAPHUS is then lowered into the water with the submersible hoist, unhooked by the swimmer, or "ZODIAC" crew, and maneuvered away quickly by the pilot.

RETRIEVAL PROCEDURES

Retrieval is basically the reverse, except that much longer nylon tag-lines on two air-tuggers are attached to the submersible while it is 5—10 m aft of the anchored ship. The submersible is then pulled sideways towards the stern of the ship and the lifting cargo hook is attached as quickly as possible. The moment the air-tuggers have the submersible directly behind the stern of the ship, the submersible hoist and "A" frame lift it out of the water and into its cradle (Fig. 1).

After launch, the submersible is guided at the surface to the desired dive site by a communicator on the ship. Tracking of the submersible at the surface is by visual range and bearings taken from the deck of the ship and/or by radar. Upon reaching the dive site DIAPHUS' 225-kg capacity main ballast tanks are vented and flooded. The dive begins, underwater telephone communication is checked and a proper rate of descent is obtained until the bottom is reached. For the duration of the dive, pilot functions include maneuvering the submersible according to a pre-determined dive plan. This plan is subject to modifications by the observer inside the submersible, or by the scientist-in-charge on the deck of the ship. Routine checks are made by the pilot of (1) depth, (2) oxygen, (3) CO_2 levels, (4) cabin pressure, (5) leak detectors for the battery pod and main pressure hull, (6) battery voltage, (7) magnitude of electrical grounds, (8) compressed air pressure, and (9) the functioning of various fans, lights, gauges, etc. If communications are lost or unintelligible for more than 30 minutes during a dive the pilot will surface the submersible. This is accomplished by blowing water out of the main ballast tanks with compressed air.

TRACKING PROCEDURES

Tracking of the submersible while it is on a dive has been done by taking visual or radar ranges and bearings from the ship on a floating tether ball attached to the submersible by a byoyant 1-cm polypropylene line (fig. 2 of

Chapter 6). This rather simple system is only as accurate as estimates from the ship and, therefore, leaves much to be desired in establishing exact positions for scientific purposes. Moreover, the tether line has twice parted from the submersible during dives. As soon as the budget permits, an acoustical underwater range and bearing device to improve our tracking capabilities will be obtained. Other methods for tracking and navigation are described in Chapters 7 and 12.

There is much more to operating the submersible than may be apparent from this description. Each activity mentioned could be expanded considerably. Details of preparation, maintenance and repair between dives often result in the pilots and deck crew spending twelve to sixteen hours a day working to accomplish four to eight hours of dive time. Between dives it is necessary to (1) recharge batteries, (2) recharge the high-pressure air system, (3) rinse and lubricate external components of the submersible, (4) check all systems, (5) repair any minor electric or mechanical breakdowns, (6) adjust lead ballast for next dive, (7) remove scientific equipment not to be used for the next dive, (8) add that which will be used, and (9) inform the U.S. Coast Guard of major changes in dive schedule, etc.

The operating cost of a submersible designed specifically for the Department of Oceanography at Texas A & M University which has both teaching and research commitments is of the order of $1,600 a day, not including the support vessel.

ADVANTAGES OF USING DIAPHUS AS COMPARED WITH CONVENTIONAL METHODS

Where scientific observations are concerned DIAPHUS has provided a major methodological breakthrough for some of the biological and geological oceanographers on the Texas A & M faculty. For years prior to purchase of the submersible studies of the Texas Outer Continental Shelf fishing banks were conducted using (1) corers, (2) grabs, (3) dredges, (4) hook-and-line fishing, (5) spearfishing, (6) rotenone poisoning, (7) observation and photography by SCUBA divers and (8) underwater television. The diver operations, although extensive, were restricted generally to depths less than 30 meters. It also has been demonstrated in Chapter 13 that these are not cost-effective. Results of observations made by Bright and Rezak in 1972 from the submersible NEKTON GAMMA to depths of 150 m at the West Flower Garden Bank convinced them of the usefulness of submersible operations for their purposes. Lateral and vertical zonation of the banks became immediately clear. Unexpected details of biotic community distribution were revealed and very important selective samples were taken in a short period of three days' diving. This has been the pattern of the results using DIAPHUS on the other banks since 1974 (Bright et al., 1976).

Basically, the unique advantages are the opportunities to (1) view the substrata and biota directly, (2) to photograph, (3) describe, (4) record them on audio and video tape; and (5) selectively sample them at a more or less leisurely pace down to depths at which divers are not effective. In pursuit of these objectives, the instruments most used have been (1) a portable black-and-white video recording system, (2) various audio cassette recorders, (3) various standard cameras, (4) hydraulic manipulator arm and (5) sample basket (Fig. 2). Video recordings are used to document wholly or in-part transects made with the submersible across a study area. Video and audio records have ultimately provided a bulk of the data. They also serve as later references to the time and position of samples and color photographs taken, thereby more-or-less tying together all of the observations. Color photographs are a necessary complement to video records, because much visual information is lost in black-and-white video and, comparatively, resolution on the portable video systems is much inferior to that of photographs.

INSTRUCTIONS FOR DATA DOCUMENTATION

The greatest difficulty from a scientific standpoint has been that of determining later exactly where and at what depth the submersible was, at a certain point. This fact is reflected in the following basic instructions devised for correlation of observations made by scientists using the submersible:

The critical elements to be correlated in order to derive a "picture" of the study area are: (1) the observation (verbal, television, photographic, sample collected); (2) the depth; and (3) the position at which the observation was made. Items (1) and (2) can be linked directly by the observer while in the process of taking and recording observations inside the submersible. Item (3), however, can only be correlated with the other two on the basis of time synchronization. Many hours of analysis and a great deal of uncertainty concerning positions and depths of observations can be avoided if the observer and pilot follow the procedure outlined below in recording observations and data on audio and video tape.

(1) Label tape with date, area of operation, cruise number, dive number, time, pilot's name, observer's name.

(2) Report time leaving surface.

(3) Report times and depths of observations made during descent.

(4) Report time reaching bottom, depth, bottom type and observations.

(5) Report time, depth and pertinent observations about every five minutes.

(6) Report time and depth for each underwater telephone communication.

(7) Report time and depth for each stop.

(8) Report time and depth when getting underway each time.

(9) Report time and depth for each series of verbal descriptions.

(10) Report time, depth, and label designation for each frame or sequence of frames of still photographs taken and for each scene of movie film taken.

(11) Report time and depth for each scene of TV tape taken.

(12) Report time, depth and adequately describe each sample taken.

It has been found that audio taped, video taped and still and motion picture photographed observations are complementary and all very useful. The video and audio tapes, however, prove to be most significant in later correlation of all types of observations. Use the audio tape as continually as possible, even if it is left running the entire time.

Records of observations taken as described above can be readily correlated with timed position fixes for the submersible by the support vessel. These fixes will be acquired every five or ten minutes depending on the requirements of the observer for scientific purposes.

Some very useful standard observations should be entered on the video and audio tapes along with the time-depth reports. These include: (1) water temperature, (2) visibility, (3) current direction and approximate magnitude, (4) bottom type and (5) color.

A submersible program at a university is not a highly subsidized operation and funds are limited. The Texas A & M program is, however, viable for the following reasons:

(1) The submersible is totally paid for.

(2) The pilots do not depend on the submersible program for their basic salaries, insofar as they devote only part-time to submersible operations.

(3) The system is relatively simple and easily maintained.

(4) Operating expenses are low.

(5) No profit-motive exists.

Accordingly, the operating charge of $1,600/day plus mobilization, demobilization, and ship costs has been sufficient to allow the submersible to more than pay its way within the Oceanography Department's overall ship and equipment operation. After its scheduled 1976 operation a few useful items will be added to the inventory of accessory equipment. There are, however, no outstanding debts to be liquidated and the policy is not to overextend financially in pursuit of immediate improvements to the submersible system. In this way the services of DIAPHUS can be offered to the faculty and students without imposing a fiscal burden on the Oceanography Department's overall ship and equipment operations.

SAFETY PROCEDURES

It is important to emphasize again that safety must be a paramount consideration in the design and operation of a submersible by a university because of its teaching responsibility. Students especially must be protected from the hazards of this type of operation in this hostile environment, because their participation is not always voluntary. In addition to the usual safety precautions found on a submersible for returning it to the surface in an emergency, the life-support system for DIAPHUS has been expanded to provide for a maximum of 96 hours. A back-up communication system with the support ship is also available. Diving plans similar to air-craft flight plans are filed with the proper branch of the U.S. Coast Guard, well in advance of actual operations, and a constant plot is maintained aboard the support vessel of movements of DIAPHUS while submerged.

SPECIFIC USES OF DIAPHUS

DIAPHUS, named after a deep-sea lantern fish which has a similar appearance, was completed in the spring of 1974 and carries the PC-14 designation. Since then, more than 200 dives have been made at several dozen locations in the Gulf of Mexico and in the channel off the Catalina Islands in California. The variation in duration of individual dives has been between 50 minutes and 8 hours of continuous operation for a cumulative total of more than 200 hours. The major portion of this time was used in research projects. A detailed discussion is presented in Chapter 6 of the methods used and the results obtained by one of the major research programs using DIAPHUS extensively. This research was and is directed primarily toward obtaining information helpful in establishing ecological base lines on the continental shelf off the coast of Texas and Louisiana (see fig. 1 of Chapter 6 for principal areas surveyed). The results are incorporated in preparing environmental impact statements. A sufficient amount of submersible time was also made available to several advanced graduate students to obtain material from surveys of the continental shelf of the Gulf of Mexico for use in their theses and dissertations.

Abbott (1975) used underwater television tapes and direct observations while making dives in DIAPHUS (1974) and NEKTON-GAMMA (June 1972). These provided an accurate description of the faunal composition of the Algal Sponge Zone of the Flower Garden Banks in the northwest Gulf of Mexico. Submersible surveys were also conducted by Abbott (1975) over the area surrounding Stetson Bank. These were hampered occasionally by poor visibility, but a gradational surface sediment pattern was still evident. Several surveys extended over 300 m beyond the base of the bank revealed a soft bottom of fine, flat-laying sediments. On a traverse toward the bank, an increase in the coarseness of the texture of the surface sediments was observed. The debris consists apparently of coarse gravelly rubble eroded from the bank. The flat, soft bottom surrounding the bank was devoid of any noticeable plant life. Several ill-defined trails, patterned burrows, sea urchins, and an occasional shell were observed as transects were made toward the bank. Nearer the bank, sparse sponge colonies were observed increasing in abundance towards the base of the bank and slopes. Other biological activity including the number of fish, increased in this direction.

The runners at the base of the submersible were allowed to drag in the soft sediment on the more distant submersible transverses. Penetration was approximately 12 cm and a firmer sediment below the overlying fine layer was revealed. A study of the cores showed this firm sediment to be a conglomerate. Water content of the fine layers was determined to be high, based on the response of the sediment during the plowing action of the submersible, but well-defined tracks were not formed. Visual observations of the sediment distribution from the submersible indicated no apparent differences between the two sides of the bank.

Sampling by DuBois (1975) over Stetson Bank was conducted primarily with the use of DIAPHUS. A total of ten dives were made resulting in six transects across the Bank. Transect paths were designed to cover as much previously unexplored area as possible in the time available. Where echinoderm species seen from the submersible were identifiable in situ, their presence, location and abundance were recorded on either audio or video tape. Specimens were collected, either because no preserved representative of the species was available or for confirmation for identification purposes. These were picked up with the submersible's hydraulic manipulator arm and placed in the sample basket beneath the observation dome. Additional sampling was done with SCUBA, rock dredges, box cores and van Veen grab samplers.

Documentation of events, phenomena and processes from within the submersible was made using a Sony video recorder, and still, as well as movie cameras. A Sony tape recorder was used to provide verbal descriptions. The purpose of this research was to compare the distribution of a coral community with that of a nearby rock on the Texas Continental Shelf.

Algal nodules were collected by Hogg (1975) from various depths in the algal nodule zones of the East Flower Garden Banks using the manipulator arm and sample basket of DIAPHUS.

Observations of in-situ distribution of the nodules were made in four ways:

(1) Direct observations made from DIAPHUS. This was by far the most effective means of observation.

(2) Audio-visual video tapes taken from DIAPHUS.

(3) Audio-visual video tapes taken from the NEKTON-GAMMA.

(4) An underwater TV camera towed over the banks at a very slow speed by a surface vessel.

It was concluded from these studies that the algal nodules of the Flower Garden Banks develop and are distributed on the Banks primarily because of activities of organisms in the nodule zone which are in turn influenced by geological processes and physical forces. All of these factors are interrelated and interdependent as are the organisms which make up the nodule community.

Another major scientific use is for a program to be conducted by Bryant in 1976 (personal communications) to study the settling rate of pipelines involving research on marine geotechnical properties of the bottom sediments near the Mississippi River.

STUDY OF NATURALLY OCCURRING HYDROCARBON SEEPAGES IN THE GULF OF MEXICO

Many oil slicks have been reported off the Texas and Louisiana coasts by various government agencies in bulletins and shown on charts as far back as the beginning of the twentieth century. Archaeological reports and pottery from kitchen middens of the Karankawa Indians of Texas offer evidence that since pre-Colombian times tar was used to make their pottery waterproof or for decorative purposes. Also, early Spanish explorers caulked their ships with tar found along the beaches of Padre Island, Texas before starting on the long voyage back to Spain.

DIAPHUS has been active in an extensive inter-disciplinary research program to study naturally occurring hydrocarbons in the Gulf of Mexico. Surface vessels carrying seismic sub-bottom profilers and gas "sniffers" are used for the reconnaissance portion of this program. An illustration of naturally occurring gases emanating from a seep on the floor of the ocean, obtained by a seismic sub-bottom profiler, is shown in Fig. 4. [The subsurface geologic structure associated with the seep may be seen at the bottom of the record.] A chart of the northwest Gulf of Mexico showing locations of a number of shallow topographic features and reefs investigated during this study using the submersible appears in fig. 1 of Chapter 6.

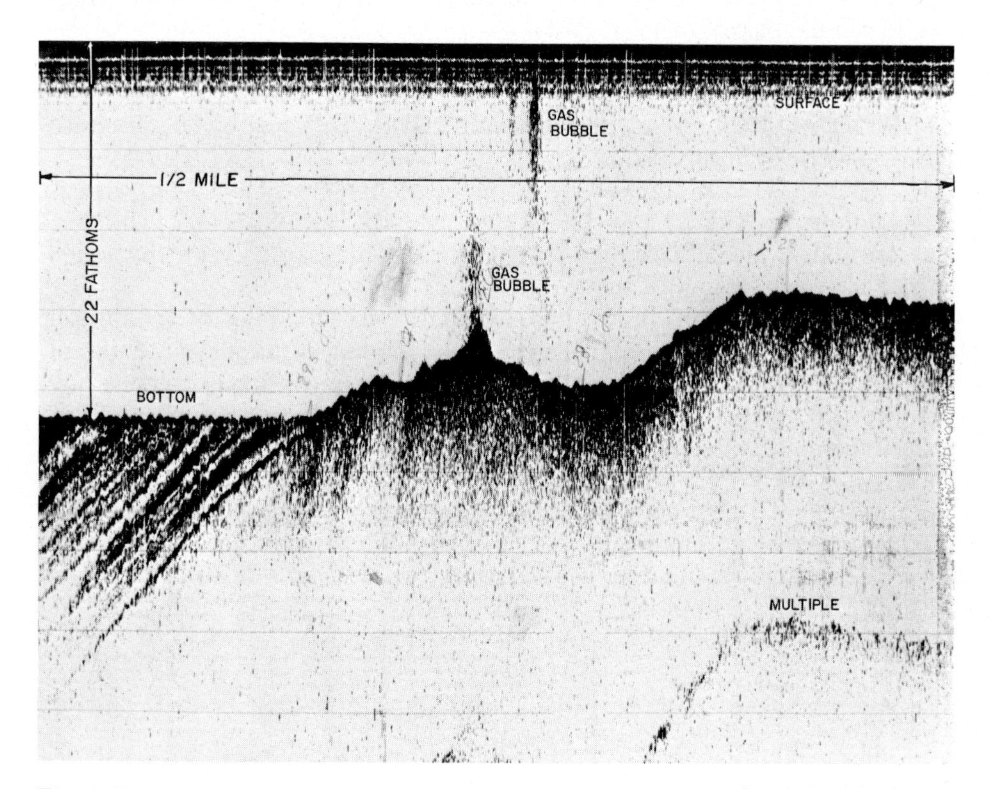

Fig. 4. Evidence of gas emanating from a naturally occurring seep on the bottom of the Gulf of Mexico on a seismic sub-bottom profiler record.

Details of the results are presented in a series of technical reports as well as publications in the scientific literature. These include a report by Bright et al. (1976) describing the biota of drowned reefs associated with natural gas seeps in the Gulf of Mexico: and a publication by Geyer and Sweet (1973) reviewing the results of the first three years of this research program. Evidence for a naturally occurring oil seep based on a seismic sub-bottom profiler record appears in Fig. 5. It is located 60 km southeast of Galveston, Texas in the Gulf of Mexico at a depth of 22 m.

In areas where evidence for oil or gas seeps is found by these means, detailed observations (Fig. 6) are then made using DIAPHUS; and some idea can then be obtained as to the areal extent of the naturally occurring hydrocarbon seepages. It was determined from the submersible, that the streams of gas bubbles emanating from the subsurface formation are aperiodic rather than continuous. This is a very important fact affecting the accuracy of computations estimating the amount of gas escaping.

Another important advantage of using a submersible in this study is to facilitate documenting with still and motion picture cameras the reaction of fish to these upward migrating gas bubbles. It became evident that they are not affected adversely by them; and in fact, they appear to be oblivious

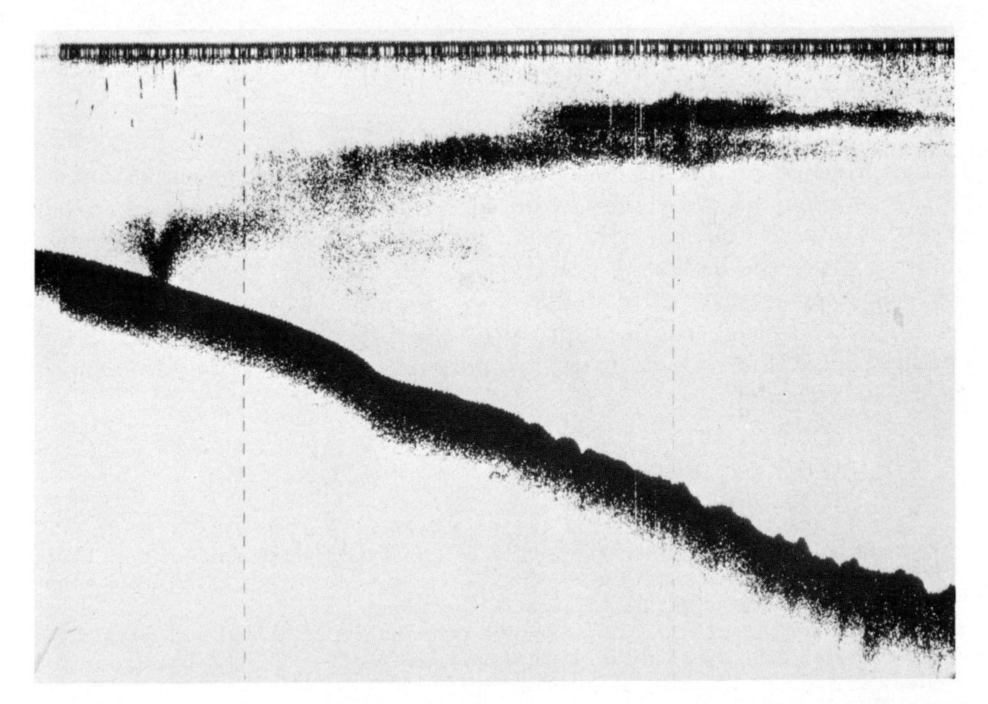

Fig. 5. Evidence of a naturally occurring oil seep in the Gulf of Mexico on a seismic sub-bottom profiler record.

Fig. 6. Visual evidence of a naturally occurring gas seep in the Gulf of Mexico in the upper Antipatharian Zone (55 m).

to the presence of the bubbles. Attempts were made in the earlier phases of this program to study the gas bubbles using a television camera traling over the side of the ship. However, the results obtained in this manner did not yield the definitive results achieved using DIAPHUS nor could any significant quantitative results be obtained. These results can be used to reach meaningful decisions with regard to promulgating realistic oil and gas production regulations as they pertain to the effect of hydrocarbons on the environment.

REFERENCES

Abbott, R.E., 1975. The Faunal Composition of the Algal—Sponge Zone of the Flower Garden Banks, Northwest Gulf of Mexico. Thesis, Department of Oceanography, Texas A & M University, College Station, Texas, 205 pp.

Abbott, R.E. and Bright, T.J., 1975. Benthic communities associated with natural gas seeps on carbonate banks in the Northwestern Gulf of Mexico. Technical Report for Study of Naturally Occurring Hydrocarbons, 191 pp.

Bright, T.J., Rezak, R., Bouma, A., Bryant, W. and Pequegnat, W., 1976. A biological and geological reconnaissance of selected topographical features of the Texas continental shelf. Final report to the U.S. Dept. Interior, Bureau of Land Management, OCS Office, New Orleans, La., 342 pp.

DuBois, R., 1975. A Comparison of the Distribution of a Coral Community with that of a Nearby Rock Outcrop on the Texas Continental Shelf. Thesis, Department of Oceanography, Texas A & M University, College Station, Texas, 153 pp.
Geyer, R.A. and Sweet, W.E., 1973. Natural hydrocarbon seepage in the Gulf of Mexico. Trans. Gulf Coast Assoc. Geol. Soc., pp. 158—169.
Hogg, D., 1975. Formation, Growth, Structure and Distribution of Calcareous Algal Nodules on the Flower Garden Banks. Thesis, Department of Oceanography, Texas A & M University, College Station, Texas, 59 pp.

CHAPTER 6

RECONNAISSANCE OF REEFS AND FISHING BANKS OF THE TEXAS CONTINENTAL SHELF

Thomas J. Bright and Richard Rezak

INTRODUCTION

Extensive scientific studies were conducted during 1974—1976 over the major commercial fishing banks and coral reefs found at considerable distances offshore on the Texas—Louisiana Outer Continental Shelf (Fig. 1). The results of these survey and research efforts in which the submersible DIAPHUS (Fig. 2) played a significant and critical role are described in this chapter. The design characteristics of this submersible are discussed in some detail in Chapter 5, as well as operating techniques, including launch and retrieval methods, and documentation of the visual and tangible data obtained.

The purpose of this research was to conduct "baseline" studies to obtain a better understanding of existing ecological conditions at these fishing banks and coral reefs. This information in turn is to be used in developing criteria to promulgate realistic regulations pertaining to the drilling for and production of oil and gas, as well as transporting these hydrocarbons to terminals and refineries along the Gulf Coast.

BENTHIC MACROBIOTA AND FISHES

Biota of the reefs and hard banks in the northwestern Gulf of Mexico (Fig. 1) are easily distinguishable into at least four assemblages, all of which are faunally linked (Table I):

(1) The sparse Claypile Bank biota (35—55 m) consists of predominantly low-growing filamentous and leafy algae and sponges with occasional "meadows" of high-standing leafy algae occupied by numerous fish.

(2) The more diverse Stetson Bank and Three Hickey Rock biota (28—56 m) dominated by the hydrozoan fire coral *Millepora alcicornis* and sponges.

(3) The highly diverse and abundant Flower Gardens and Twenty-eight Fathom Bank biota with coral reefs (22—49 m) algal-nodule and sand-covered platforms (45—76 m) and drowned reefs (76—100+ m).

(4) The deep-water biota of the South Texas Fishing Banks (56—78 m) and Fishnet Bank (61—82 m).

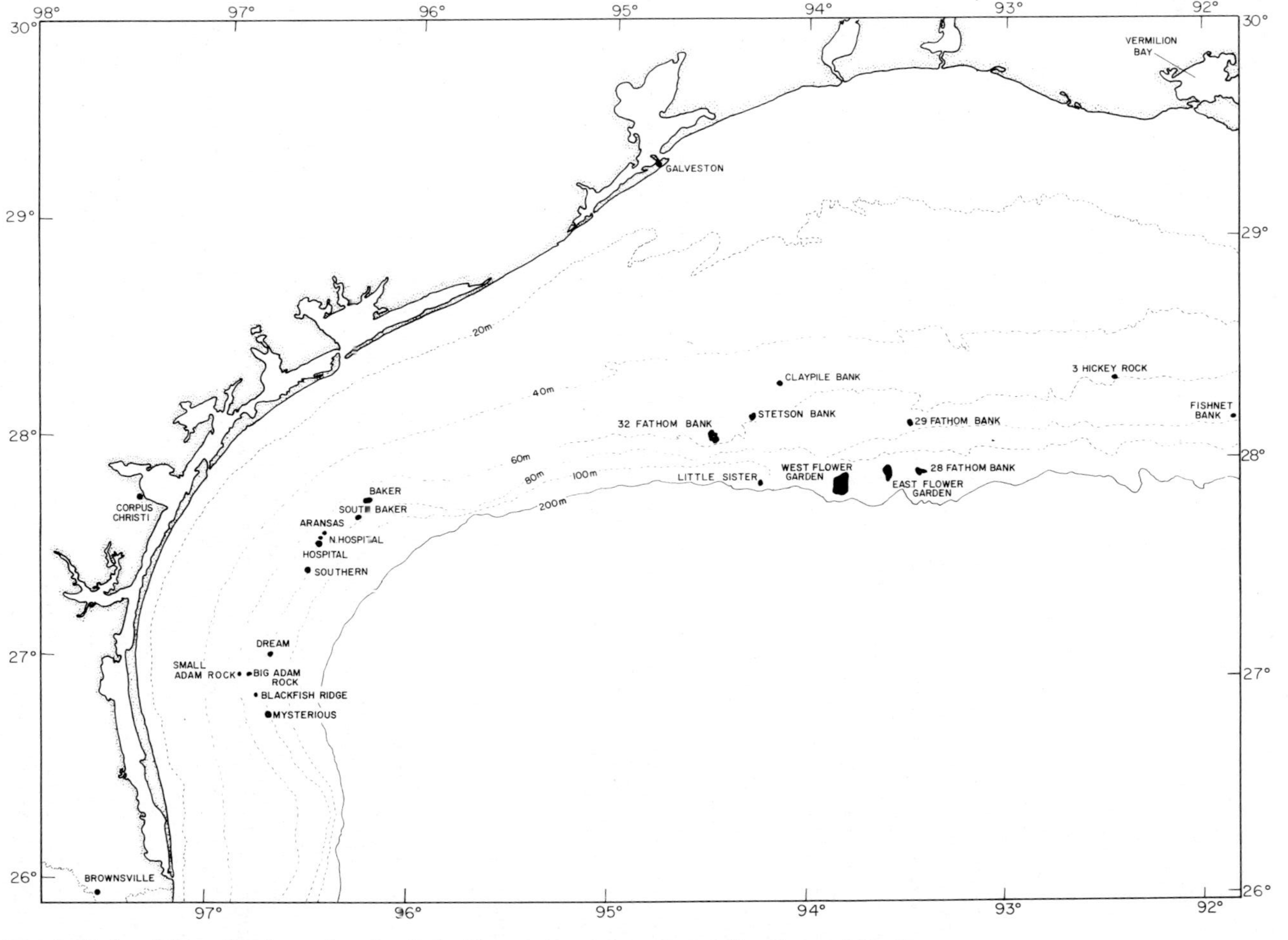

Fig. 1. Major fishing banks and reefs of the Texas—Louisiana Outer Continental Shelf.

Fig. 2. Research submersible DIAPHUS. Note polypropylene line attached to the submersible and leading to a floating "tether buoy" at the surface.

FLOWER GARDEN BANKS

Bright and Pequegnat (1974) listed over 250 species of benthic invertebrates and more than 100 fishes from the West Flower Garden, the distinctive biotic zonation of which is basically the same as that of the East Flower Garden (Fig. 3). Above 45—49 m both banks are covered with thriving submerged coral reefs which, except for their total lack of shallow-water alcyonarians, are good examples of the *Diploria—Montastrea—Porites* community which is so common on reefs in the Caribbean and southern Gulf of Mexico (Figs. 3—7).

Unlike the West Flower Garden the East Flower Garden harbors sizeable knolls occupied almost entirely by populations of the small branching coral *Madracis mirabilis* (*Madracis* Zone) (Figs. 3, 8, 9). Finger-sized remains of dead *Madracis* are extremely important components of the sediment on and adjacent to the reef. In some cases the coarse carbonate sand which typically occurs between coral heads in the *Diploria—Montastrea—Porites* Zone is entirely supplanted by *Madracis* rubble.

Other knolls at the East Flower Garden are covered completely by lush growths of leafy algae including *Caulerpa*, *Chrysymenia*, *Halymenia*, *Gloiophlaea*, *Lobophora*, *Microdictyon*, and others (Figs. 3, 10, 11). The presence at the East Flower Garden of this Leafy Algae Zone and the *Madracis* Zone,

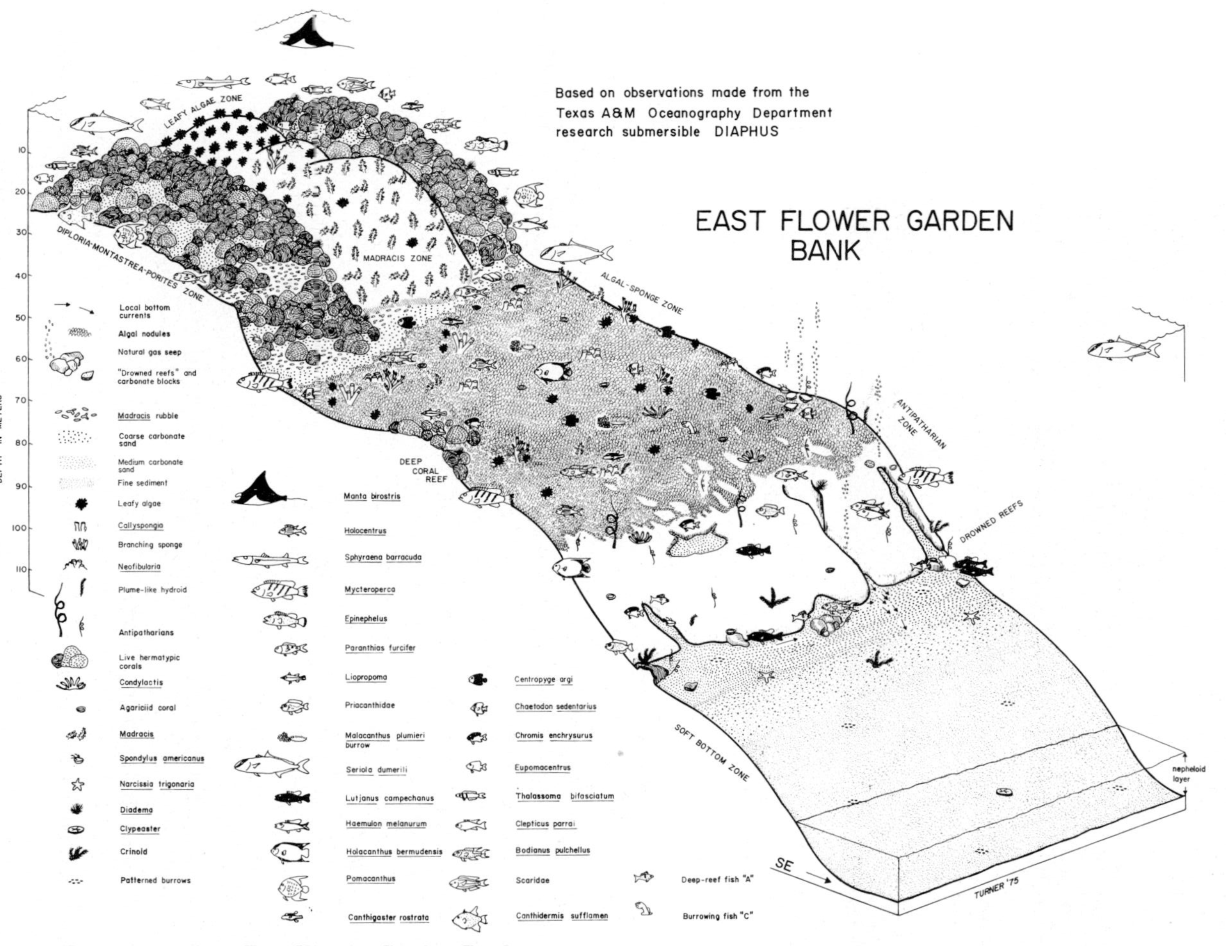

Fig. 3. Biotic zonation, East Flower Garden Bank.

Fig. 4. Coral reef at West Flower Garden Bank (25 m depth).

Fig. 5. Brain coral, Spanish hogfish (*Bodianus rufus*), Bluehead (*Thalassoma bifasciatum*), Brown chromis (*Chromis enchrysurus*) and other reef fishes at West Flower Garden (25 m depth).

Fig. 6. Extreme closeup of expanded polyps of the coral *Montastrea* sp. at night at the West Flower Garden (25 m depth).

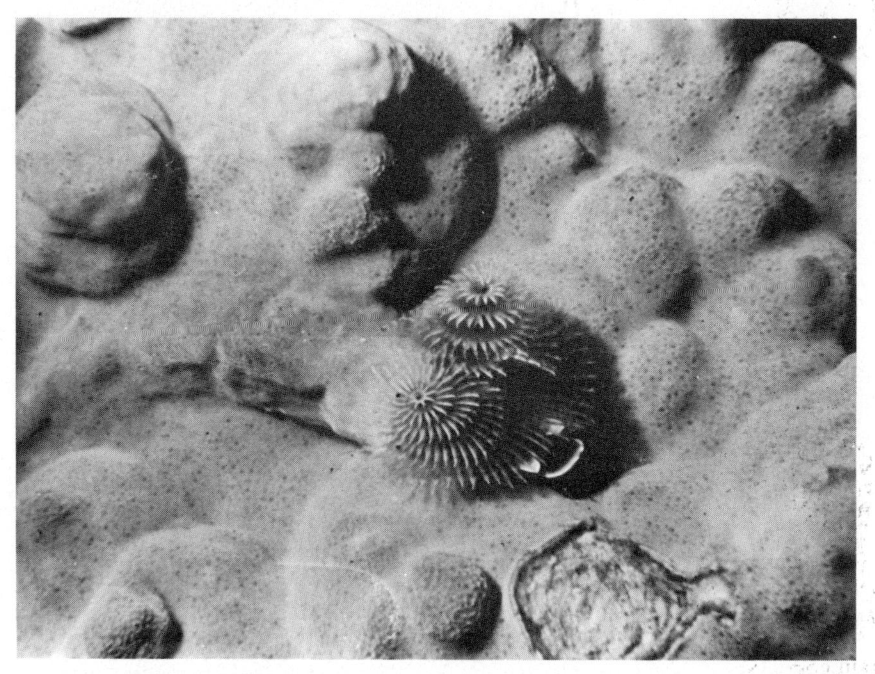

Fig. 7. Feather duster worm, *Spirobranchus giganteus* in tube embedded in fire coral *Millepora alcicornis* at West Flower Garden (25 m depth).

Fig. 8. *Madracis* field at top of East Flower Garden (27—30 m depth).

Fig. 9. *Madracis* in the field pictured above (27 m depth).

Fig. 10. Field of leafy algae (26 m depth).

Fig. 11. Leafy algae in field pictured above (26 m depth).

Fig. 12. The branching sponge *Verongia* (just below manipulator arm claw) leafy algae and *Madracis* clumps (25 m depth).

Fig. 13. *Neofibularia* and *Madracis* clumps (25 m depth).

Table I. Organisms encountered by us at various fishing banks, with indications of relative abundances.
**** very abundant, *** abundant, ** moderate population, * known to be present, p - presumed present.

	Claypile	Stetson	East Flower Garden and West Flower Garden				South Texas Fishing Banks	Fishnet
Depths in meters	35-55	28-56	22-49	49-76	76-92	92-107	56-78	61-82
CORALLINE ALGAE								
encrusting		*	****	***	**	*	*	*
nodules				****	*			
Lithothamnium spp.			****	****	**			
Lithophyllum spp.		*	**	**	*		*	*
CALCAREOUS GREEN ALGAE			*	**				*
LEAFY ALGAE	**		****	***	*		*	*
FORAMINIFERS (encrusting)								
Gypsina plana		*	*	***	*			
SPONGES	**	***	**	***	**	*	**	**
Agelas sp.		*	**	*	*	*	*	
Callyspongia spp.	**	*	**	**				
Ircinia campana							**	
Neofibularia nolitangere	*	**	*	***	**		*	
Verongia spp.	*		**	***				
PLUME-LIKE HYDROIDS		*		**	**	**	**	
ALCYONARIAN WHIPS								
Ellisella sp.					*	*		
ALCYONARIAN FANS					**	**	**	
Hypnogorgia sp.	*						*	*
Scleracis sp.				*	*		**	
Thesea sp. "B"					*			
Thesea sp. "A"							*	
ANTIPATHARIANS			*	***	***	**	***	***
ANEMONES								
Condylactis sp.			*	***	**			
HYDROZOAN CORALS								
Millepora alcicornis		***	***	*				
ANTHOZOAN STONY CORALS								
Stephanocoenia intersepta		*	*					
Madracis decactis		*	**					
Madracis mirabilis			****					
Madracis asperula				*				
Madracis brueggemanni				*			*	
Agaricia agaricites agaricites			**					
Saucer-shaped agariciid			*	**	*		*	
Helioceris cucullata			*	*				
Siderastrea sidera	*		*					
Porites astreoides			****					
Diplurla strigosa			****					
Diploria spp.			****	**				
Colpophyllia natans			***					
Colpophyllia spp.			***					
Montastrea annularis			****					
Montastrea cavernosa			****	**				
Scolymia sp.			*					
Mussa angulosa			*					
ahermatypic solitary sp. "A"							*	
ahermatypic solitary sp. "B"						*		
ahermatypic solitary		*						

as well as knolls of intermediate biotic composition which bear various types of sponges, *Madracis* clumps, patches of leafy algae and extensive encrustations of coralline algae (Figs. 12, 13), is indicative of a greater degree of

	Claypile	Stetson	East Flower Garden and West Flower Garden				South Texas Fishing Banks	Fishnet
Depths in meters	35-55	28-56	22-49	49-76	76-92	92-107	56-78	61-82
GASTROPODS								
Busycon spp.							*	*
Cypraea spp.		*	**				*	
PELECYPODS								
rock borers	***	****						
Jouannetia quillingi		****						
Lithophaga bisculcata		*						
Lima sp.			***					
Spondylus americanus			***	***			***	
BRACHIOPODS								
Argyrotheca barrettiana					***	***	**	
POLYCHAETE WORMS								
Hermodice sp.			**	**			***	****
Spirobranchus giganteus		**	****	**				
CRABS								
Carpilius corallinus			*					
Stenorynchus seticornis		**	**	**	**	*	*	
LOBSTERS								
Panulirus sp.			*					
Scyllarides sp.			*				*	
MANTIS SHRIMPS								
Gonodactylus spp.		*	*	***				
STARFISH								
Narcissia trigonaria				*	*			
BASKET STARS								
Gorgonocephalidae					**		**	**
SEA URCHINS								
Clypeaster sp.		**				**		
Diadema antillarum		****	****		*		*	**
SEA CUCUMBERS								
Isostichopus sp.	*	**	*				**	
CRINOIDS				*	****	****	****	****
PATTERNED BURROWS	****	****		**	****	****	****	****
FISHES								
Ginglymostoma cirratum								
Nurse shark			*					
Manta birostris								
Manta ray			**					
Gymnothorax moringa								
Spotted moray			*	*				
Gymnothorax spp.								
moray eels		*	*	*				
Aulostomus maculatus								
Trumpetfish			**					
Holocentrus ascensionis								
Longjaw squirrelfish			**					
Holocentrus spp. (large species)	*	**	***	**	**		**	
Holocentridae (small species)		**	**	*			**	*
Myripristis jacobus								
Blackbar soldierfish		*	*					
Sphyraena barracuda								
Great barracuda		***	***	*			***	

lateral biotic variability on the approximately 28-hectare crest of this bank than is found at the West Flower Garden where the *Diploria—Montastrea—Porites* Zone predominates everywhere above 45—49 m (approximately 40

Depths in meters	Claypile 35-55	Stetson 28-56	East Flower Garden and West Flower Garden 22-49	49-76	76-92	92-107	South Texas Fishing Banks 56-78	Fishnet 61-82
Epinephelus adscensionis Rock hind	**	***	***	**				
Epinephelus cruentatus Graysby			**	*				
Epinephelus spp. hinds	**		***	**				*
Dermatolepis inermis Marbled grouper			*	*	*			
Mycteroperca spp. groupers		**	**	**	**		*	**
Paranthias furcifer Creolefish		***	***	**			*	
Serranus annularis Orangeback bass				**				
Liopropoma sp. basslet		**	*	**			**	**
Priacanthidae bigeyes	**		*	**	*		**	**
Priacanthus arenatus Bigeye		*	*					
Apogon spp. cardinalfishes		*	***				**	
Malacanthus plumieri Sand tilefish			*	**	**			
Sand tilefish burrows			*	**				
Amblycirrhitus pinos Redspotted hawkfish			**	*				
Rachycentron canadum Cobia							*	
Caranx spp. jacks		**	**	*			**	
Caranx ruber Barjack			**					
Seriola dumerili Greater amberjack	**	**	**	**	*		**	**
Selene vomer Lookdown							*	
Scomberomorus spp. mackerels		*	*				*	
Lutjanus campechanus Red snapper	p	***	*	***	***	***	***	***
Lutjanus spp. snappers (not Red)		*	*				*	
Rhomboplites aurorubens Vermilion snapper	p	***	*				***	***
Haemulon melanurum Cottonwick		**	*	**	**		**	
Calamus spp. porgys		**	**	**			*	
Equetus spp. drums	*	*	*	*	*	*	*	
Equetus acuminatus High hat		*		*			*	
Equetus lanceolatus Jackknife-fish			*	*	*	*	*	
Mulloidichthys martinicus Yellow goatfish		*	**	*				
Pseudupeneus maculatus Spotted goatfish	**	**	**	*				
Chaetodon aculeatus Longsnout butterflyfish			*	*				
Chaetodon ocellatus Spotfin butterflyfish		**	***					
Chaetodon sedentarius Reef butterflyfish	*	***	***	***	*	*	***	***
Pomacanthus spp. angelfishes		***	***	*			*	
Pomacanthus paru French angelfish	**	***	***	*				
Pomacanthus arcuatus Gray angelfish		*	*					

	Claypile	Stetson	East Flower Garden and West Flower Garden				South Texas Fishing Bank	Fishnet
Depths in meters	35-55	28-56	22-49	49-76	76-92	92-107	56-78	61-82
Holacanthus ciliaris Queen angelfish		**	**	**			*	
Holacanthus bermudensis Blue angelfish	*	**	*	**	*	*	**	**
Holacanthus tricolor Rock beauty			**	**	*	*		*
Centropyge argi Cherubfish			*	***				
Eupomacentrus spp. damselfishes		***	***	*			**	
Eupomacentrus partitus Bicolor damselfish			***	*				
Chromis cyaneus Blue chromis		*	***					
Chromis multilineatus Brown chromis		**	***					
Chromis enchrysurus Yellowtail reeffish	*	****	***	****	***	***	****	****
Bodianus rufus Spanish hogfish		**	***					
Bodianus pulchellus Spotfin hogfish		**	**	***	*	*	***	***
Halichoeres garnoti Yellowhead wrasse			*					
Thalassoma bifasciatum Bluehead		****	****	*				
Clepticus parrai Creole wrasse			****					
Scarus spp. parrotfishes			**	*				
Sparisoma viride Stoplight parrotfish			**	*				
Gobiosoma sp. sharknose goby		***	****	***			*	
Acanthurus spp. surgeonfishes		**	**	*				
Acanthurus coeruleus Blue tang		*	*					
Balistes capriscus Gray triggerfish		*	*					
Balistes vetula Queen triggerfish		**	**	*			*	
Canthidermis sufflamen Ocean triggerfish			***					
Melichthys niger Black durgon			***					
Lactophrys triqueter Smooth trunkfish			**					
Canthigaster rostrata Sharpnose puffer		***	***	***			*	
Ogcocephalus vespetilio Longnose batfish				*				
Deep-reef fish "A"					***	***	***	*
Fish "B"							**	
Burrowing fish "C"	****	**		**			*	

hectares). It is evident from Table I that the coral reefs at the East and West Flower Gardens (22—49 m) house more species of fishes and epifauna, particularly corals, than do zones deeper on the same banks or on other banks at any depth. It is interesting, however, that the authors have rarely encountered commercial snappers (fishes of the family Lutjanidae) on the coral reefs, although the Red snapper, *Lutjanus campechanus*, is abundant on the lower reaches of the Flower Garden Banks around rocks and drowned reefs. Large groupers, *Mycteroperca* spp., are more uniformly present at all depths. However, due to a general reduction in fish abundance and numbers of species in the deeper zones compared to the coral reefs, large groupers appear more conspicuous around topographical irregularities below 50 m. On the other hand, the smaller groupers and hinds, *Epinephelus* spp., although common on the coral reef, are not often seen at greater depths.

STETSON, THREE HICKEY ROCK AND CLAYPILE BANKS

In comparison to the biotic populations of the Flower Gardens above 49 m, those of the banks occupying similar depth ranges elsewhere on the shelf (Stetson, Three Hickey Rock, Claypile) are less diverse and numerically smaller. Stetson Bank (Fig. 14), with a crest depth of about 28 m, supports an epifaunal community dominated by the hydrozoan stony coral *Millepora alcicornis*, sponges, and the rock-boring pelecypod *Jouannetia quillingi* (Figs. 15—17) (Neumann, 1958; Bright et al., 1974). The substratum at Stetson is siltstone and claystone in various stages of induration, most of the outcrops being soft claystone and are easily perforated by the abundant rock borers (Fig. 17). A majority of the surface area of the rock is bare (Figs. 18, 19). Where epifauna occurs it is generally restricted to the upper halves of the outcrops, the degree of coverage varying from zero to 100% (Fig. 15). The authors have never visited Three Hickey Rock but have viewed photographs which indicate obvious similarities to Stetson Bank in the nature of the biota. *Millepora* and sponges appear to predominate. In general, it is felt that Stetson Bank and probably Three Hickey Rock are manifestations of a hard-bank assemblage composed primarily of a limited number of the species occurring on the coral reefs at the East and West Flower Gardens with notable deficiencies in the populations of anthonzoan corals and fishes. With regard to the important commercial and sport fishes, however, Stetson Bank appears to be comparable (Table II).

Possibly because the crest of Claypile Bank is somewhat deeper, approximately 35 m, a *Millepora*—Sponge assemblage such as that occupying Stetson has not developed. The fact that Claypile's outcrops are much lower in relief than those at Stetson may also be significant in this respect. The benthic community which has developed at Claypile Bank is a rather limited one, composed primarily of several species of leafy algae and a sparse population of sponges. The presence of numerous rock borers, possibly pelecypods, reflects the similarity of the outcropping rocks to those of Stetson. In places on Claypile Bank sizeable meadows of leafy algae resembling *Sargassum* were recorded on videotape but none was collected and the identification is speculative. The greatest concentrations of fishes were seen in and over these meadows. There are also unconfirmed verbal reports of other *Sargassum* bearing banks off Louisiana.

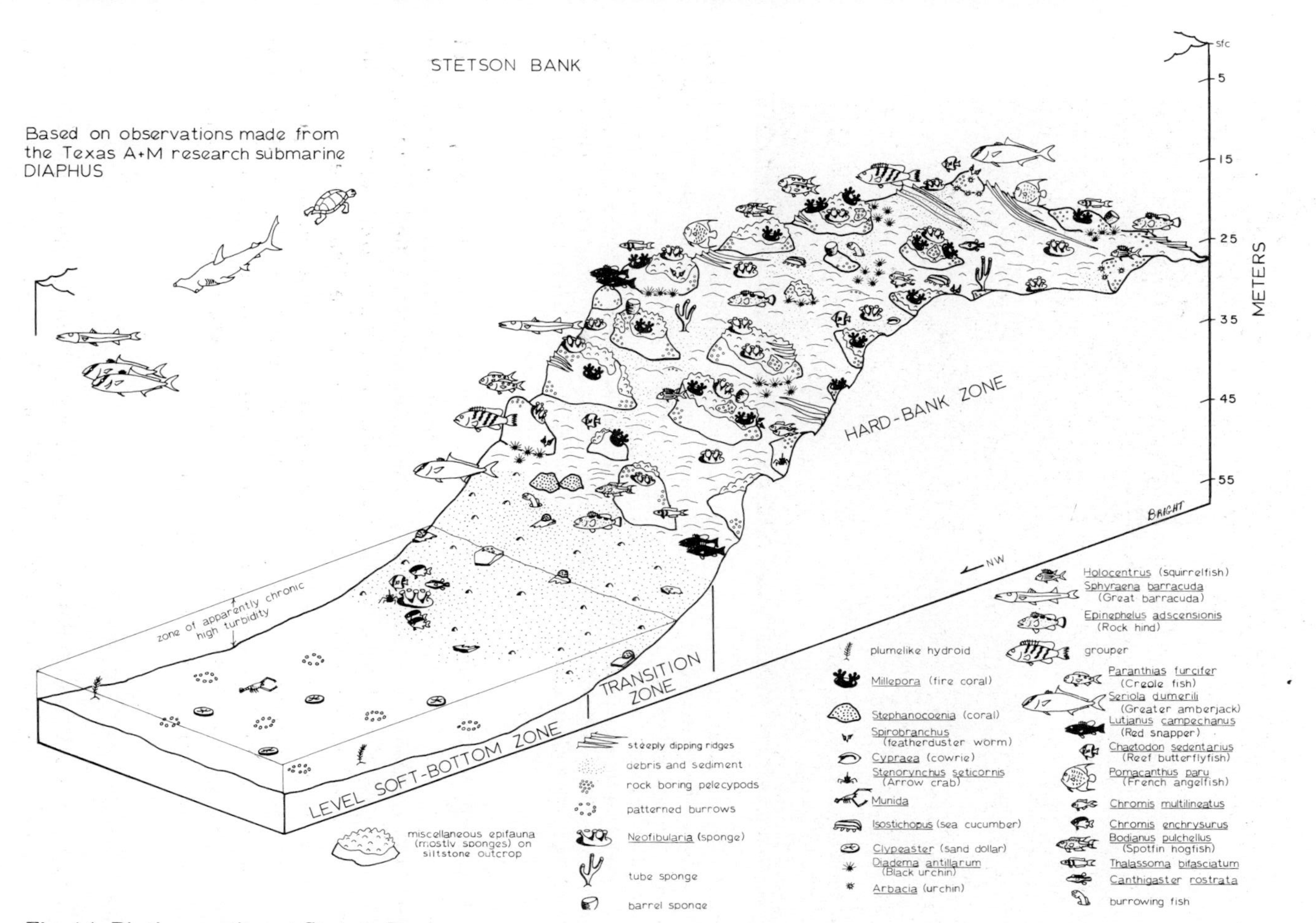

Fig. 14. Biotic zonation at Stetson Bank.

Fig. 15. Top of a claystone outcrop almost completely covered with epifauna. *Millepora*, which covers the crest and central part of this outcrop, can be distinguished by its short upward directed pillars and branches which are normally white-tipped. Sponges occupy the right side of the outcrop below the crest. *Diadema* spines can be seen in the upper left. Several spiral-shaped branchae of the feather duster worm, *Spirobranchus*, appear in the lower half of the picture. The fishes seen, *Thalassoma bifasciatum* in yellow phase (upper right) and *Canthigaster rostrata* (below and to left of first fish) are about eight cm long.

Information obtained by the study concerning Claypile Bank is sparse, but the bank is obviously occupied by a benthic assemblage which must be categorized separately from those of the other banks studied. Corals of any kind are insignificant at Claypile Bank, although *Siderastrea siderea* occurs rarely in small knobs several inches in diameter. The fishes seen there all occur also on the Flower Garden Banks, but the most conspicuous species, at Claypile Bank, remain to be identified, and therefore has been designated as burrowing fish "C". It occurs only on the deeper parts of the Flower Gardens below the coral reefs. There are verbal reports from sport divers of a large mollusk population at Claypile Bank.

TWENTY-EIGHT FATHOM BANK

Twenty-eight Fathom Bank, unlike Stetson, Three Hickey, and Claypile is comparable to those parts of the East and West Flower Garden Banks

Fig. 16. Tube sponge, *Callyspongia*, attached to bare siltstone bottom extensively pitted by boring pelecypod holes. The sponge, approximately 0.5 m high, is surrounded by a cluster of *Diadema antillarum*.

designated as Algal—Sponge Zone (Fig. 3). A comparison of Table III with the two columns in Table I covering the 49—92 m depth range at the Flower Gardens shows that the diverse populations on all three banks are extremely similar within that depth range. Twenty-eight Fathom Bank, however, lacks the coral reef and other communities which cap the Flower Garden Banks above 50 m.

The bottoms in the Algal—Sponge Zones of all three banks are covered primarily with first-size nodules composed of encrustations of coralline algae, mostly *Lithothamnium* with some *Lithophyllum*, and lesser amounts of the encrusting foraminifer *Gypsina plana* (Edwards, 1971; Hogg, 1975;

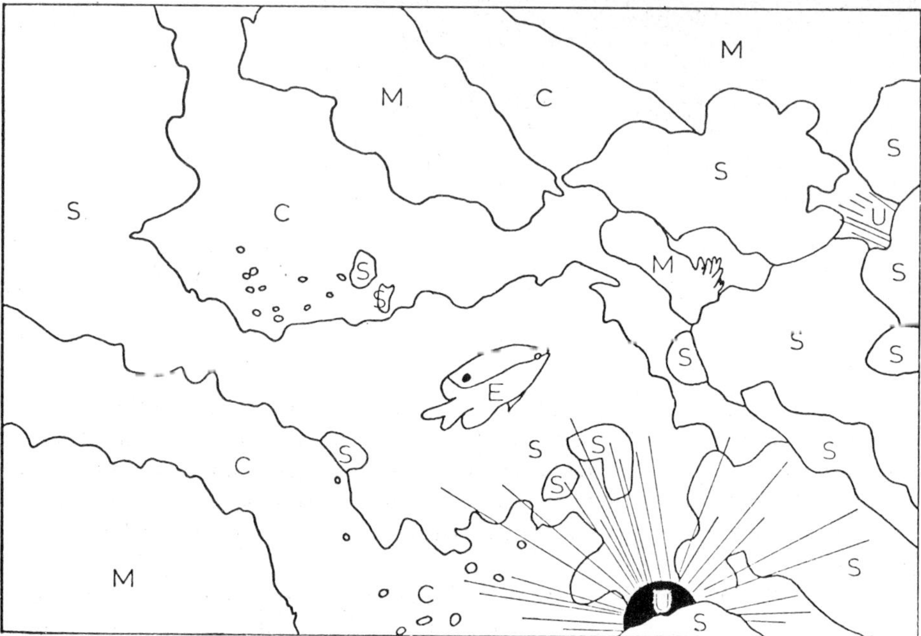

Fig. 17. Typical epifauna encrusted siltstone outcrop surface. *C* = mostly bare claystone with boring pelecypod holes; *E* = *Eupomacentrus* sp. (damselfish); *M* = *Millepora*; *S* = sponges; *U* = *Diadema antillarum* (black urchin). The fish is about 8 cm long.

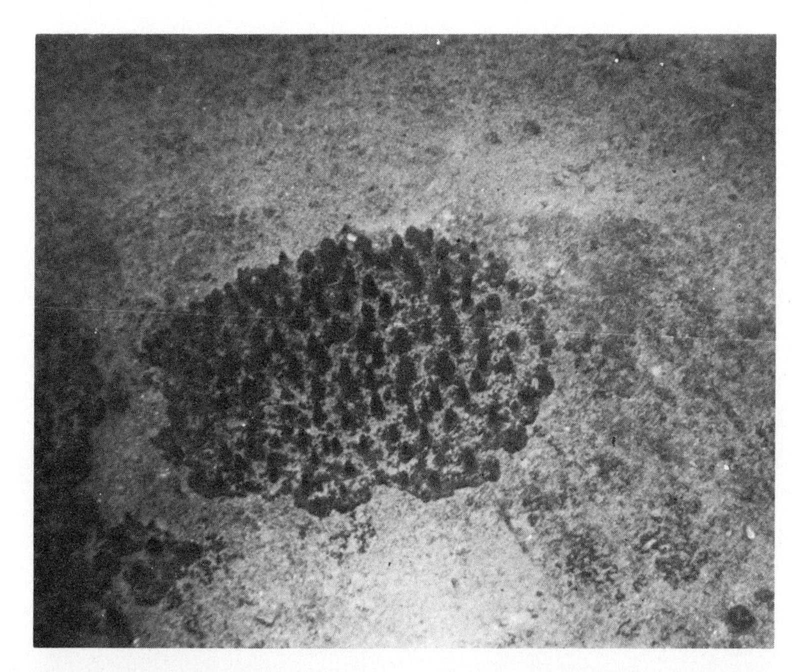

Fig. 18. Typical *Neofibularia* sponge growth on low relief hard-bottom. Traces of high dip bedding can be seen on the bottom.

Fig. 19. Siltstone outcrops crossed by high dip bedding upon which branching encrustations of *Millepora* are growing. Rock boring pelecypod holes are evident in the foreground just below the sponge, and can be seen on most of the outcrops visible. Below the fish, *Thalassoma bifasciatum*, sediment can be seen in the depression between two outcrops. The fish is approximately 10 cm long.

Table II. Summary of results of hook-and-line fishing from our research vessels, 1972-1975. *** most frequently caught,** often caught, * sometimes caught.

	EFG + WFG	Stetson	South Texas Fishing Banks
Carcharhinid sharks	*		**
Gymnothorax spp. morays	*		*
Enchelycore nigricans Viper moray	*		
Holocentrus spp. squirrelfishes	***	**	**
Sphyraena barracuda Great barracuda	**	**	***
Epinephelus guttatus Red hind	*		
Epinephelus adscensionis Rock hind	**	**	
Epinephelus cruentatus Graysby	***		
Mycteroperca spp. groupers	**	*	*
Dermatolepis inermis Marbled grouper	*		
Paranthias furcifer Creole fish	*	*	*
Priacanthidae bigeyes	**	**	
Malacanthus plumieri Sand tilefish	**		*
Rachycentron canadum Cobia			**
Seriola dumerili Greater amberjack	***	***	***
Caranx spp. jacks	**	**	*
Selene vomer Lookdown			*
Coryphaena hippurus Dolphin	*		* juveniles
Scomberomorus spp. mackerels	*	*	*
Lutjanus campechanus Red snapper	***	***	***
Lutjanus spp. snappers	*	*	*
Rhomboplites aurorubens Vermilion snapper	***	***	***
Haemulon melanurum Cottonwick	***	***	***
Calamus spp. porgies	**	*	*
Pomacanthus spp. angelfishes	*	*	
Acanthurus spp. surgeonfishes	*	*	
Balistes vetula Queen triggerfish	**	*	
Balistes capriscus Gray triggerfish		*	
Canthidermis sufflamen Ocean triggerfish	*		
Melichthys niger Black durgon	*		

Table III. Conspicuous benthic organisms and groundfishes seen at Twenty-eight Fathom Bank. Depths given indicate our observations only and do not preclude presence of the species at other depths.

	Depths of observation (meters)	Comments
Algae		
Coralline algae	52-91	Probably *Lithothamnium* and *Lithophyllum*. Forming nodules. Encrusting outcrops and rubble.
Soft algae	52-67	Probably extend somewhat deeper.
Sponges		
Neofibularia	52-61	
Agelas	61	
Anemones		
Condylactis	61	
Antipatharians	52-85	
Echinoderms		
Sea cucumber	55	Probably *Isostichopus*.
Comatulid crinoids	67-88	
Fishes		
Holocentrus spp.	67	
Mycteroperca spp.	52-67	
Paranthias furcifer	67	Dense schools.
Epinephelus spp.	52-67	
Malacanthus plumieri	52-61	
Seriola dumerili	55-67	
Lutjanus campechanus	61	
Equetus spp.	52	
Chaetodon sedentarius	67	
Holacanthus sp.	67	Either *H. bermudensis* or *H. ciliaris*.
Pomacanthus paru	52-67	
Centropyge argi	61	
Chromis enchrysurus	52-88	Very abundant.
Bodianus pulchellus	67-88	
Balistes capriscus	67	
Balistes vetula	52-67	
Xanthichthys ringens	67	Sargassum triggerfish.

Fig. 20. *Lutjanus* (snapper) (55 m depth).

Abbott, 1975) (Fig. 3). The coralline algae are important and abundant on the coral reef as well as in the Algal—Sponge Zone and they extend significantly onto the drowned reefs to depths exceeding 90 m. In the lower reaches of the Algal—Sponge Zone the nodules give way to coralline algal crusts adhering to the hard carbonate substratum. The coralline algae decrease downward in percent cover but are still moderately abundant in depths of 80 m or more. Among and attached to the nodules is a sizeable population of leafy algae, generally the same organisms which occur in the Leafy Algae Zone at the East Flower Garden. Sponges are very conspicuous, particularly the encrusting *Neofibularia nolitangere oxeata*, the tube sponge *Callyspongia vaginalis* (Fig. 16) and the branching *Verongia* sp. (Fig. 12). Other particularly conspicuous invertebrates of this zone are small saucer-shaped growths of agariciid stony corals and a large anemone, *Condylactis* sp. The expected fishes are seemingly as abundant at Twenty-eight Fathom Bank as in similar depths at the Flower Gardens, the commercial species being well represented (Fig. 20).

NATURAL-GAS SEEP OBSERVATIONS

Natural-gas seeps issue abundantly from Twenty-eight Fathom Bank and the East Flower Garden below the coral reef (see fig. 6 in Chapter 5). These seeps are intermittent in nature and characteristically emit repeated short bursts of several to hundreds of bubbles, each less than a few centimeters in diameter. There is no evidence that such seeps have had any effect on the

Fig. 21. White slimy covering on bottom of "Canyon" (76 m depth).

benthic populations. We have observed very small amounts of white mucus-like material at the points from which gas escapes the rock. No such "deposits" have been detected where gas escapes through sand, although the bottom of a large surge channel at 70—80 m at the East Flower Garden was found to be totally covered with a similar-appearing substance (Fig. 21). Speculation by fishermen that snappers and groupers are attracted to gas seeps has not been confirmed by observations of this study. The fishes are nearly always inclined to position themselves over or beside rocks, outcrops or other bottom irregularities. Where they occur, the gas seeps happen also to be associated with these features. Fishes congregating near bottom irregularities bearing seeps, however, seem to be oblivious of the gas, showing no behavior which would indicate either an affinity for or aversion to it. In addition to the East Flower Garden and Twenty-eight Fathom Banks, gas seeps have been seen by us at Fishnet, Claypile and Baker Banks.

ANTIPATHARIAN ZONE

The Antipatharian Zone is the deepest hard-bank assemblage examined in this study. It occurs on the drowned reefs and adjacent parts of the Flower

Garden Banks (Figs. 3, 22—25), as well as Fishnet Bank and all of the South Texas Fishing Banks visited in this study (Baker, South Baker, North Hospital, Hospital Rock, Southern, Dream, and Big Adam) (Fig. 26). It is presumed that it also occurs at Aransas Bank, but no observations were made there.

The Antipatharian Zone represents a transition downward from the shallow-water benthic biota to a truly deep-water assemblage (Table I). Interestingly, whereas the assemblage is developed at the crests of the South Texas Banks at about 56 m (Figs. 27, 28), truly comparable deep-water populations at the Flower Gardens usually start at depths greater than 70 m (Figs. 3, 22—25). The generally clearer water at the Flower Gardens may be a factor here, particularly in influencing the lower limit of lush coralline algal and soft algal growth. Missing from the zone proper are any stony corals except sparse populations of the saucer-shaped agariciid (Fig. 29), a small species of *Madracis*, and other small ahermatypic varieties.

Lithophyllum is present in reduced quantities and leafy algae are sparse. Abundant populations of comatulid crinoids (Fig. 30), deep-water alcyonarian fans (Fig. 31), deep-reef fish "A" (Fig. 32) and fish "B" are present.

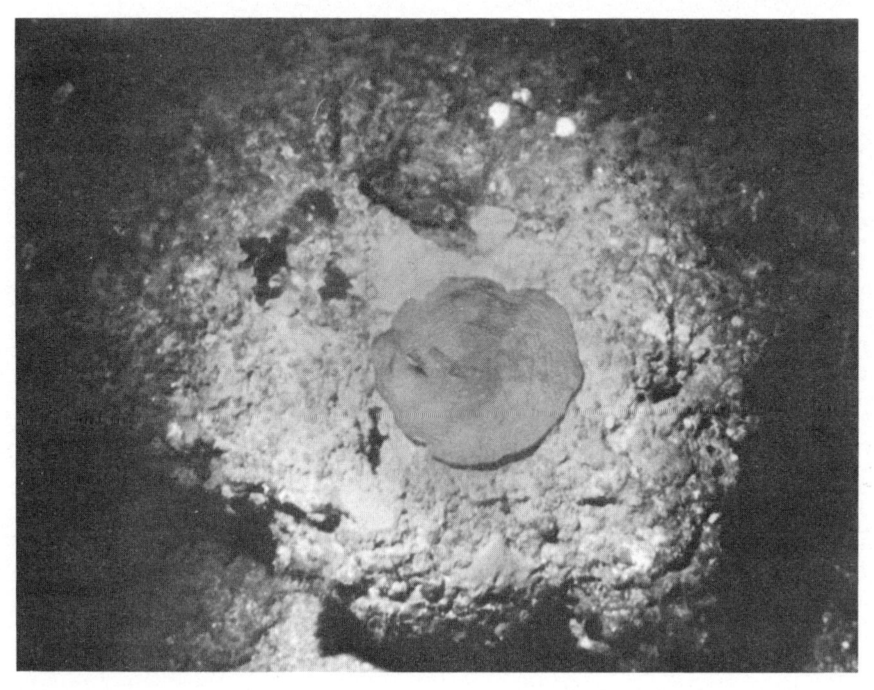

Fig. 22. Saucer-shaped agariciid coral and encrusting coralline algae in lower Antipatharian Zone (84 m depth).

Fig. 23. Deep-reef fish species "A" (84 m depth).

Fig. 24. Deep-water alcyonarians on drowned reef at the West Flower Garden (125 m depth).

Fig. 25. Deep-water corals on drowned reef at the West Flower Garden (130 m depth).

The most conspicuous organisms in this zone are the bedspring-shaped white antipatharian "sea whips" (Figs. 27, 30, 33). Antipatharians are very rarely seen shallower than 55 m at the Flower Gardens (Fig. 3). On the South Texas Fishing Banks and Fishnet Bank they are abundant from the crests down, and thin-out with increasing depth. The South Texas banks apparently differ from the other banks in their possession of conspicuous populations of the large white sponge *Ircinia campana* (Fig. 33).

The deep-reef fish "A" (Fig. 32) is a particularly reliable indicator of the Antipatharian Zone assemblage. It has not been seen shallower than 80 m at the Flower Gardens but occurs from the crests downward at Fishnet (61+ m) and the South Texas banks (56+ m). The Yellowtail reef fish, *Chromis enchrysurus*, is undoubtedly the most abundant fish of its size, namely 5—10 cm, on the Texas—Louisiana banks below 50 m and within the Antipatharian Zone particularly. It frequents all of the banks in schools of up to several hundred, although it also occurs in smaller groups or singly. At least in the spring, *Chromis enchrysurus* occupies territories and engages in antagonistic behavior toward other fishes in which it changes temporarily from its typically dark-above light-below coloration to a dusky gray. Although we have no evidence to indicate it, the species would seem to be an ideal

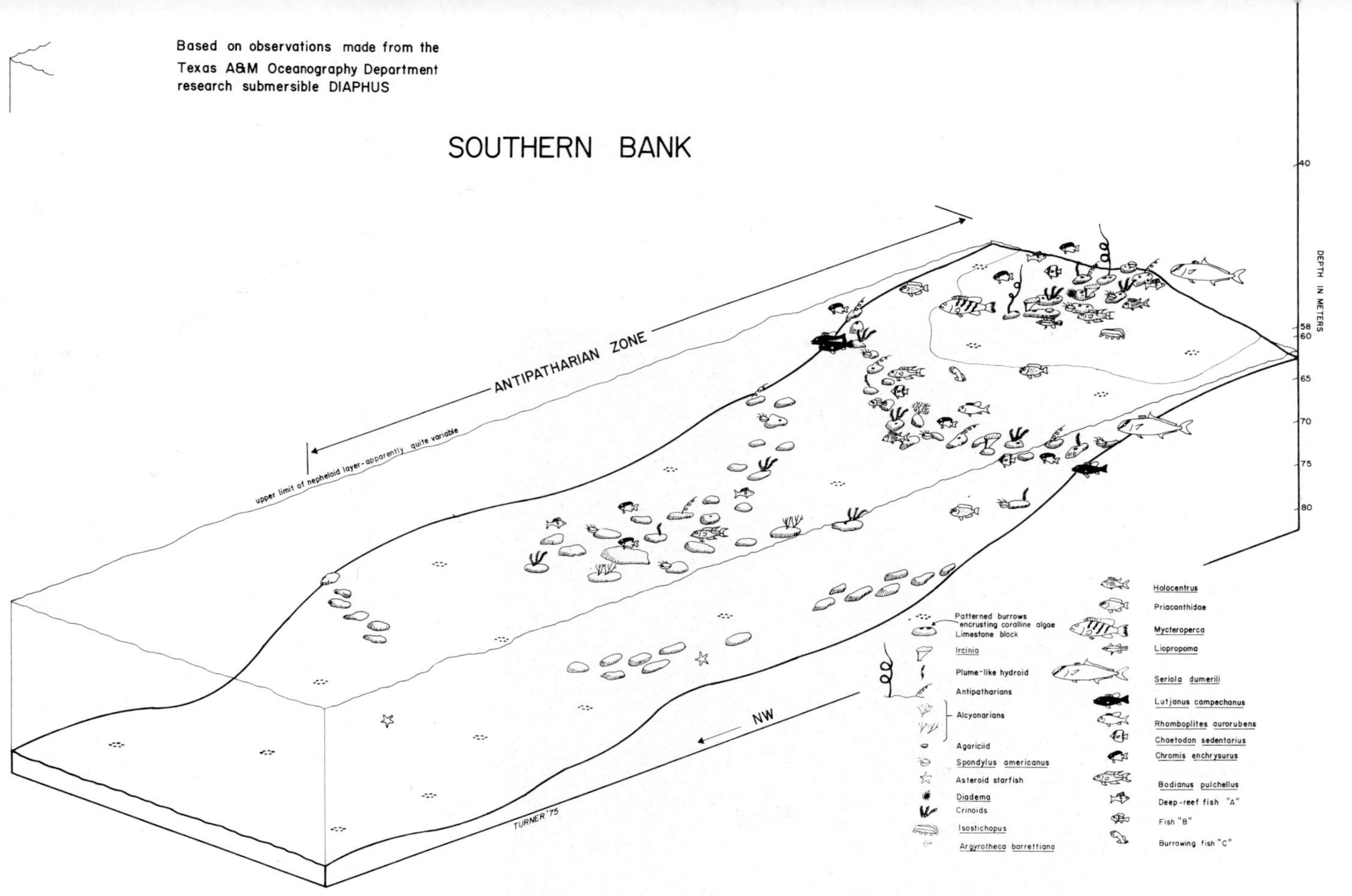

Fig. 26. Biotic zonation at Southern Bank, representative of basic zonation on all South Texas Fishing Banks and Fishnet Bank off Louisiana.

Fig. 27. Epifauna on rocks at crest of Southern Bank, bedspring-shaped antipatharians, crinoids in foreground, *Ircinia* near center (58 m depth).

forage fish for snappers, groupers and other large fishes frequenting the banks (Fig. 34).

The South Texas banks are particularly subject to nearly total inundation by thick nepheloid layers (turbid water layers) which overlie the predominantly soft bottom of the Texas—Louisiana Outer Continental Shelf (Figs. 3, 14, 26, 35). Off South Texas the difference between relief of the hard-banks and thickness of the nepheloid layers is so small that, generally, only the top few meters of the banks are in relatively clear water (Fig. 26). It is strongly suspected that during storms or prolonged heavy weather the banks are entirely covered by turbid water. Even the rocks at the tops of these carbonate banks are covered with thin veneers of fine sediment wherever the sparse epifauna and coralline algae do not occur. It is believed that the epifauna and coralline encrustations are best developed at the crests of

Fig. 28. Pinnacle-like structures about 0.75 m depth at 58 m depth on Southern Bank.

Fig. 29. Saucer-shaped agariciid coral and some coralline algae attached to rock at Southern Bank (61 m depth).

Fig. 30. Antipatharian and crinoid at Southern Bank (61 m depth).

Fig. 31. *Hypnogorgia* (73 m depth, South Baker Bank).

Fig. 32. Deep-reef fish species "A" (64 m depth at Southern Bank).

these banks and tend to decrease in abundance downward into the nepheloid layer.

The nepheloid layer observed at Stetson Bank was well down toward its base (Fig. 14), those at the Flower Gardens were well off the hard-banks altogether (Fig. 3), and that at Fishnet started at 80 m (20 m below the bank's crest). The Flower Gardens are, therefore, because of their greater relief and position at the edge of the continental shelf, bathed perpetually by clear oceanic water. Stetson Bank is probably subject to only occasional heavy doses of turbid neritic water, whereas the South Texas banks must frequently be covered by the nepheloid layer. It is speculated, therefore that the Antipatharian Zone assemblages of the South Texas banks are rather adapted to turbid water conditions, whereas the assemblages occupying other biotic zones at Stetson, the Flower Gardens and Twenty-eight Fathom Bank require predominantly clear water conditions.

There seem to be indications that even the biota of the Antipatharian Zone thrive better in clear water. Certainly, Antipatharian Zone populations are more extensive on drowned reefs at the Flower Gardens than on the South Texas banks, and they appear to be better developed at the tops of the South Texas banks than on their more turbid flanks when we visited it. Big Adam Rock, which has relatively little relief above the surrounding

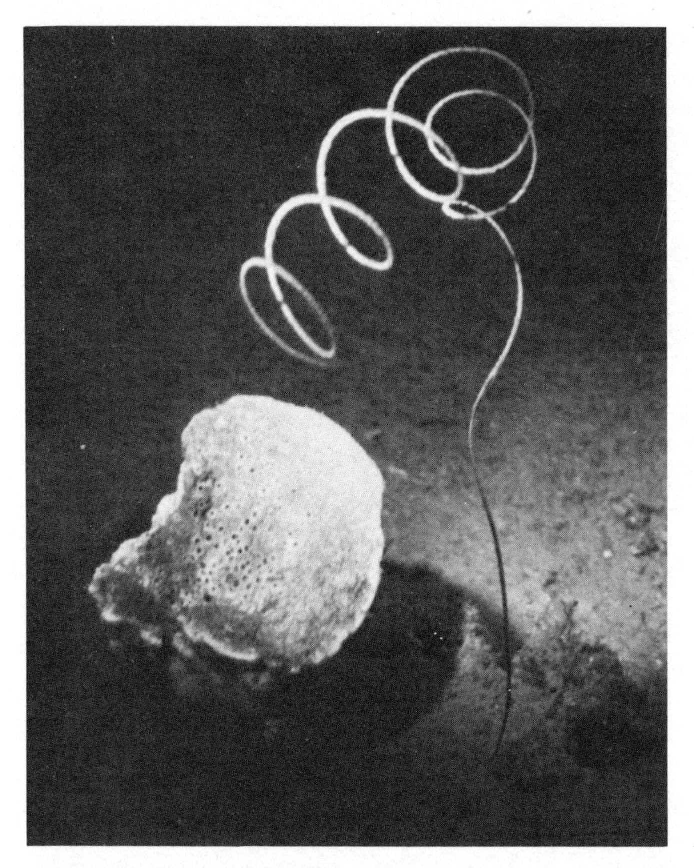

Fig. 33. *Ircinia* and antipatharian at Southern Bank (64 m depth).

soft bottom, was entirely covered by the nepheloid layer. It has a much sparser benthic population than its neighboring banks a few miles north. Fishnet Bank, on the other hand, with a nepheloid layer somewhat farther down on its sides, appeared to harbor a more diverse and abundant Antipatharian Zone population than any of the South Texas banks. However, speculations concerning the significance of the nepheloid layer as a controlling environmental factor remain to be confirmed.

Where commercial or potentially commercial finfish are concerned, the snappers *Lutjanus campechanus* and *Rhomboplites aurorubens* were about as abundant on the South Texas Fishing Banks as on the nothern banks, but there were few sightings and hook-and-line captures of groupers of the genus *Mycteroperca* in the south. This had not been expected, nor was the absence of sightings or captures of hinds of the genus *Epinephelus* on the South Texas banks. Fishnet Bank, on the other hand, harbors at least moderate populations of *Mycteroperca* and *Epinephelus*. If indeed the

Fig. 34. *Mycteroperca*, bigeye, antipatharians, *Ircinia* and *Hypnogorgia* at 73 m depth on South Baker Bank.

Fig. 35. Patterned burrow cluster in soft sediment adjacent to South Baker Bank (80 m depth).

groupers and hinds are less abundant on the South Texas banks, the reasons are not apparent unless overfishing is a cause. On the banks from Stetson north large schools of good-sized Creolefish, *Parathias furcifer*, were observed and the species is sometimes caught on hook-and-line. The Cottonwick, *Haemulon melanurum*, is abundant and easily caught on all the banks.

GEOLOGY OF THE BANKS

The banks of the Texas Outer Continental Shelf may be divided into two main groups. Those banks north of 27°46′N are associated with salt domes in the subsurface and their distribution is generally the same as the distribution of shallow salt domes. The banks south of 27°46′N are not associated with any shallow subsurface structures and their distribution is most probably controlled by an ancient shoreline at a depth of approximately 60 m during the Late Pleistocene.

The relief on the banks is quite variable with those banks in the northern area generally having greater relief. Twenty-eight Fathom Bank has the greatest amount of relief with a maximum of 118 m in a distance of 670 m. The least amount of relief is on Thirty-two Fathom Bank with a total of 6 m in a distance of 3,201 m.

Banks such as those described here occur on the outer shelf eastward to the head of the Mississippi Canyon. The crests of these banks increase in depth towards the east, the deepest one is in the Mississippi Canyon at a depth of 179 m. This increase in depth of crests is due to downwarping of the shelf caused by the weight of the Mississippi Delta.

All of the banks are covered by a heavy growth of coral and coralline algae except for Stetson and Claypile Banks. These two banks are the only ones known to have outcrops of Tertiary bedrock exposed at the surface of the bank. Some of the banks such as the West and East Flower Gardens are living coral reefs. Most of the banks are covered by dead reefs (drowned reefs) that were living from 6,000 to 18,000 years ago at times when sea level was considerably lower than it is at present.

NORTHERN BANKS

Direct geological observations using submersibles have been made at West Flower Garden, East Flower Garden, Twenty-eight Fathom, Stetson, and Claypile Banks. Typical of the larger banks on the northern shelf is the occurrence of gently sloping terraces covered with sediment bounded by steep rocky cliffs. These terraces and associated cliffs are especially obvious on the West Flower Garden and Twenty-eight Fathom Banks. The rocky cliffs represent drowned reefs that are now dead but were flourishing during a lower stand of sea level (Figs. 24, 25). Scattered over the terraces are isolated patch reefs that developed as sea level rose. These features are well illustrated on the chart of the West Flower Garden Bank. The rocky cliffs, patch reefs, as well as the irregular parts of the hard substrate of the Algal—Sponge Zone are places where large schools of snapper, grouper, Creolefish, barracuda and jacks seem to congregate. There are three drowned reef levels at the West Flower Garden Bank. They occur at 56, 91 and 128 m. At the East Flower Garden Bank there is one large drowned reef that occurs from about 63 m to a depth of 85 m (Fig. 3). At Twenty-eight Fathom Bank drowned reefs occur at 52, 56, 80 and 90 m on the north side and a single reef from 100 to 170 m on the south side.

The sediments that surround the actively growing reefs are coarse sands and gravels grading into finer sediments with increasing depth of water. The distribution of sediment types on the West Flower Garden is typical of the actively growing reefs. At the crest of the reef, between the large coral heads, a coarse coral-molluscan sand covers the bottom (Fig. 4). This sand is moved by severe storms into chutes that carry it to the base of the reef where it is spread by currents into a narrow band immediately adjacent to the base of the reef at depths of from 45 m to 49 m. Close to the base of the reef are large blocks of reefrock that have been torn loose by storms and tumbled down the steep slopes. Beginning at a depth of about 49 m and extending to a depth of about 73 m the bottom is covered by a coarse gravel composed of nodules of coralline algae comprising the substrate of the Algal-Sponge Zone. From 85 m to a depth of 106 m the sediment consists of a foraminiferal—coral—coralline algae sand. Below 106 m the sand gives way to the sandy, silty clays that are the normal deposits of the outer continental shelf.

Thirty-two Fathom Bank

Thirty-two Fathom Bank has a very low relief but the nature of the record on the precision depth recorder indicates a hard bottom. No direct observations nor sampling have been conducted at Thirty-two Fathom Bank. However, experience gained from the study of other banks at comparable depths indicates that the bottom there should be covered with the hard, coralline algae nodules typical of the Algal—Sponge Zone,but this has not been verified.

Stetson Bank

Submersible observations show Stetson Bank to be a hogback of Tertiary claystone and poorly cemented siltstone. Fig. 15 illustrates a typical outcrop of claystone that has been extensively bored by sponges and clams. Massive outcrops such as this are common as can be seen on Fig. 14. The swales between the massive outcrops are underlain by siltstones. In some areas, thin stringers of well-cemented siltstone stand above the bottom as small ridges several centimeters in height and one or two centimeters thick (Fig. 19). As can be seen (Figs. 15—19), large areas of the bank are bare rock with little or no cover by encrusting organisms.

SOUTHERN BANKS

Direct geological observations have been made on Baker, South Baker, North Hospital, Hospital Rock, Southern, Dream, and Big Adam. The greatest relief on the southern group of banks is found on Southern with a maximum of 22 m, the average relief on the other banks being about 10—12 m. As mentioned earlier, these banks differ from the northern banks in that they are not associated with salt domes. The banks are all drowned reefs that were thriving coral-algae reefs during lower stands of sea level.

Southern Bank is typical of this group and the diagram (Fig. 26) illustrates the nature of the bottom topography. Four levels of reef development are shown at depths of 72, 68, 63 and 58 m. These are identical to the massive rocky, drowned reefs of the northern banks but have very little relief. The substrate between these reef levels is a pavement of dead coralline algae covered by a thin film of fine clay and silt-size sediments (Fig. 3) deposited from the nepheloid layer which more or less continuously covers all but the crest of Southern Bank.

The crest of the bank is quite rough with a relief on the surface of approximately one meter (Fig. 27). In places, the surface gives the impression of a solution karst (Fig. 28) similar to those commonly seen in some low, supratidal and high, intertidal areas. Exposure of the bank during early Holocene time is probably the cause of the solution karst.

As these reefs are no longer actively growing, the coarse, gravelly and sandy sediments that are found surrounding actively growing reefs to the north are not present at the surface. Sediment cores taken adjacent to the drowned reefs show the sandy and gravelly sediment to be covered by about a foot of sandy and silty clay.

The reason for the demise of these once flourishing reefs is probably the development of the nepheloid layer in the area. This layer of highly concentrated suspended sediment in the bottom waters of the area attains a thickness of approximately 20 m. Consequently, only those reefs that stand higher than 20 m above the surrounding bottom protrude above the upper surface of the nepheloid layer. During times of storms, even the crests of these larger banks must be immersed in the turbid waters of the nepheloid layer. Most corals and coralline algae are not capable of surviving in such turbid waters.

CONCLUSIONS

Since 1974 the use of the research submersible DIAPHUS has enabled oceanographers from Texas A & M to visit, first hand, sea-bottom structures and biotic habitats they had been studying for years by more conventional methods. The opportunity to traverse, visually inspect, photograph and selectively sample fishing banks and reefs has resulted in a significant improvement in the amount and quality of information available concerning the biology and geology of these important features. Moreover, the impressions left in the scientist-observer's minds have permitted a much more efficient synthesis of data gathered than would be possible had they not actually "seen" the study areas.

Biotic assemblages of reefs and hard-banks of the Texas—Louisiana Outer Continental Shelf can be distinctly grouped according to their natures into four categories: (1) the sparse Claypile Bank biota (35—55 m) of predominantly low-growing filamentous and leafy algae and sponges with occasional "meadows" of high-standing leafy algae occupied by numerous fish; (2) the more diverse Stetson Bank and Three Hickey Rock biota (28—56 m) dominated by the hydrozoan fire coral *Millepora alcicornis* and sponges; (3) the highly diverse and abundant Flower Gardens/Twenty-eight Fathom Bank biota with coral reefs (22—49 m), algal nodule and sand-covered platforms (45—76 m), and, drowned reefs from 76 to 100+ m, bearing an assemblage of organisms directly comparable to the deep-water biota of category 4 described below; (4) the deep-water biota of the South Texas Fishing Banks (56—78 m) and Fishnet Bank (61—82 m) characterized by the presence of antipatharian whips, deep-water alcyonarian fans, comatulid crinoids, certain species of deepdwelling fishes and sparse populations of encrusting coralline algae.

Geologically, the banks are of two types: (1) those associated with salt domes and; (2) those not associated with tectonics of any kind. The relief on the banks varies considerably with the northern banks (type 1) generally having the greatest amount of relief. The southern banks (type 2), once flourishing coral reefs, are now dead. The development of a nepheloid layer probably was the cause of their death.

ACKNOWLEDGEMENTS

We thank the following people for their help in identifying specimens: Robert Abbott, Bart Baca, Dr. Elenor Cox, Charles Giammona, Arthur Leuterman, Glen Lowe, Dr. Linda Pequegnat, Jack Thompson (all from Texas A & M University), and Joyce Teerling (New Orleans, Louisiana).

The pertinent projects received support from the U.S. Department of the Interior, Bureau of Land Management (Contract No. 08550-CT5-4);

Signal Oil and Gas Company; and the Texas A & M Oceanography Department Study of Naturally Occurring Hydrocarbons in the Gulf of Mexico.

REFERENCES

Abbott, R.E., 1975. The Faunal Composition of the Algal—Sponge Zone of the Flower Garden Banks, Northwest Gulf of Mexico. Thesis, Texas A & M University, College Station, Texas, 205 pp.

Bright, T.J. and Pequegnat, L.H. (Editors), 1974. Biota of the West Flower Garden Bank. Gulf Publishing Co., Houston, Texas, 435 pp.

Bright, T.J., Pequegnat, W.E., DuBois, R. and Gettleson, D., 1974. Baseline Survey, Stetson Bank, Gulf of Mexico, Biology. For Signal Oil and Gas Co., 59 pp.

Edwards, G.S., 1971. Geology of the West Flower Garden Bank. Texas A & M Sea Grant Publ., TAMU-SG-71-215: 199 pp.

Hogg, D.M., 1975. Formation, Growth, Structure, and Distribution of Calcareous Algal Nodules on the Flower Garden Banks. Thesis, Texas A & M University, College Station, Texas, 57 pp.

Neumann, A.C., 1958. The configuration and sediments of Stetson Bank, northwest Gulf of Mexico. Texas A & M Univ., Dep. of Oceanogr., Tech. Rep., Ref. 58-5T: 125 pp.

CHAPTER 7

THE ROLE OF MANNED SUBMERSIBLES IN SEDIMENTOLOGICAL AND FAUNAL INVESTIGATIONS ON THE UNITED KINGDOM CONTINENTAL SHELF

John B. Wilson

INTRODUCTION

As a result of the extensive investigations during the 19th century the U.K. continental shelf is, scientifically speaking, the best known in the world. Therefore, to assess the advances made in the last six years or so by the use of manned submersibles, it is appropriate to begin with a brief historical review of the different techniques used and of the impact that they had on these shelf investigations over the past 150 years.

Most of our knowledge of continental shelf and slope faunas and sediments has come from observations on specimens collected with dredges and grabs operated from surface vessels. The use of dredges for scientific purposes goes back to at least the mid-eighteenth century. The naturalist's dredge (introduced by O.F. Müller in 1799) was first used on the British continental shelf by Edward Forbes in the early 1830's. His successful use of it in the Irish Sea and off the Shetland Islands between 1832 and 1839 resulted in the British Association for the Advancement of Science forming a series of Dredging Committees to support the work around the British shelf. This work continued through the mid-nineteenth century and a detailed knowledge of the fauna was built up. Larger-scale investigations by H.M.S. LIGHTNING (1868) and H.M.S. PORCUPINE (1869, 1870) on the outer part of the continental shelf, slope and into deeper water added considerably to this body of knowledge (Thomson, 1874). Following the first dredgings on Rockall Bank by H.M.S. PORCUPINE in 1869, the next major investigation was the S.S. GRANUAILE expedition by W.S. Green and others in 1896 (Green, 1897).

Small samples of the sea-floor sediments had been collected on a routine basis for many years during Admiralty Hydrographic surveys of the continental shelf. Initially a simple lead line "armed" with soft tallow was used to obtain very small samples of the sea floor. Later, considerably larger samples were obtained using a variety of more sophisticated sounding instruments (Thomson, 1874). Data collected during these surveys and published as notations (S. Sh. Co. etc.) on Admiralty charts is still the basis of our knowledge of the nature of sediments over large parts of the British shelf in spite of recent work.

The first largely undisturbed grab samples from the shelf were collected in the 1920's and 1930's following the pioneer work of Petersen and others (Petersen, 1918). These gave the first reliable indications of density of particular species, together with information on mode of life and sediment/faunal relationships for several groups of animals.

The next major advances — underwater photography, underwater television and SCUBA diving techniques — came in the late 1940's and 1950's. Improvements in cameras and housings, film, underwater lenses and electronic flash systems have continued up to the present day. Over the last few years both colour and stereophotography have been adapted for underwater use.

The advent of underwater television enabled scientists to observe processes on the sea floor as they happened (Barnes, 1963). Early systems were clumsy, had poor resolution, required elaborate lighting and were unsuitable for detailed sedimentological or ecological work. Some of these difficulties have been overcome recently with the development of the image intensifier.

The impact of SCUBA diving on inshore, shallow water, marine biology and geology in the United Kingdom and elsewhere has been dramatic. Much useful work has been done by those scientists who are competent divers and also by divers with little or no scientific training who collect specimens and make observations for non-diving marine scientists. In British waters much of this work is described in the reports of the Underwater Association, the British Sub Aqua Club and in volumes edited by Flemming (1973) and Drew et al. (1976).

SCUBA diving, although ideal for inshore or near surface work, does have severe limitations. The deeper the diver attempts to work, the shorter the time that can be spent on the bottom collecting data. Scientific operations at depths greater than 50 m are largely impracticable. The deeper parts of the shelf further offshore are thus not accessible to scientific SCUBA divers.

Until the advent of the manned submersible in the 1960's, therefore, there was no way in which direct scientific observations could be made in situ in deeper waters. Although rapid progress in the design and building of submersibles was made in the United States preliminary design work on a British submersible was not commenced until 1963. SURV (Standard Underwater Research Vehicle) was built in 1967 and underwent sea trials in 1968 and 1969 (Jack, 1969). It was found, however, to be inadequately designed for work in conditions such as those encountered on the U.K. continental shelf (Flemming, 1968).

Also in 1968, a joint project between Vickers Ltd and the Canadian company International Hydrodynamics Ltd, resulted in the building of the submersible PISCES II. Evaluation trials in the autumn of 1969 demonstrated the potential of the submersible to British marine scientists (Mott, 1972). The system enables non-diving scientists to go to all depths on the shelf, in comparative comfort, for periods of several hours to make detailed, first-hand visual observations on sediments and faunas. For the first time, the scientist can choose exactly where a sample should be collected to answer a particular question. The doubts and uncertainties as to precise bottom conditions that arise every time an interesting dredge sample is collected from a surface ship are thus eliminated. Rare and delicate species can be observed and photographed in situ, collected and transported to the surface in a virtually undamaged condition. In those areas of the U.K. continental shelf where the nature of the sediments varies rapidly, representative samples of the various types present may be collected relatively easily. This again eliminates the uncertainties in the interpretation of grab samples collected by surface ships in such areas.

It is interesting to note that these first submersible trials in U.K. waters took place just 100 years after the cruises of H.M.S. PORCUPINE.

SEDIMENT AND FAUNAL OBSERVATIONS

The purpose of this chapter is to present examples of the different types of observations on sediments and faunas that have been successfully made from manned submersibles on the U.K. continental shelf. These demonstrate the great potential of the submersible as a research tool for the sedimentologist and palaeoecologist. It is not appropriate here to present a detailed account of the actual submersible operations, since they are discussed in other chapters, nor to present the conclusions.

These investigations were undertaken using the Vickers Oceanics submersibles PISCES II and III by the National Institute of Oceanography (now known as the Institute of Oceanographic Sciences, Wormley) in the English Channel; and in collaboration with the Institute of Geological Sciences, on the west of Scotland shelf and Rockall Bank between 1970 and 1973. Work on the solid geology of the United Kingdom continental shelf and Rockall Bank using PISCES between 1969 and 1973 is the subject of Chapter 10 by Eden, McQuillin and Ardus in this volume. During most of these dives, sediment and faunal observations were also made, samples collected and photographs taken.

Methods

Much experience in the use of external handling tools in the collection of samples from submersibles has been gained by the ALVIN team at Woods Hole Oceanographic Institution (Winget, 1969).

Experience with PISCES showed that cobbles and small boulders with epifauna on them can be collected relatively easily using the hydraulically operated telechiric manipulator which has 6 planes of movement and can grip up to 68 kg.

Sediment samples can be collected using a scoop (Fig. 1) held in the manipulator. The scoop is drawn over the surface of the sediment at precisely the point from which a sample is required. The scientist can observe the actual collecting procedure and note exactly what happened. Should the first sweep not yield sufficient material a second sweep can be taken either alongside the first or, if deeper penetration is required, within the groove left after the first sweep. The sample accumulates at the bottom of the jute bag. On completion of the sampling operation the metal scoop frame is rotated by the manipulator so that the entrance to the bag is folded against the metal, thus ensuring minimal loss of the fine fraction during subsequent manoeuvres. The sample is then deposited by the manipulator in the basket (Fig. 2) fitted to the front of the submersible. During the 1970 and 1971 operations with PISCES II, 6 separate scoop samples could be carried. Each scoop is clearly marked with a different letter or number so that samples can be easily identified. At the end of each dive the samples are all removed from the basket by divers and placed in the standby dinghy used for the recovery

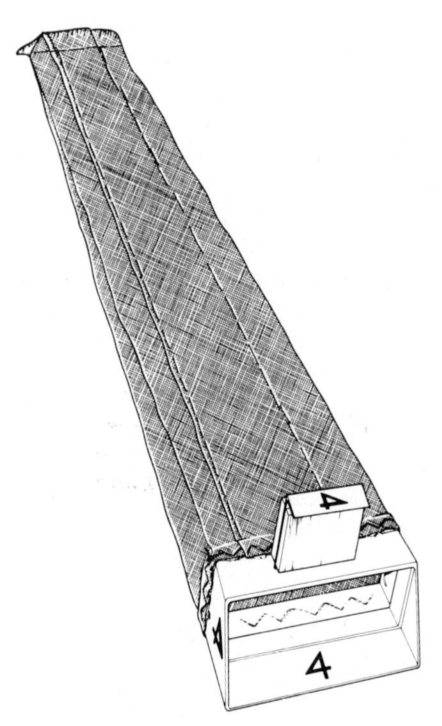

Fig. 1. Sediment scoop. The frame is made of 16 gauge mild steel and is 20 cm wide by 7 cm in height. The jute bag is 45 cm long. When the sample — up to 5 kg of sediment — has been collected, it normally occupies the bottom third of the bag. The plywood blocks on the handle enable the telechiric manipulator to grip the scoop rigidly during the sampling operation.

Fig. 2. PISCES III as fitted for sediment and faunal sampling on Rockall Bank, June 1973. Note large sample basket with several sample scoops attached by means of strong rubber bands to the side (scoops *L* and *M* are clearly visible). The rubber bung on the rod attached beside scoop *L* is for stoppering core tubes. The telechiric manipulator can be seen above the basket. Its claw is directly above the small sample basket. When required, the scoops are detached by the manipulator breaking the rubber band. A still camera, flash, television camera and flood light are mounted on the pan and tilt head on the port side. At the top of the picture a second still camera, flash and the submersible's flood lights can be seen.

operations before the towline is attached. This is done to prevent serious damage to the samples during the towing of the submersible back towards the mother ship and at the critical transition through the surface waves as it is winched on board the mother ship.

Underwater navigation

Position fixing has always been a problem. During the 1970 operations the best that could be achieved was to obtain the position of a surface float attached to the submersible. The mother ship manoeuvred until the float was approximately 100 m on the beam. The ship's position was then determined and the range to the float was measured. The range was then plotted perpendicular to the ship's head and the position plotted on an X—Y grid. This position was then relayed to the submersible using the underwater communications system. The use of simple X—Y coordinates eliminated the need to transmit long complicated numbers. Position fixes were taken at intervals while the submersible was moving and also each time it stopped to take samples.

In 1971 SPATE II (Submersible Position And Tracking Equipment) developed by Vickers Oceanics was used. With this system the bearing of the submersible relative to the mother ship and its slant range were determined using sonar to interrogate a transponder mounted on the submersible. Knowing the depth, the horizontal distance of the submersible from the mother ship could be calculated.

Numerous errors were present in both these methods. With the float, it could never be guaranteed that it floated directly above the submersible. Indeed, the effects of midwater and surface currents almost certainly ensured that it was rarely directly above the submersible. With the SPATE system, complications arise from spurious echoes and from the difficulty of identifying the precise bearing of the strongest signal. These difficulties increased in deeper water and the resulting fixes were much less reliable.

In practice, it is rarely essential for most sedimentological and faunal work to know exactly where the submersible is at all times while actually on the bottom. What is important is to be able to determine the relative positions

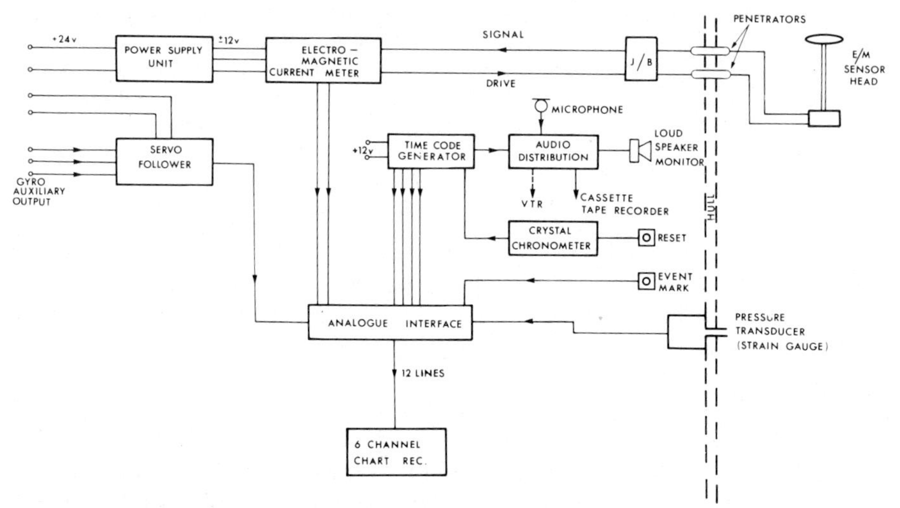

Fig. 3. Block diagram of the navigational logging system as used in PISCES III, June 1973, showing the major components of the system. It was built in modules which could be fitted with ease into the limited space available within the submersible. *J/B* = junction box; *E/M* = electromagnetic; *VTR* = video tape recorder.

Fig. 4. Rock outcrop showing isolated patch of epifauna consisting of tetractinomorph sponges, alcyonarians, hydroids, starfish, the echinoid *Echinus* sp. and encrusting bryozoans and serpulids. Over most of the outcrop, the fauna consisted only of encrusting bryozoans and serpulids. Depth 116 m. Top of Rockall Bank (57°39′N, 13°39′W).

of features observed, photographs taken, or samples collected, with a high degree of accuracy after the completion of the dive.

With these requirements in mind, a system was developed by C.A. Hunter of the Institute of Oceanographic Sciences for logging essential data during the dive (Fig. 3). The data recorded on the six track pen recorder included forward and athwartships components of velocity measured from an electromagnetic log (Tucker et al., 1970), gyro-compass bearings of the submersible's head, depth as measured by a strain gauge pressure transducer, time, and events such as the taking of photographs or the collection of samples.

Time was measured using a crystal controlled chronometer recording hours, minutes and seconds of elapsed time from the start of the dive. As the paper record logging the data only lasted for four hours, the practice during long dives was to start the elapsed time from the moment of departure from the landing site rather than from the time of arrival on the bottom. Further delays of fixed intervals of time could be built in throughout the dive at points where long stops were made should this be found to be

Fig. 5. Coarse carbonate detritus accumulating in the spaces between cobbles and boulders. Note the encrusting epifauna of bryozoans and serpulids on the boulders. Depth 96 m. Top of Rockall Bank (57°37′N 13°43′W).

necessary. Time signals were fed into both the videotape recorder and the cassette tape recorder.

On returning to the laboratory the records from the data logger were digitised. The data were processed using a modified version of the navigational program developed for R.R.S. DISCOVERY to derive the track of the submersible along the bottom from the starting position of the dive. Values of the bottom currents at times when the submersible is known to have been stationary can be obtained from the electromagnetic log data. These values, as well as providing useful current data, may be used to make further adjustments in the track.

Sediments and faunas in rocky areas and boulder fields

The collection of samples in rocky areas from surface vessels is virtually impossible. Equipment, such as dredges, grabs and cameras, that is lowered into such areas is very prone to damage or loss. Any results obtained can never be considered as widely representative of the bottom conditions. With

Fig. 6. Boulders and cobbles with epifauna. The larger boulders support a fauna of tetractinomorph sponges, anthozoans and bryozoans in addition to the encrusting bryozoans and serpulids. The smaller cobbles and pebbles have a much less diverse encrusting fauna and rarely support larger forms. Depth 155 m. West side of Rockall Bank (56°58′N, 14°35′W).

a submersible, however, outcrops such as shown in Fig. 4 can be examined in detail and systematic changes in the faunas observed, photographed and sampled where this is feasible. In well-jointed rocks it is often possible to prise out a joint-bounded block carrying epifauna seen to be typical of large areas of outcrop.

From the viewpoint of the carbonate sedimentologist these areas of high carbonate productivity are of the utmost importance. Carbonate detritus derived from fauna living on the adjacent rocks accumulates in small pockets within the outcrop area (Fig. 5). With care, these can be sampled from the submersible, and provide valuable information on the early processes of breakdown of the coarse shell debris into material more easily transported by wave and current action.

Although dredging from the surface can be attempted in boulder fields, the delicate epifaunal material collected is frequently badly damaged and the orientation of the boulders in the sample with respect to each other can only be guessed at. The fauna on smaller pebbles frequently differs from that

Fig. 7. Boundary between gravel wave and the encroaching sand patch. For details see Eden et al., 1973. Depth 86 m. Blackstones Bank, Sea of the Hebrides (56°5′N, 7°9′W). I.G.S. photograph UW/P71/4AF/4.

on larger boulders. Particular examples of both types seen from the submersible (Fig. 6) can be photographed in situ and then collected.

Sediment boundaries

In studies of sedimentary processes, much of our knowledge of sediment boundaries comes from the interpretation of sonographs (Belderson et al., 1972). Detailed information on the precise nature of boundaries and of the sediments involved is most easily obtained from a submersible. Such boundaries (Fig. 7) can be followed for distances of several hundred metres if bottom conditions (currents, visibility, etc.) permit.

Sedimentary structures

Bottom photographs can often provide much useful data on sedimentary structures. Samples collected by surface vessels from strongly rippled areas

Fig. 8. Symmetrical gravel ripples composed almost entirely of bioclastic debris. Note the concentration of coarse debris in the trough of the ripple. Wavelength ca. 75 cm, amplitude ca. 24 cm. Depth 183 m. Top of Rockall Bank (57°37′N, 13°45′W).

however cannot give a true picture of the grain-size characteristics of the sediment, as it is not usually known from which part of the ripple the sample came. The observer working from a submersible can collect samples from particular parts of the crests and troughs and thus obtain an accurate picture of the variation present (Fig. 8).

Sometimes unexpected structures are observed which, although covering a very small area, provide vital information. Features such as the current lineations seen on Fig. 9 would rarely be detected on a discontinuous series of bottom photographs.

As a submersible can remain on the bottom for periods of several hours, qualitative observations on the nature of sediment movement in response to tidal currents can be made. For example, on a dive in the Western English Channel 16.7 km southeast of St Mary's, Scilly Isles, observations were made at the time of maximum spring tidal current activity and the sand was not seen to move when the currents were at their strongest. Movement did, however, take place quite suddenly for a period of 20—30 sec during which plumes of sand were seen to rise from the crests of the ripples and a "carpet"

Fig. 9. Current lineation in sandy gravel. Note large area of scour associated with the boulder in the immediate foreground with the shadow behind it. Depth 158 m. North side of Rockall Bank (57°55′N, 13°52′W).

of sand moved in the direction of the current (N.C. Flemming, personal communication, 1971).

Observations on particular species

The ability to observe particular carbonate-producing species in situ and to examine their relationships to other species and to the sediment is probably the most important use that a palaeoecologist can make of a submersible. Dispersal or aggregation of species can be studied (Grassle et al., 1975; Wilson et al., 1977). The areal extent of the groups of dispersed individuals may be quite small (e.g. Fig. 10) and would most likely be missed on bottom photographs. Local aggregations within these dispersed populations, any changes of density towards the edges of the population, and the nature of the edge can be investigated with comparative ease from a submersible. The echinoderms — particularly echinoids and ophiuroids — have been found to be particularly suitable for such investigations (Grassle et al., 1975; Wilson et al., 1977).

Fig. 10. Shelly mud floor with population of crinoids (*Antedon* sp.). Note *Turritella communis* Risso shells on the surface. Depth 100 m. For details see Eden et al., 1971. Sea of the Hebrides (56°53'N, 7°25'W). I.G.S. photograph UW/P70/3B/3.

In species such as *Ophiotrix fragilis* (Abildgaard) known from diving investigations (Warner, 1971) the spectacular displays of aggregation (Fig. 11) can be studied more extensively for longer periods of time and in deeper water.

Submersible observations of the faunas associated with particular environments can generally give much information not available any other way. An example of this was a study made of faunas associated with iceberg plough marks (Belderson et al., 1973) on the west side of Rockall Bank in 1973 using the submersible PISCES III. The sediment floor of the plough mark was seen to be populated by the echinoid *Spatangus raschi* Lovén (Fig. 12). The individuals were irregularly distributed over the sediment. Occasionally groups of three or four could be seen. Isolated "patches" of the deep water coral *Lophelia prolifera* (Pallas) were also present (Fig. 13). Those "patches" observed range in size from approximately 10 m to nearer 40 m across. The boundaries of each "patch" are quite distinct and are usually marked by the outward growing face of the living coral colonies (Fig. 14). The central parts

Fig. 11. Aggregation of brittle stars (*Ophiothrix fragilis* (Abildgaard)) on a rock ledge. Depth 55 m. English Channel (50°14.3′N, 4°25.7′W). Photograph by D. White.

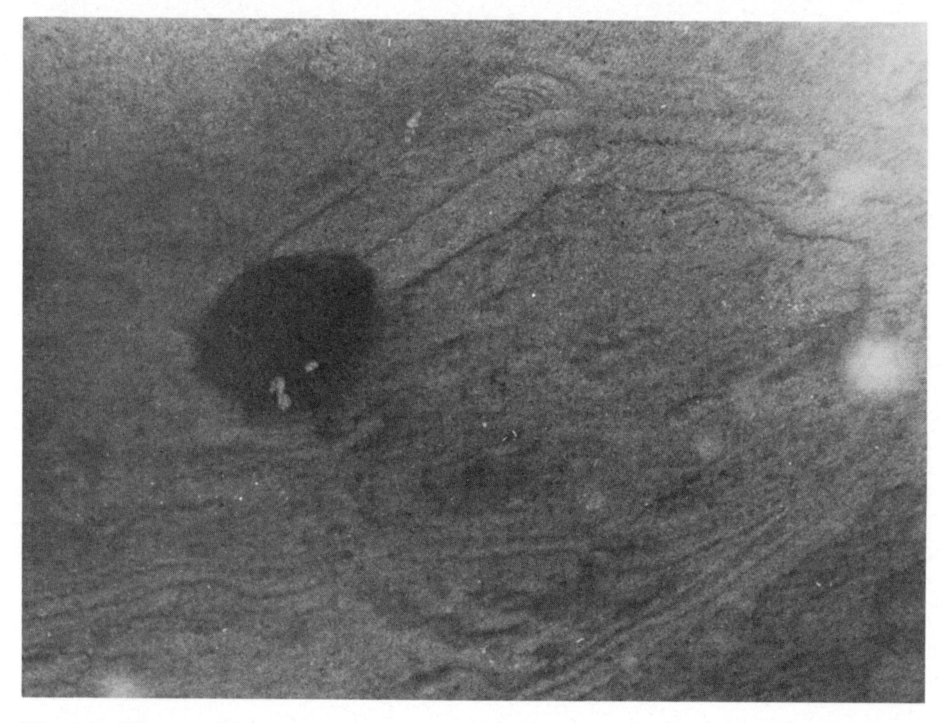

Fig. 12. Silty sand floor of iceberg plough mark showing the echinoid *Spatangus raschi* Lovén moving slowly across the surface. Depth 256 m. West side of Rockall Bank (57°36′N, 14°29′W).

Fig. 13. Edge of "patch" of the deep-water coral *Lophelia prolifera* (Pallas). The living colonies are approximately 1 m in height. Note the large anemone attached to the dead coral. Depth 256 m. West side of Rockall Bank (57°35'N, 14°29'W).

Fig. 14. Edge of living colony of *Lophelia prolifera*. Note expanded polyps. Depth 256 m. West side of Rockall Bank (57°35'N, 14°29'W).

of the "patch" are generally occupied by dead corals. Portions of living coral that break off the growing face fall onto the sediment in front of it. Sometimes quite large portions of the colony break off and land on the sediment upside down. These broken portions generally do not survive, but they become substrate for the growth of new colonies. Thus the "patch" increases in size.

DISCUSSION

The above few examples demonstrate the value of the manned submersible as a tool for making direct observations and collecting samples on the continental shelf. Indeed, as most of these results could not be obtained in any other way, the advent of the manned submersible in the 1970's is as significant an advance as was the introduction of the naturalist's dredge in the 1830's.

Most of the initial difficulties have now been overcome and the experience gained should enable any future scientific submersible operations in British waters to be successful. In the last three years or so the commercial demand for submersibles in European waters has increased dramatically, and many more are now operational in connection with the oil industry in the North Sea. This increased demand coupled with the effects of inflation has priced the submersible out of the reach of most scientific users. These increased costs have stimulated the development of unmanned submersibles such as CONSUB (see Chapter 2) attached by an umbilical cable to the mother ship. Although these can never take the place of the manned submersible it will be necessary to utilize such craft in the immediate future. It may be several years, therefore, before further sedimentological and faunal work can be done from manned submersibles on the U.K. continental shelf.

ACKNOWLEDGEMENTS

Figs. 7 and 10 are reproduced by permission of the Institute of Geological Sciences. Fig. 11 was taken by D. White Esq, formerly of Vickers Oceanics Ltd. Dr N.C. Flemming, Messrs R.A. Eden and D.G. Roberts are thanked for collection of additional sedimentological and faunal data during their dives.

REFERENCES

Barnes, H., 1963. Underwater television. In: Oceanography and Marine Biology, An Annual Review, 1: 115–128.

Belderson, R.H., Kenyon, N.H., Stride, A.H. and Stubbs, A.R., 1972. Sonographs of the Sea Floor. Elsevier, Amsterdam, 185 pp.

Belderson, R.H., Kenyon, N.H. and Wilson, J.B., 1973. Iceberg plough marks in the northeast Atlantic. Palaeogeogr., Palaeoclimatol., Palaeoecol., 13: 215—224.

Drew, E.A., Lythgoe, J.N. and Woods, J.D. (Editors), 1976. Underwater Research. Academic Press, London, 430 pp.

Eden, R.A., Ardus, D.A., Binns, P.E., McQuillin, R. and Wilson, J.B., 1971. Geological investigations with a manned submersible off the west coast of Scotland, 1969—1970. Inst. Geol. Sci. Rept. No. 71/16: 49 pp.

Eden, R.A., Deegan, C.E., Rhys, G.H., Wright, J.E. and Dobson, M.R., 1973. Geological investigations with a manned submersible in the Irish Sea and off western Scotland, 1971. Inst. Geol. Sci. Rep. No. 73/2: 27 pp.

Flemming, N.C., 1968. SURV trials April—May 1968. Nat. Inst. Oceanogr. Int. Rep. No. A 31: 15 pp.

Flemming, N.C. (Editor), 1973. Science Diving International. British Sub Aqua Club, 282 pp.

Grassle, J.F., Sanders, H.L., Hessler, R.R., Rowe, G.T. and McLennan, T., 1975. Pattern and zonation: a study of the bathyal megafauna using the research submersible ALVIN. Deep-Sea Res., 22: 457—481.

Green, W.S., 1897. Notes on Rockall Island and Bank, with an account of the petrology of Rockall, and of its winds, currents, etc.: with reports on the ornithology, the invertebrate fauna of the bank and on its previous history. I. Narrative of the cruise. Trans. R. Ir. Acad., 31: 39—47.

Jack, R.L., 1969. SURV — A practical report. Underwater Assoc. Rep., 4: 35—37.

Mott, G.G., 1972. The PISCES submersible system. Trans. Inst. Mar. Eng., 84: 234—242.

Petersen, C.G.J., 1918. The sea bottom and its production of fish food. Rep. Dan. Biol. Stn. Board Agric., No. 21: 62 pp.

Thomson, C.W., 1874. The Depths of the Sea. Macmillan, London, 527 pp.

Tucker, M.J., Smith, N.D., Pierce, F.E. and Collins, E.P., 1970. A two-component electromagnetic ship's log. J. Inst. Navigation, 23: 302—316.

Warner, G.F., 1971. On the ecology of a dense bed of the brittle-star *Ophiothrix fragilis*. J. Mar. Biol. Assoc. U.K., 51: 267—282.

Winget, C.L., 1969. Hand tools and mechanical accessories for a deep submersible. WHOI Tech. Rept. No. 69—32: 180 pp. (unpublished manuscript).

Wilson, J.B., Holme, N.A. and Barrett, R.L., 1977. Population dispersal in the brittle-star *Ophiocomina nigra* (Abildgaard) (Echinodermata: Ophiuroidea). J. Mar. Biol. Assoc. U.K.

CHAPTER 8

SUBMARINE GEOLOGY FROM SUBMERSIBLES

Bruce C. Heezen

INTRODUCTION

The visual observation of contemporary geological processes in the deep sea is the principal use of the submersible in geology. Of course, if erosional or tectonic processes have exposed unique sections of ancient rocks the submersible may also be employed in the study and mapping of stratigraphy and structure.

I have long been interested in the visual aspects of the sea floor. This interest arose abruptly in the summer of 1947 when Maurice Ewing placed a small Woods Hole research ship at my disposal for three months and suggested that I might make a photographic investigation of the New England continental margin. We took 200 stations with our primitive single-shot flash-bulb camera. It took years to understand and interpret our results, which were finally summarized in a book Charles Hollister and I published a quarter of a century later (Heezen and Hollister, 1971). It was not until that book was nearly completed that I had the opportunity of directly observing the sea floor from a submersible. In each of those first instances I was a grateful guest of another scientist. I shall long be indebted to Robert Dill for giving me my first direct view of the sea floor from DEEP STAR 4000 in the San Diego Trough.

In recent years a group of us have been quite active in submersible work during which we have employed a number of different submersibles. This work, initiated and supported by the Office of Naval Research, Ocean Science and Technology Division (Code 480), is now supported by their successor the Naval Ocean Research and Development Activity (Code 400).

In the present chapter I quite arbitrarily limit the discussion to a few investigations which we have made in the eastern Pacific between Hawaii and Central America. The subjects range from the initial form of volcanic relief created by submarine volcanism near Hawaii, through seamount volcanic forms, oceanic-plateau sedimentation and erosion, tectonic dislocation, and on to sea-floor subduction in the Middle America Trench.

The dive series conducted in 1974 and 1975 were arranged by Dr. Alexander Malahoff of ONR and Lt. Cmdr. George Verd of Submarine Development Group 1. The Officer in Charge of DSV SEACLIFF in 1974 was Lt. Cmdr. L. Bliss and in 1975 the Officer in Charge was Lt. Wm. Schotts. The Officer in Charge of DSV TURTLE in 1975 was Lt. Cmdr. John Cameron.

Cmdr. V.K. Nield of ONR (now NORDA) assisted in making arrangements and participated in one cruise. The research reported here was supported under Contracts N-00014-67-A-0108-0036 and N-00014-76-C-0264. I wish to acknowledge with thanks the collaboration of Alexander Malahoff, Michael Rawson and Daniel Fornari, scientific observers on the 1974 and 1975 dive series in the Pacific. In preparing this review I have drawn upon more detailed reports prepared in collaboration with the above scientists.

VOLCANIC FOUNDATIONS

The oceanic crust is created at the Rift Valley of the Mid-Oceanic Ridge by injection of mantle materials. This injection is accompanied at the juvenile sea-water—sea-floor interface by sea-floor volcanism which creates (together with associated tectonism) the complex initial topography of the sea floor. To the ship-borne oceanographer this initial topography manifests itself as a faint pattern of overlapping hyperbolae on the echogram, a lack of penetration on the seismic profiler, bent, gouged and empty coring tubes, remote photographs of broken rocks and pillows and dredge hauls of pillow basalt. Although all the above are indicative of the sea-floor geology and provide a quite reliable guide to the situation, the complexity defies a detailed analysis of the remotely obtained data in the more precise terms of features and processes.

Of the three existing syntactic foam submersibles two are at present limited to 2000 m and only one can reach 3600 m. This situation is soon to be modified when DSV TURTLE is modified to reach 3000 m and DSV SEACLIFF is modified to reach 6000 m. In the summer of 1974 one syntactic foam submersible (ALVIN) had been modified for use in 3600 m. It was employed in the exploration of submarine volcanic processes in the Rift Valley of the Mid-Atlantic Ridge (Ballard et al., 1975). During the same weeks we were using the 2000-m DSV SEACLIFF in the investigation of submarine volcanic features near the "big island" of Hawaii (Fig. 1). The feature of prime interest was the Puna Ridge, a steep-sided submarine volcanic ridge which extends for more than 70 km east-northeast from Cape Kumahaki.

Puna Ridge

The Puna Ridge is the submarine continuation of the East Rift zone of the active Kilauea Volcano. Previous investigations from surface ships had established that the ridge crest is devoid of sediments. A number of remote bottom photographs had revealed the presence of pillow lavas (Moore and Reed, 1963) and a detailed bathymetric survey revealed steep-walled linear elevations and depressions at the ridge axis which parallel the general northeast trend of the ridge (Fig. 2). There was some evidence that submarine erup-

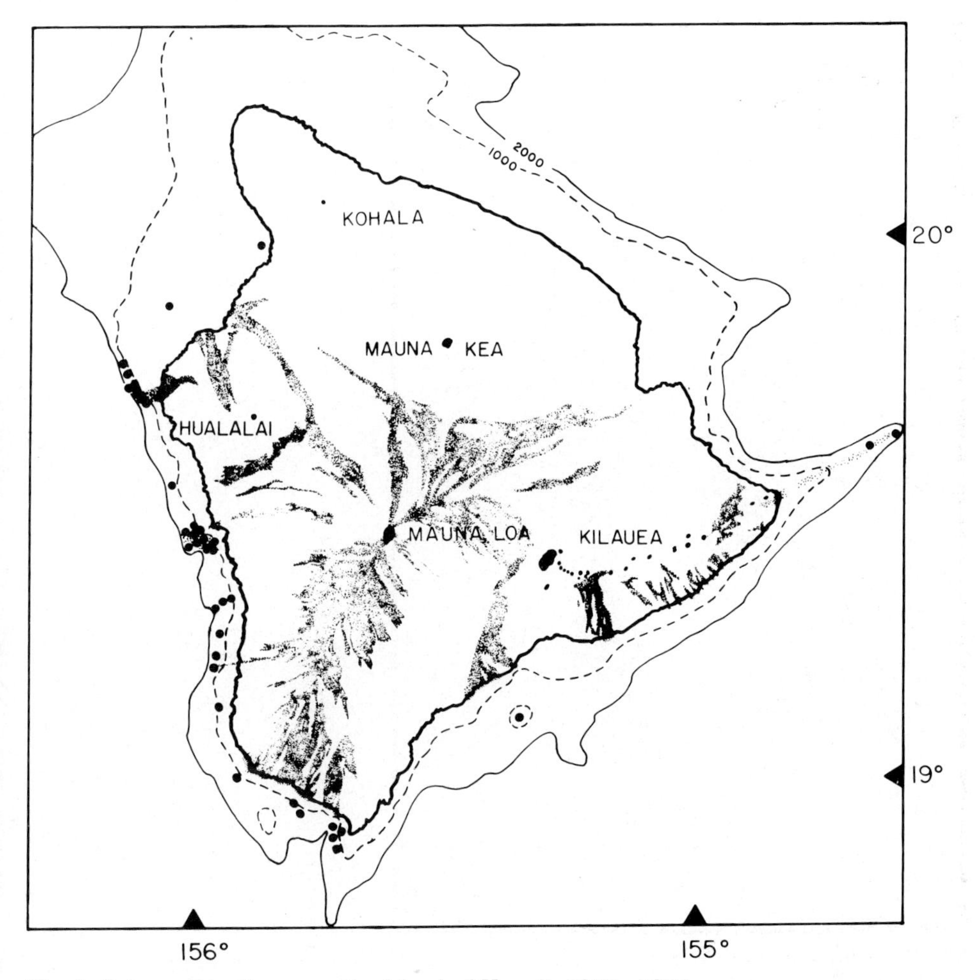

Fig. 1. Submersible dives near the Island of Hawaii, 1974—1975.

tions had occurred along the ridge crest in historical time. The Puna Ridge seemed, therefore, an ideal location to study submarine rift volcanism. Several miles of the ridge crest lay within the 2000-m depth capability of SEACLIFF. It had to be anticipated that significant differences would be found between the volcanic features on the flank of a large central volcano and those in axial Mid-Oceanic Rift Valley, particularly when the scales and a vastly different tectonic style are considered. One of our disappointments was that due to persistent heavy northeast trades we managed to get only two dives to the Puna Ridge. These were, however, well worth the effort. The two dives were made in June 1974. On the first (Dive 108), Dr. Alexander Malahoff was scientific observer (Fig. 3). On the second (Dive 115), I was

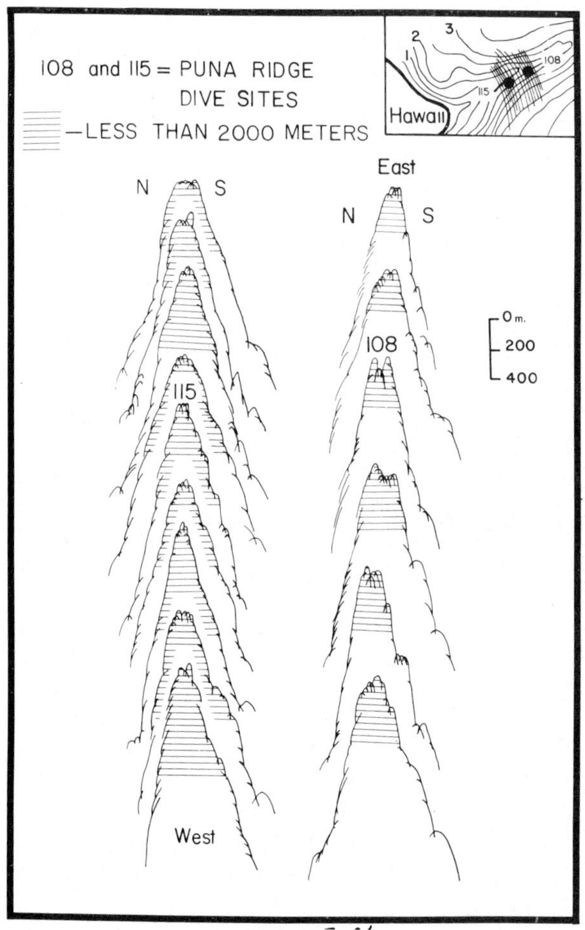

Fig. 2. Echogram profiles of the crest of Puna Ridge in the area investigated with DSV SEACLIFF during July 1974.

scientific observer. On both dives the SEACLIFF was piloted by Lt. Cmdr. L. Bliss with Lt. M. Smith as co-pilot (Fornari et al., 1977).

Both dives encountered a nearly sediment-free pillow-lava bottom nearly devoid of bottom life (Fig. 4). The sparsity of attached filter feeders was especially striking in view of the appreciable currents encountered. Both dives began on the southern margin of the crest zone and proceeded north or northwest to the crest. On both traverses a dominant feature seen was linear fissures parallel to the ridge axis. These long fissures attained widths up to 10 m and depths of the same order (Fig. 5). We descended into some of the widest ones but passed over the more awesome narrow ones which looked like potential submersible traps. In numerous cases fissures terminated only to be replaced in a few meters distance by a new subparallel fissure. The

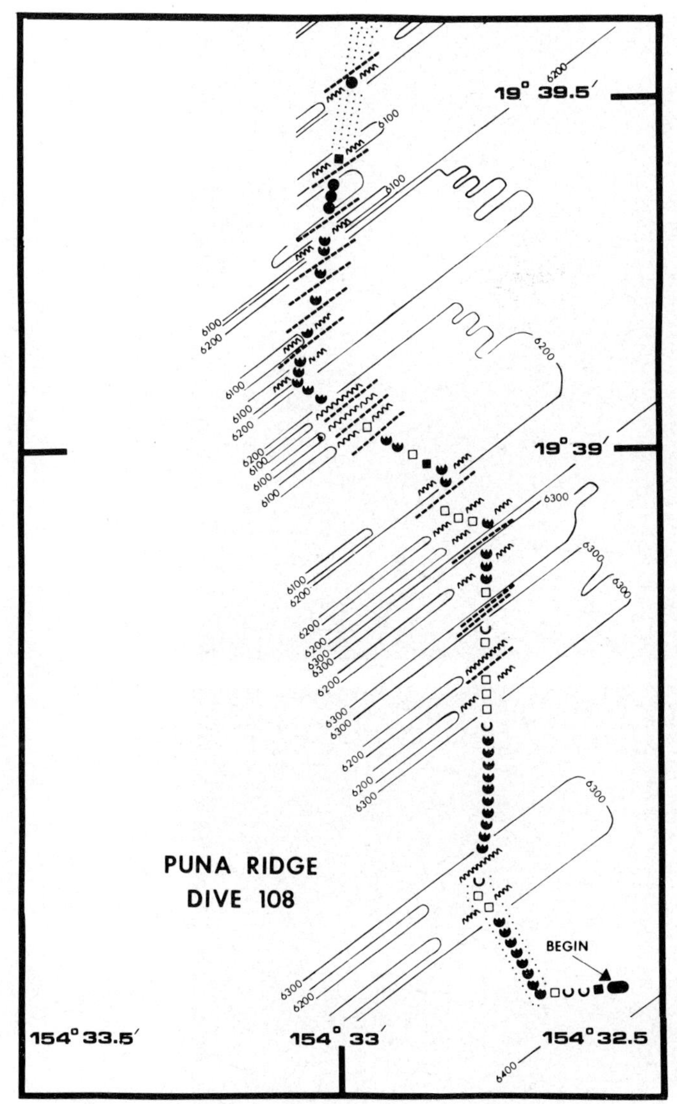

Fig. 3. Lava forms and fissures on the Puna Ridge crest observed along submersible traverse during DSV SEACLIFF Dive 108. See Fig. 8 for legend.

fissured areas are associated with a somewhat greater abundance of broken pillows and collapsed lava tubes. A particularly striking aspect of several of the fissures is that extrusive forms on opposite sides match across the chasm indicating that the fissure post-dates the latest outpouring of lava.

Although the mechanism of pillow-lava formation has long been correctly understood, the dramatic films obtained by Lee Tepley (Tepley, 1974),

Fig. 4. Pillow lavas and fissures of the crest of the Puna Ridge. Upper: Small fissure observed on Dive 108. Lower: Pillows and lava tubes at Dive Site 115.

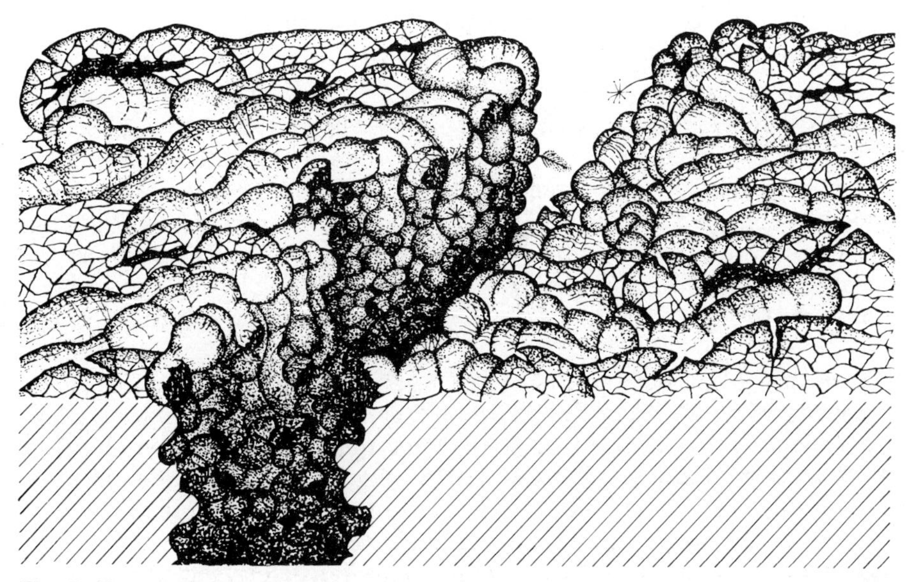

Fig. 5. Fissure at the crest of the Puna Ridge. From an assemblage of photographs (Dive 108).

when Mauna Ulu lavas poured into the sea, provide such a detailed demonstration that we should be spared future attempts to provide novel explanations. We had, however, expected that pillows at 2000 m might have shown rather obvious differences to those formed in 10 m, since the hydrostatic pressure at 2000 m by suppressing the formation of large vesicles, might cause a discernible change in flow characteristics of lava. If there is such a difference we have yet to identify it. Volcanologists have noted that vesicularity is an important factor in the creation of different flow forms in subaerial flows. We, therefore, expected to see some differences between 10 m and 2000 m in the large flow forms.

The small prominences which occur along the crest of the Puna Ridge had been interpreted (Malahoff and McCoy, 1967) as the submarine analogs of the subaerial cones which occur in the Puna district of Hawaii along the East Kilauea Rift Zone. Although our observations do not preclude that interpretation, neither do the observations on the two dives support it. The features observed seemed all rather linear and large equidimensional features were not observed.

Broken pillows occurred in at least two types of situations. The long parallel fissures which rent apart the pillow fields, caused increased breakage of the fragile shells of thin-walled pillows, but had little effect on the more solid variety of pillow. Thus the occurrence of collapsed pillows seems to have more to do with the thickness of the outer shell and whether or not the pillows are hollow or full, than with the occurrence of fissures.

In our dives on the Puna Ridge and on other dives in the vicinity of Hawaii we noticed a range of pillows and cylinders, from those with a perfectly smooth surface to slightly cracked to an elephant-like hide which in extreme cases develops a deep radial pattern separating surface polygons. Deductively one would expect that absolutely smooth cylinders and pillows (Fig. 6) must be hollow, for otherwise how could cooling have occurred so evenly as to avoid the creation of shrinkage joints? On numerous occasions we have determined that smooth cylinders are hollow (Fig. 7). In the case of broken pillows the situation is even more obvious. We have looked through the collapsed tops into the hollow interiors of hundreds of smooth-surfaced pillows.

Another class of submarine flow form resembles subaerial features found close to vents. Thin 1—3 cm thick crusts form over fast-flowing lava streams only to collapse into fields of broken plates minutes later when the flow passes. We observed similar features on the Puna Ridge. In some cases the surfaces of the broken thin plates were as smooth as a pie crust, while in other cases it had been warped and crumpled in a plastic state before collapse. In the former case the scene resembled plaster fallen from some ceiling above, while in the latter case the scene resembled a litter of gnarled weather-beaten wood on a mountain slope.

Pillow wall. On the second dive (Dive 115) a remarkable wall was observed and traced for more than 2 km along the axis of the ridge (Fig. 8). The top of the wall is less than 5 m wide. When we drove the submersible along the wall during the last two hours of the dive the sea floor was only visible out of the forward view port. Neither the observer (on the port side) or the co-pilot (on the starboard side) could see the bottom through the side ports. The pilot would make some distance along the wall, then stop and turn the boat, so that the observer could see the bottom. From experience in approaching the bottom in water of similar clarity, we know that we can see the bottom (with the lights employed in 1974) from altitudes of 7—10 m. Thus we conclude that for the entire traverse along the crest of the pillow wall the height of the wall exceeded 10 m. Echograms from the surface ship and several submersible transits up the sides of the wall suggest that the wall maintains a height of 20—30 m in the area of investigations (Dive 115).

Where examined on their near-vertical sides, the walls were composed predominately, but not exclusively, of unbroken radially fissured solid pillows (Fig. 9). None of the pillows were of the smooth-surfaced hollow type (Figs. 10 and 11). Even where the pillows were broken off there was no hollow interior exposed. A narrow talus slope, which lies at the base of the wall, is composed predominately of the radial segments of solid pillows with an occasionally complete pillow still in tact (Fig. 12). The scattered broken pillows seen on the near-vertical sides of the pillow wall are the obvious source of the talus. At first we considered the possibility that the wall is a narrow

Fig. 6. Smooth-surfaced lava cylinders at the crest of the Puna Ridge. Upper: Dive Site 108. Lower: Dive Site 115.

Fig. 7. Collapsed lava tubes at crest of Puna Ridge. Upper and Lower: Dive 108.

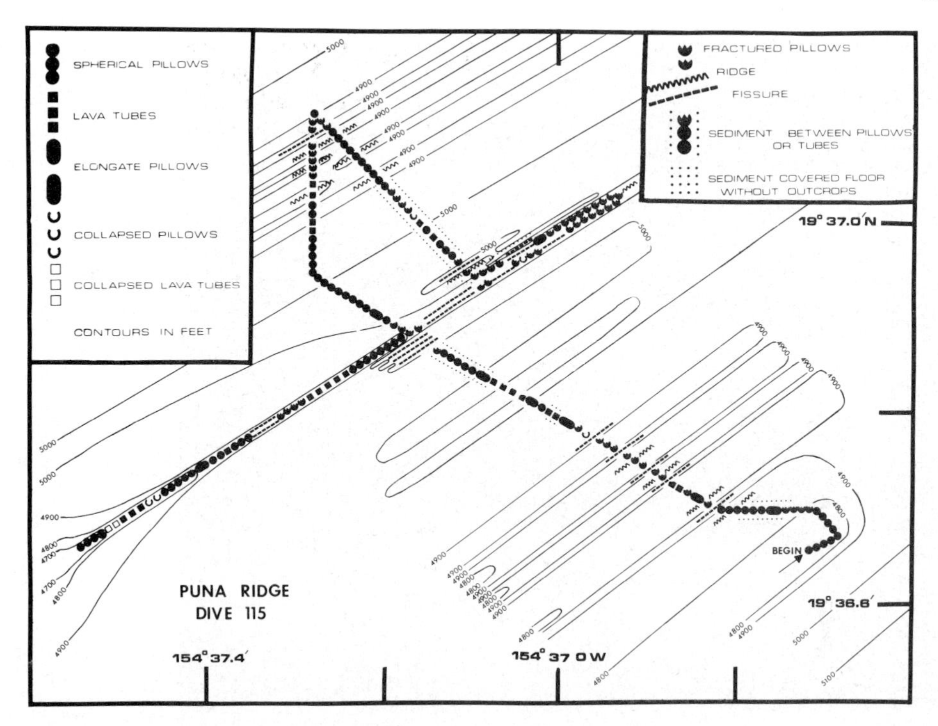

Fig. 8. Lava forms, pillow walls and fissures on the Puna Ridge crest observed along submersible traverse during DSV SEACLIFF Dive 115.

horst, but the predominance of unbroken pillows on both faces of the wall made this explanation seem unlikely. The feature seemed more likely to be constructional. The linearity suggests a fissure eruption of perhaps somewhat cooled lavas, which were rapidly quenched after upwelling from the fissure and with each successive pulse of the eruption producing an additional tier of pillows on either side of the feeding fissure. The lavas did not spill out on the adjacent ocean floor as in the case of those which flowed down slope leaving the collapsed covers of transient lava lakes, hollow pillows and hollow cylinders. The erupting lavas which built the wall, cooled within 2 or 3 m of their vent. Although one might suspect that pillow walls may be present in the Rift Valley of the Mid-Atlantic Ridge, the FAMOUS divers did not report this phenomenon (ARCYANA, 1975).

The pillow wall at the crest of the Puna Ridge is, in a way, a small model of the whole Puna Ridge. The Puna Ridge is narrow and sharp crested, because lavas emanating from axial-fissure eruptions flow only a short distance over the sea floor before being arrested by cooling. The distance lavas reach from submarine fissures is much less than in the case of subaerial eruptions due to the presence of sea water.

The crest zone of the Puna Ridge, in the vicinity of Dive 108, was marked

Fig. 9. Pillow lavas on the pillow wall. Upper and Lower: Dive 115.

Fig. 10. Pillows and lava cylinders observed on the crest of Puna Ridge south of the pillow wall. Upper and Lower: Dive 115.

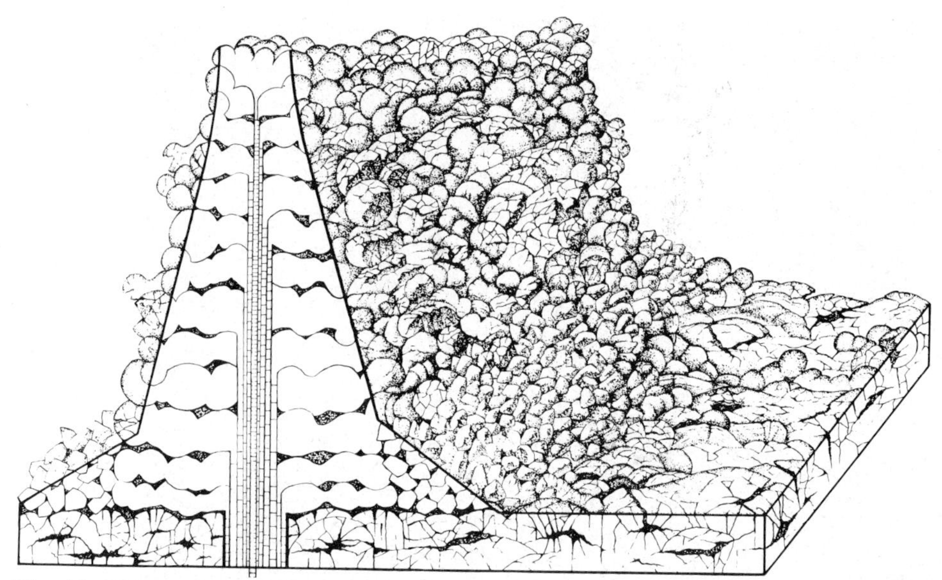

Fig. 11. Pillow wall at the crest of Puna Ridge. From an assemblage of photographs, Dive 115.

by deep fissures; a pillow wall was not observed. The lavas of the latest eruption had been deeply fissured, perhaps as a forerunner of a new eruption. Lava cylinders drooped down the sides of the ridge, some solid, some hollow. In depressions between the upturned margins of fissures and between rounded linear pillow mounds at the crest of the ridge, we observed the broken plates and gnarled lava sheets of the collapsed surface of temporary lava lakes, as well as collapsed pillows whose lava had flowed out before the entire mass was chilled. The pillow walls represent the extreme case of cooling arrest where lavas reach only a couple of meters distance from the vent before cooling into solid pillows. The more mobile, farther-reaching flows, which produced the lava cylinders, drooped down the flanks of the Puna Ridge, and probably reached no more than a few hundred meters from their feeding vent. Pillow walls are relatively dense and presumably are underlain by an even denser feeder dike. The flows which traverse a few hundred meters of sea floor with their hollow pillows and hollow lava cylinders produce a much less dense accumulation. Although the very thin-shelled pillows will collapse under a few meters of overburden and the thinnest-shelled pillows have already collapsed of their own weight, there are many voids between the massive pillows and within the thick-walled hollow pillows and cylinders which will survive under an appreciable overburden. Here lies the explanation of the long-perplexing observation of gravimetry that the average density of volcanic pedestals is far below the average density of the constituent volcanic rock (Woollard, 1954). Low-density weathering products

Fig. 12. Pillow wall and pillow talus. Upper: Pillows on the near vertical pillow wall, Dive 115. Lower: Talus at the foot of the pillow wall, Dive 115.

also play a role and the scoriaceous nature of subaerially erupted lavas plays a minor role. However, the principal explanation lies in the submarine eruption process itself, with the predominance of hollow pillows and cylinders of intermediate wall thickness. Neither those solid pillows which are found in the pillow walls, or the thin broken eggshells which are found near vents, predominate.

Ka Lae Canyon. The Hawaiian volcanoes are characterized by rift zones which radiate from the central vent (MacDonald and Abbott, 1970). The Puna Ridge represents the northeast submarine continuation of the east rift zone of Kilauea. The southern tip of Hawaii lies along the south rift zone of Mauna Loa. Four dives were made along the submarine continuation of this zone in the vicinity of Ka Lae Point encountering a vastly different environment (Fig. 13).

Scientific observers on these dives were Alexander Malahoff, Donald Peterson, John Lockwood and myself. A fault cuts off the west side of Ka Lae Point and extends as a steep cliff for several kilometers north into the island as the Kahuka Pali. It also continues to the south as the eastern slope of a submarine canyon which must owe its origin principally to the fault. The age of the fault is unknown but is comparatively young, on an island where old features are less than thousands of years old.

Although we presume that Ka Lae Canyon is principally a result of faulting, erosion has occurred since the faulting. Dikes were observed standing 10—20 m above the sea floor on the eastern side of the canyon (Malahoff et al., in prep.). The polygonal jointing in a horizontal plane was so dramatically expressed that there is absolutely no doubt that the features are dikes formed in some more erodable material which had subsequently been removed (Fig. 14). The dikes are 1—2 m thick and occur in swarms of 3—5. They lie parallel to the wall in N—S orientation.

Although there was little or no sign of pillows on the eastern fault-controlled wall of Ka Lae Canyon, a dive to the floor of the canyon (107) revealed a pillow and cylinder terrain thinly blanketed by ooze. Current effects were observed on the walls, but on the dive to the floor no current evidence was seen. Although sediment has been eroded to expose the dikes, and recent sediment is not being deposited in the current-swept environment of the canyon walls, there is no evidence that transport is occurring along the axis of the canyon below about 1800 m depth. Turbidity currents are not and apparently never have been an important factor here (perhaps never having occurred at all). The erosion seems to be caused by ocean currents sweeping around Hawaii augmented by the tides.

Sediment-covered slopes. Elsewhere around the pedestal of Hawaii in areas remote from the submarine extension of rift zones and removed from those

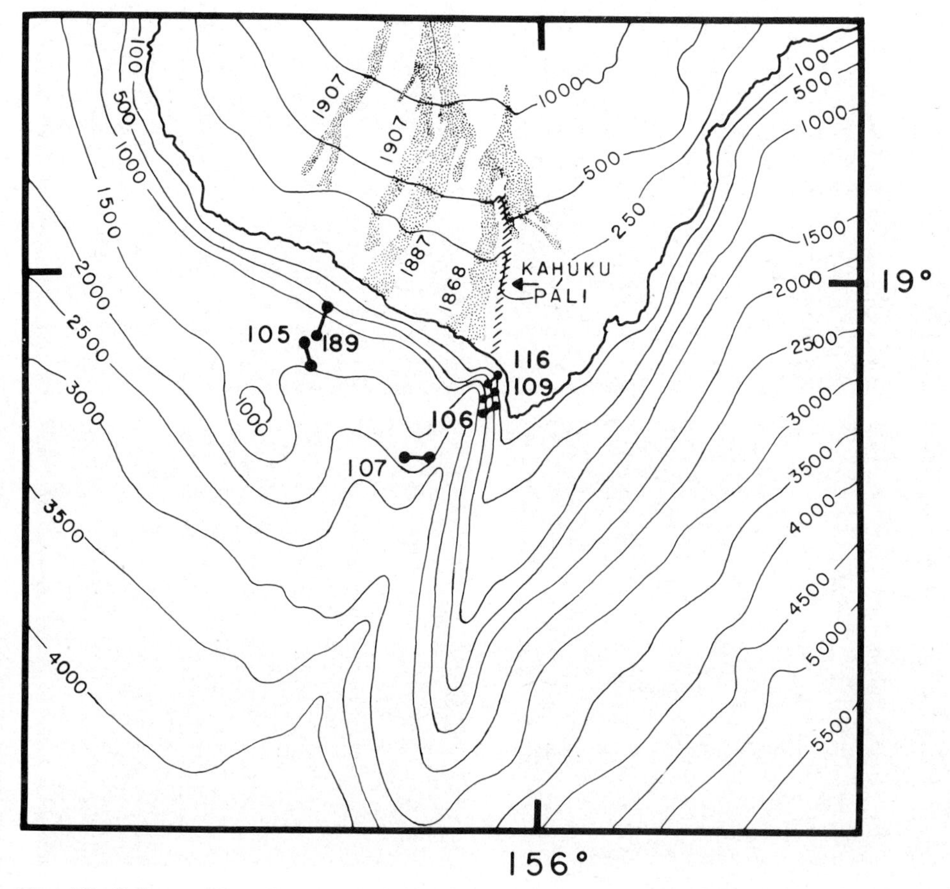

Fig. 13. Submersible dive sites in Ka Lae Canyon near the southern most point of the Island of Hawaii. Stippled areas on land represent historical flows.

points where historical flows have entered the ocean, the sea floor is generally covered with sediment. This sediment generally is quite clearly current smoothed and shaped. Talus blocks are not infrequent. The most striking observation is that sedimentation has been able in a few centuries or millenia to completely mask the underlying volcanic relief.

Submerged terraces. Along the northwest Kona Coast broad terraces were observed down to depths of about 400 m (Heezen et al., in prep., a). On a terrace at about 350 m mangrove mats were observed. Samples gave radiocarbon ages of 4590 yrs. ±260 B.P. Here the lava terrain is eroded and over wide areas partially covered by thin calcareous deposits.

Fig. 14. Polygonal jointing in eroded dikes on eastern flank of Ka Lae Canyon. Upper left: fractured lava with abundant growth of filter feeders — Dive 106, 1100 m. Upper right: top of excavated dike — Dive 116, 350 m. Middle left: fractured lava on east bank of Ka Lae Canyon — Dive 106, 1100 m. Middle right: dike swarm — Dive 116, 1000 m. Lower left and right: polygonal columns of jointed dike scattered near base of exposed dike — Dive 116, 300 m.

SEAMOUNTS

Once oceanic volcanoes die their pedestals sink slowly with the underlying cooling oceanic crust. The pillows can be observed in many cases for a long period before carbonate deposition and manganese encrustations obscure the volcanic forms. We have observed the crests and flanks of several oceanic volcanoes. Here we will limit our discussion to four seamounts on or near the Cocos Ridge (Heezen and Rawson, 1977a).

Of the four conical seamounts to be discussed, Tutu Seamount rises from the crest of the Cocos Ridge, Lorraine Seamount rises from the south flank of the Cocos Ridge in the Panama Basin; and Tortuga and Bielicki Seamounts are located on the north flank of the Cocos Ridge in the Guatemala Basin. All are extinct volcanoes which presumably erupted after the creation of the Cocos Ridge in mid-Miocene time. Pillows and lava cylinders dominated the visual scene on each of the five dive transects. Significant differences in the thickness of manganese, the presence or absence of ripple fields, variations in the amount of sediment accumulation on benches and differences in the degree of fracturing were the principal variations observed in an otherwise continuous expanse of lava pillows and cylinders.

Photographs of ripple marks at 1000 m on an Atlantic seamount (Heezen et al., 1959), ripples at 1370 m on a Pacific guyot (Menard, 1952) and the dredging of mid-Cretaceous rudistids from Pacific guyots (Hamilton, 1956) long ago demonstrated that such isolated features, seemingly of rather trivial cross-section as compared to the broad ocean basins, are subject to denudation by the motions of oceanic water masses. Certainly on empirical evidence alone, any conical seamount which rises above 2000 m in an ocean basin can be expected to be nearly devoid of sediment.

Since the ocean is, in general, a highly stratified medium, motions in the horizontal plane are normal and vertical motions are exceptional. The reduction in cross-section imposed by a seamount on the motion of ocean waters is more important than one might at first realize, for a water mass is compelled to pass around the obstruction without significant vertical excursions. If the moving waters are to continue their flow it is necessary, because of the longer path, for the waters passing around an ideal conical mountain to increase velocity by a factor of $\sim\pi/2$ or ~ 1.6.

Since a velocity of about 15 cm/sec is necessary for the erosion of foraminiferal sand, we may conclude that the maximum velocities of mid-oceanic water motions must frequently reach at least 10 cm/sec. An increase by a factor of 1.6 would raise that velocity to above the minimum required for erosion.

Ripple marks in small fields of foraminiferal sand were observed near the summit of Lorraine Seamount (Dive 16). These moving sands appeared to be a permanent aspect of the area, for they have extensively undercut ledges of manganese-encrusted lithified sediment (Fig. 15). Ripples were not observed

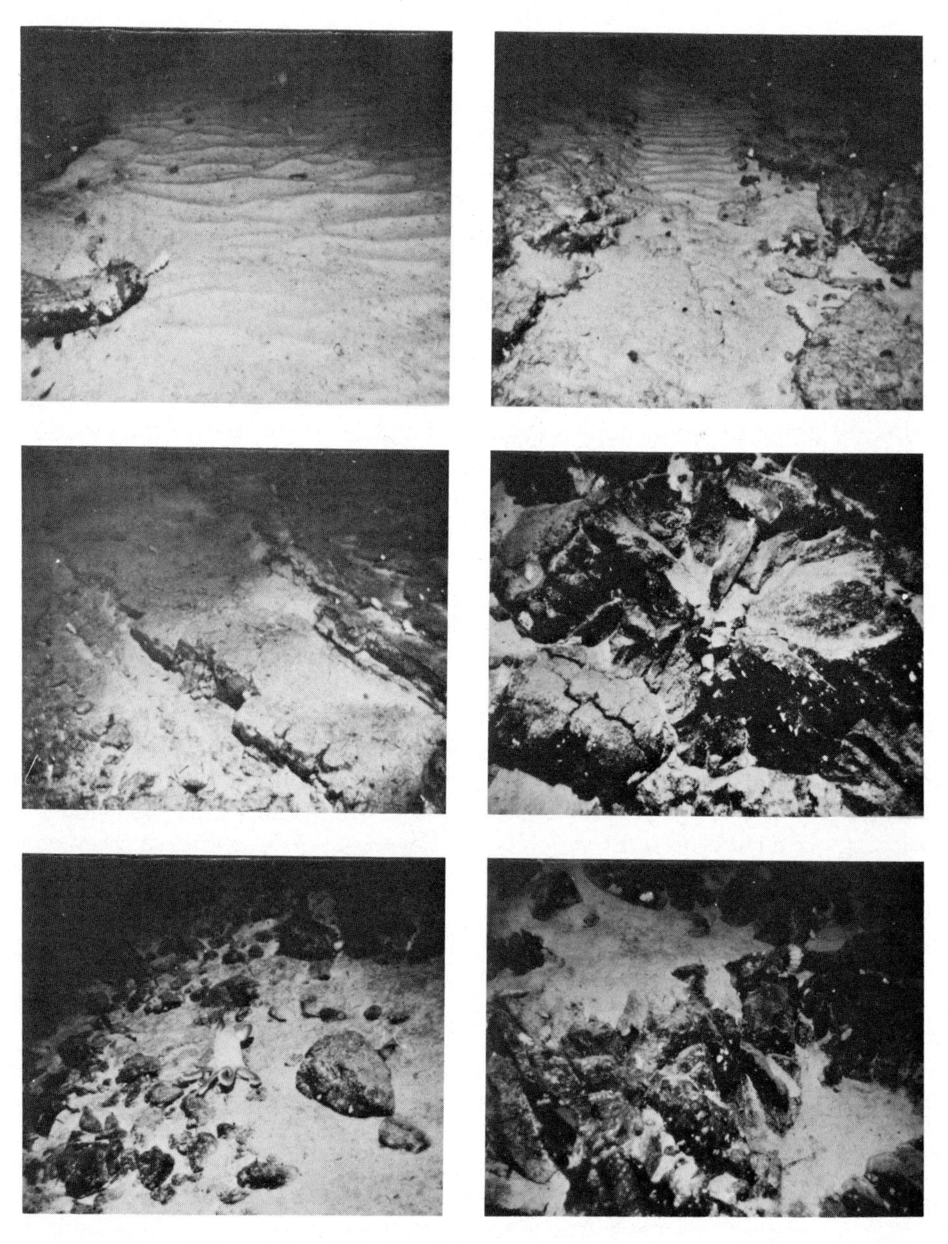

Fig. 15. Views of the crest and flanks of three volcanic seamounts. The top photographs show small patches of rippled sand which are often found at seamount summit areas. The middle photographs illustrate outcrops of bedded rock (left) and pillow lavas (right). The lower photographs show small patches of smooth sediment which accumulate on the less steep areas of the seamount flanks. Upper left: Lorraine Seamount (16—175): 05°00′N, 85°02′W; 1127 m (604 tau). Upper right: (16—173): 05°00′N, 85°03′W; 1127 m (604 tau). Center left: (16—119): 04°59.8′N, 85°03.2′W; 1143 m (612 tau). Center right: Tortuga Seamount (13—22): 05°00′N, 87°30′W; 1036 m (555 tau). Bottom left: Lorraine Seamount (15—155): 05°00′N, 85°07′W; 1371 m (735 tau). Bottom right: (13—31): 05°00′N, 87°30′W; 1005 m (536 tau). 1 tau = 1/400 sec.

on the lower flanks of any of the four seamounts investigated. Ripple marks appear to be characteristic features of summit areas and slope-break regions but do not normally occur on the flanks of seamounts.

The 1500-m summit of Ita Maitai Guyot in the western equatorial Pacific is capped by 200 m of post-Paleocene foraminiferal sand and nannoplankton ooze. This cap thins towards the break in slope and unconsolidated sediment is absent from the upper flanks of the guyot. Drilling on the summit of Horizon Guyot Ridge also penetrated a cap consisting of ~300 m of post-lower Cretaceous foraminiferal sand. The break in slope of Horizon Guyot Ridge is also an area of current activity where cherts originally formed in the central sediment cap have been later exposed by erosion (Lonsdale et al., 1972). Mid-Cretaceous rudistid reefs at the break in slope of the summit plateaus of numerous western Pacific guyots are, after 100 m.y., still unburied by sediments. The central sediment cap found on guyots and the occurrence of rippled sand on seamount summits may be the result of common phenomenon. Sediments appear to become trapped in a local current system which carries sand back and forth across the summit, but does not disperse it to the seamount flanks.

Manganese is deposited on all sediment-free submarine rock surfaces. At first it merely darkens the rocks, but as growth increases, surfaces become jet black and as it thickens further shapes are increasingly altered until eventually all small irregularities are eliminated. The ability of an observer to accurately evaluate such changes, not only depends upon his experience and general familiarity with bottom features, but on the types of rocks involved. It can be, for instance, more difficult to estimate the extent of manganese coating on tabular rocks than it is on more irregular ones which become detectably rounded more quickly and often assume strangely subdued shapes. Manganese crusts were present on each of the four seamounts. On the basis of a number of geochemical measurements (Bender et al., 1970), the general deep-sea rate of manganese growth has been shown to be about 1 mm/m.y. This rate apparently varies widely with water motion and proximity to sources of manganese. A mid-Miocene rock which has been continuously exposed to sea water, can be expected to possess a crust at least 1—2 cm thick.

Of the four seamounts, Tortuga and Tutu Seamounts have the smallest thickness of manganese encrustation. Shapes are so little altered that pillow lava could be easily identified. In fact, the pillows resembled the recent submarine flows we observed on the flanks of the island of Hawaii. A greater thickness of manganese was observed on Lorraine Seamount (Dives 15—16). The undercut horizontal ledges indicate that the manganese crusts are more resistant to abrasion by the shifting foraminiferal sand than the underlying rock on which the manganese originally began accumulating. The thickest accumulation of manganese was observed on Bielicki Seamount (Dive 29) where shapes were so completely altered that the pillow lavas were only occasionally identifiable. Samples of manganese crust formed over devitrified volcanic rubble in Bielicki Seamount are about 10 cm thick and the similar subdued shapes observed throughout the dive suggest that ~10 cm is a good average thickness for the seamount. Such an accumulation suggests a long period during which currents have prevented the deposition of sediments. Bielicki Seamount rises from the lower flanks of the Cocos Ridge and might be somewhat older than the ridge itself.

So far I have noted those aspects of the visual scene on the four seamounts which contrast with the recent lava terrain of the Puna Ridge: the fields of rippled sand, the presence of manganese crusts, the sediment-filled voids, and the much greater population of attached filter feeders. But it is really the similarities between the Puna Ridge and the seamounts which are the most striking aspect. (We did not observe eggshell pillows, nor fields of pie-crust or wood-pile lava, but this is not surprising since such types are generally found in depressions, do not have much relief, and can be assumed to be easily and quickly buried by sediments.) We did observe lava cylinders draped one over the other drooping pendulously down steep slopes.

I doubt that some of the few hundred-year-old lava cylinders near Hawaii could be visually distinguished from the several million-year-old pillows and cylinders seen on the seamounts, if it were not for the noticeable manganese crust and general absence of glassy rinds on the older lavas. The more or less equidimensional pillows resembling grain bags stacked in a warehouse, were also indistinguishable, except by the above criteria, from fields of pillows observed near Hawaii. Perhaps the most striking difference is that recent lavas near Hawaii are rapidly covered by sediments, while the flanks of the Neogene seamounts, with the exception of benches, are almost devoid of sediment.

DEEP-SEA FLOOR

The vast expanse of sea floor which lies between the seamounts, islands, and ridges we have explored is at present far below the depth limit of all of the syntactic foam submersibles. Only the bathyscaphs TRIESTE and ARCHIMEDE can dive beyond the average ocean depth. A remote device, the Bottom Ocean Monitor (BOM), has been devised to obtain indirect remote visual observations of the rate of physical and biological activity on the deep-sea floor (Gerard et al., 1976). Our first deployment of the apparatus in 1975—1976 demonstrated such an active deep-sea environment that we can now anticipate very fruitful deep submersible operations in the near future when the depth capability of syntactic foam submersibles is extended to 6000 m.

To obtain a long time-series of photographs at a single location, the camera was programmed to obtain a photograph each four hours for several months. We anticipated that swimming organisms would appear, if at all, in single frames and would be gone by the time of the next exposure, but much to our surprise this was also true of the benthic creatures. A holothurian or a crab would appear in one photograph, but in every case by the time the next photograph was taken the animal had moved out of the 4-m^2 field of view.

I had assumed that life in mid-ocean on the deep-sea floor at 5000 m was at a much slower pace and that there was little to be learned from submersible dives to these environments that could not more easily be remotely obtained. Now I have a different opinion based on our remote instrument (BOM) results. Of course the bulk of investigations will continue to be made remotely but the submersible will be increasingly needed to tackle the more difficult problems, to determine what needs to be observed remotely, and to verify the validity of remote observations.

OCEANIC PLATEAU

Early in 1975 we made a submersible investigation of sedimentary, tectonic and volcanic features of the Cocos Ridge (Fig. 16) (Heezen and Rawson, 1977a).

Much of the ridge crest lies within the present 2000-m depth capability of DSV TURTLE. Before we commenced the dive series we made a study of

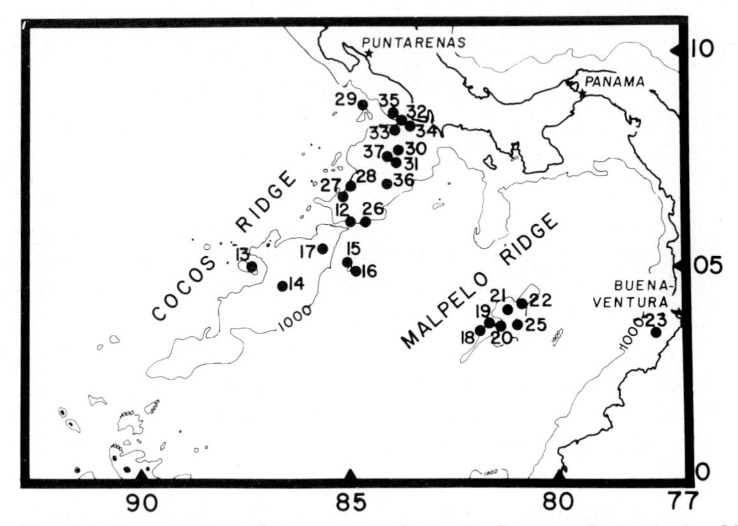

Fig. 16. Twenty-five dives were made in the Panama Basin during March and April 1975. Ten dives were devoted to the crest of the Cocos Ridge. Five dives (13, 15, 16, 27 and 29) were devoted to seamounts, six dives were made to the Malpelo Ridge (18—22, 25). One dive explored a canyon off Colombia (23) and four dives (32—35) observed contemporary subduction in the Middle America Trench.

3.5-kHz echograms previously obtained on the Cocos Ridge. We focused on those features which could be reached with a 2000-m submersible, which gave prospect of significant visual effects and which were sufficiently concentrated to ensure identifiable contrasts on ~2-km transects.

Normal crest of the Cocos Ridge

Excluding the fairly local occurrences of channels, seamounts, moats and elevated blocks, the remainder of the Cocos Ridge maintains a remarkably uniform pattern of conformable internal acoustic stratification and a smooth gentle surface topography. Neither the seismic reflection profiles (Fig. 17), the higher-resolution shallow-penetration 3.5-kHz echograms nor the still higher-resolution 12-kHz echograms suggest the existence of any widespread erosional or depositional forms. The reflectivity of the smooth crest over broad areas is remarkably strong and uniform, giving few if any hints of otherwise undetected small features at or near the limit of resolution. Thus, it was not at all remarkable that a 1100-m dive to the normal smooth crest (Dive 14) revealed a tranquil, tracked-pelagic environment without any evidence of smoothing or lineation and with a near absence of stalked filter feeders. The visual observations, in fact, confirmed what could be concluded from a study of the remote recordings: the tranquil depositional environment now prevailing at the dive site has existed unchanged since the creation of the ridge in the Middle Miocene.

Channels

Although most of the Cocos Ridge crest is covered by a continuous conformable blanket of ooze and chalk (Fig. 17), the 3.5-kHz echograms reveal several classes of erosional channels.

Deeply eroded channels. The first channel to be investigated, a prominent deeply eroded one ~50 km long, had been crossed several times by R/V VEMA (Truchan and Aitken, 1973). This channel, which crosses the Cocos Ridge from NW to SE, is joined to at least one scour moat around the base of a large seamount. The first dive (Dive 12) revealed that the channel is floored near its southeastern end by smoothed sediment and isolated small outcrops of pillow basalt (Fig. 18). Modest scour was observed around the scattered outcrops. The total lack of sediment on the outcrops confirmed the competence of contemporary currents in preventing deposition. The floor of the channel is substantially smoothed; animal "volcanoes" do not survive long in this environment and those piles present tend to be elongated with the current. A ripple field was fleetingly observed at the beginning of a second dive (Dive 26) in this channel.

A seismic reflection profile across the channel near the sites of Dives 12

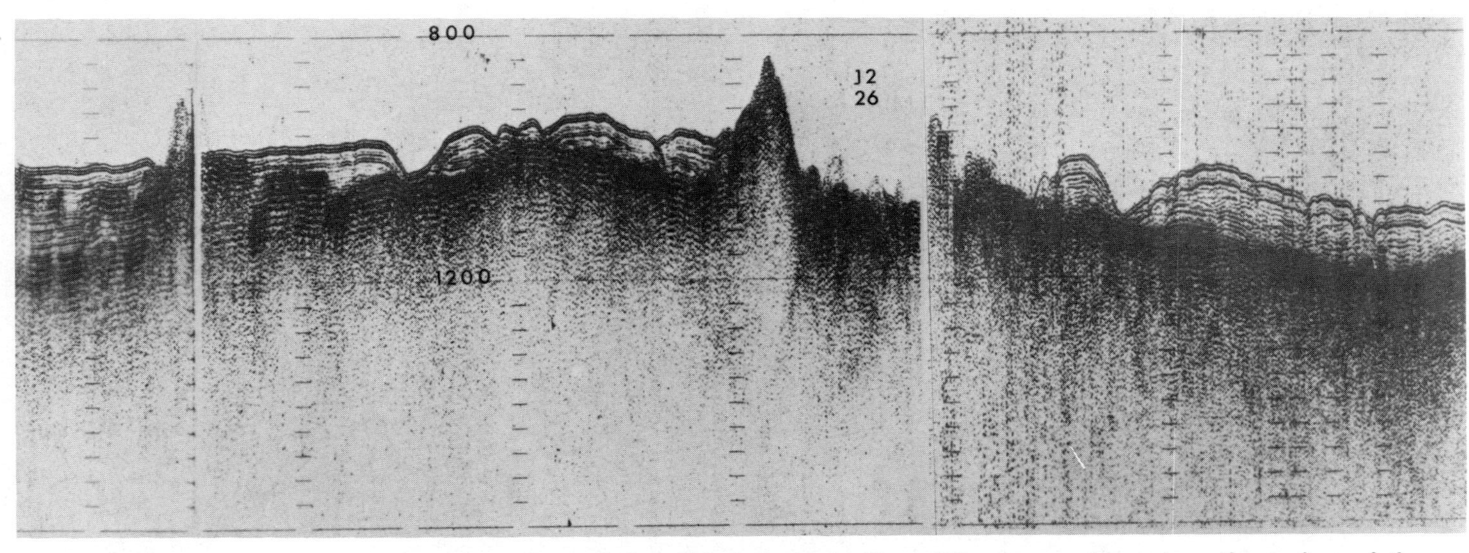

Fig. 17. Seismic reflection record across the Cocos Ridge. Relatively tranquil conditions were observed on the surface of the uneroded sediment pile. Scattered outcrops of pillow lava occur in the deeply eroded areas (Dives 12 and 26). Profile extends east to west from 0.6°02′N, 84°15′W to 06°14′N, 85°35′W. Depths in tau (1 tau = 1/400 sec).

Fig. 18. Scour around pillow lavas on the floor of a Cocos Ridge channel (based on Dives 12 and 26). Pillow lavas appear to have little manganese cover and presumably have been denuded in the recent past by current scour. At this site the seismic reflection profile indicated an absence of sediment. An extremely fuzzy pattern was detected on 3.5-kHz echograms and was identical to that associated with rock outcrops. Nevertheless the two dives traversed smoothed sediment bottom with only occasional scattered outcrops of pillow lavas.

and 26 indicates an almost total lack of sediment strata (Fig. 17). The 3.5-kHz echogram shows lacy hyperbolae from many small reflectors suggesting that surficial sediment is thin or absent in the channel. We had anticipated that dune forms, possibly produced by inflow over the Cocos Ridge from the Pacific Basin to the Panama Basin, might occur near 2000 m in this channel. Instead it was found that the level of current activity has been below the threshold velocity necessary for the establishment of major bed-forms. The sediment surface was extremely smooth. The burrow holes, tracks and feces commonly observed on tranquil bottom (Heezen and Hollister, 1971) were absent or highly subdued. The population of stalked filter feeders was greater than that seen on the normal crest. Yet at the present time the current in the channel is sufficient to prevent the mantling of outcrops even by a sprinkling of ooze. Erosion, or at least nondeposition, has removed, or prevented the deposition of, several hundred metres of sediments.

The evidence that contemporary currents are eroding and smoothing the channel renders unnecessary Wilde's (1966) hypothesis that tephra-charged ephemeral turbidity currents eroded the channels. Instead, the process of channel erosion seems to be present-day water motions involved in oceanic circulation and, perhaps more importantly, the superimposed motions induced by the tides.

The pillow lava forms observed were not deeply encrusted by manganese which would have been the case had a stable channel floor been maintained for a long time as a nondepositional environment. This may mean that erosion has only recently reached the volcanic basement.

Shallowly incised channels. Channels which cut into, but not entirely through, the sediment cover exposing acoustic reflectors are another class of features which appeared to give promise of significant visual observations. The 3.5-kHz echograms had suggested contemporary activity, but the possibility that the channels are relics of a more active recent past could not be completely excluded until such a channel was visually explored (Fig. 19).

A submersible traverse (Dive 30) was made across the axis and up the east bank of a small channel which dissects the evenly stratified oozes of the elevated Cocos Ridge crest near 84°W longitude (Fig. 20). The current observed in the channel axis flowed from N to S at about 1/3 knot. The numerous rattail fish consistently swam into the current. The comparatively robust benthic life of the sea floor consisted of stalked sponges bent over the current, and occasional crabs and echinoids. Fifty meters above the channel axis the current decreased and above that depth we entered a northerly flowing current. In the transition zone between the southerly and the northerly currents, a well-defined pattern of transverse ripples was developed on the channel bank (Fig. 20). Beneath the upper northerly current, scour and other current effects were somewhat more subdued than near the channel axis. Near the end of the dive at a depth of 1080 m, ledges of flat-lying man-

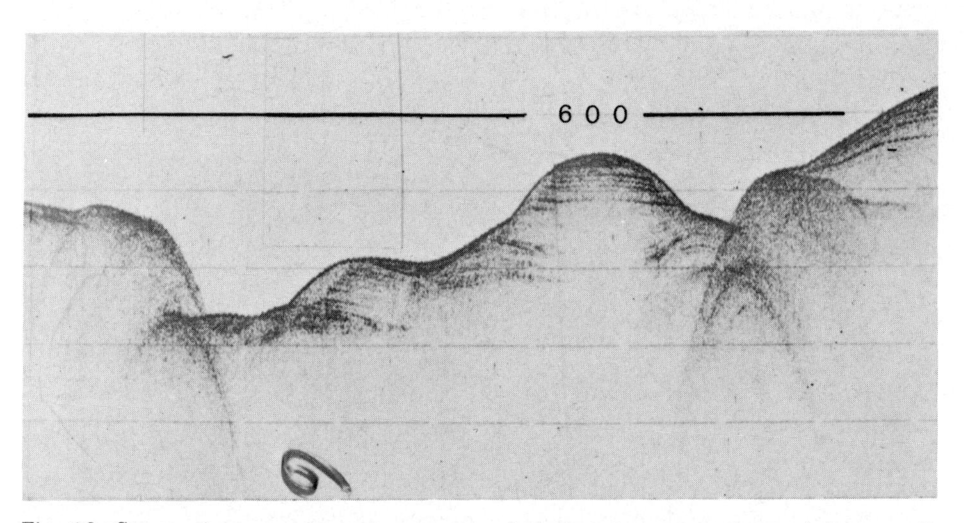

Fig. 19. Scour, ripple marks and outcrops of chalk ledges were observed during a dive (30) which traversed this Cocos Ridge Channel. Echogram (3.5 kHz) east to west from 07°29'N, 83°58'W to 07°29'N, 84°01'W.

Fig. 20. Sketch of principal features observed in an erosional channel at the crest of Cocos Ridge (Dive 30). Sketch is greatly compressed horizontally, the ledge on the right and channel axis to the west being approximately 2 km apart. In the floor of the channel a current flowed from north to south deflecting attached fauna and smoothing out bioturbations but leaving no scour marks or ripple marks. On the east flank of the channel the current diminished and changed from southerly to northerly. At this location ripple marks were prominently seen on the channel floor for a distance of a few hundred meters. Higher on the east wall current effects were more subtle than beneath the deeper southerly flowing water. The flat-lying ledges of chalk were colored by a patina of manganese but were not deeply encrusted. The ledges (at the right) were completely devoid of ooze although observers noted that the current was insufficient to carry away sediment stirred up from the adjoining ooze areas by the submersible.

ganese-stained chalk were observed. At that point in the dive the current ceased to flow and sediment stirred up by the submersible did not drift off, but settled back to the sea floor after a few minutes. The rock ledges were devoid of sediment cover indicating that the absence of current was temporary.

Scour moats

Moats surrounding seamounts and knolls are well known. For a time during the debate concerning the origin of the moats, isostatic subsidence was proposed as a general explanation. Except for the Hawaiian moat, which is accepted as a recent isostatic feature, most moats are now believed to be products of current erosion and differential deposition.

On the crest of the Cocos Ridge scour moats are also occasionally associated with fairly small topographic features which rise sharply, though not particularly high, above the ridge crest.

Several hundred meters of pelagic ooze is lacking from the seamounts which rise from the otherwise thickly sedimented ridge crest. Since this deficit does not appear as an apron surrounding the seamount, it can be assumed that currents in the scour moat and channel system transport and disperse the missing sediments.

Malpelo Ridge badlands

We made six dives (18—22, 25) to the Malpelo Ridge, an oceanic plateau which rises from the center of the Panama Basin (Heezen et al., in prep., b). I will only comment here on one aspect of that area. The flanks of Malpelo Ridge are dissected by narrow valleys which cut deeply into the sedimentary cap of the ridge.

Although erosion was not visually perceptible on most of the badland topography, the bottom being covered by recent ooze, recent erosion was clearly perceptible at the axis of several of the gullies. Erosion had exposed and undercut older oozes for about 1 m above the gulley floor (Fig. 21). The badland gulleys do not cut into the crest of the Malpelo Ridge and they disappear at depths greater than about 3000 m. The gulleys are fault controlled but appear only to develop when weak currents erode the channels.

The fauna

The principal visual features of the sea floor are the effects of animal life; the next most frequent observations are the life itself and in most areas rocks and bedforms are the rarer observations. Except on extremely steep escarpments and seamounts where rocks dominate, the majority of the sea floor is sediment covered and without dramatic bedforms; it is the bottom life and its activities which reveal the principal visual information on the physical environment.

The Panama Basin region is fairly rich in animal life at the 1000—2000 m depths to which we dove. We saw thousands of small bathypterid (tripod) fish. These are excellent creatures for the observer interested in subtleties of the current pattern for they invariably stand on three stiffened elongated fins and point directly into the current. Rattail fish swim with their mouth near the bottom and their body pointed up at an angle of 30—40°. They respond to any current by swimming directly into it, but if the current stops, they immediately swim in random patterns quickly rearranging themselves into the current when it resumes. Both fish are large enough to be easily identified, but small enough that they are often difficult to photograph, particularly the tripod fish which are of soft coloration. In the channels where current activity was rather obvious from other physical indications, there is also a greater abundance of crabs. If the tripod fish and rattail fish are the best and most reliable indicators of instantaneous current directions, it is the stalked filter feeders which give the best indication of continuing currents. In the Cocos Ridge channels there were abundant stalked sponges, less abundant gorgonians and occasional pennatulids. In the channel examined

Fig. 21. Current scour on the Cocos and Malpelo Ridges. The rocks in the picture at the upper right are lithified Miocene cherty oozes. The ripples in the upper left are in isolated patches between rock exposures near the crest of an elevated block (Dive 31). The two photographs (Dive 30) in the middle tier are from a scour channel which lies near the base of the seamount illustrated above. Robust stalked sponges are abundant, with assorted small scour moats. The ripple pattern seen in the photograph at the center right occurs between the southerly flowing deep water and northerly flowing upper water at 1128 m depth along the east flank of the channel. The two photographs at the bottom of the page illustrate erosion of "badlands" topography along the flanks of the Malpelo Ridge. Upper left (31—36): 07°33′N, 84°02′W, 579 m (305 tau). Upper right (31—195): 07°33′N, 84°02′W, 549 m (620 tau). Center right (30—156): 07°29′N, 84°00.7′W, 1151 m (618 tau). Bottom left (25—22): 03°40′N, 81°05′W, 1874 m (1012 tau). Bottom right (25—96): 03°40′N, 81°05′W, 1890 m (1021 tau).

Fig. 22. Seamounts, erosional channels and benthic life. The top pair of photographs show the manganese encrusted rocks of Bielicki Seamount which lies on the descending outer wall of the Middle America Trench. Pillow lavas exposed on the floor or an erosional channel on the Cocos Ridge are seen in the photograph at lower right. Upper left (29—19): manganese encrusted faulted rocks on Bielicki Seamount, 08°43′N, 84°45′W, 1950 m (1048 tau). Upper right (29—15): tabular manganese-encrusted rocks on Bielicki Seamount, 08°43′N, 84°45′W; 1950 m (1048 tau). Center left (13—15): vesicular pillow lavas near the crest of Tortuga Seamount, 05°00′N, 87°30′W; 1036 m (555 tau). Center right (32—198): echinoid on a smoothed portion of floor of the Middle America Trench, 08°23′N, 83°51′W; 1661 m (892 tau). Bottom left (25—110): basket star on tranquil bottom of the southern flank of Malpelo Ridge, 03°40′N, 81°05′W; 1859 m (998 tau). Bottom right: (12—133): pillows exposed on floor of erosional channel on Cocos Ridge, 02°06′N, 84°58′W; 1844 m (990 tau). (1 tau = 1/400 sec).

on Dive 30 the stalks of the sponges appeared shorter and thicker than those in other areas. Occasionally they stood erect, but more often they bent with the current, sometimes even touching bottom and scribing arcs in the sediment. The large plowers and trackers are mostly known but the makers of most of the cones, piles and holes are not certainly identified although candidates are clear from dredging results. The modification or absence of these mounds, tracks and trails can be subtle indicators of current smoothing (Heezen and Hollister, 1971).

Tilted uplifted crustal blocks

A prominent topographic high on the northern Cocos Ridge was investigated on Dive 31 (Fig. 22). Prior to the direct observation we had assumed that the 650-m peak was a volcano and that pillow lavas would be observed. This was not the case. The rocks exposed are not basalts but are tilted tabular consolidated chalks which dip 20° to the south. Except for an occasional patch of rippled sand the peak is swept bare. The chalks were certainly not deposited in an environment as dynamic as that which presently prevails on the peak. The history of this high broken region appears to include deposition of a sequence of chalks on the original basaltic basement in a tranquil environment followed by faulting and elevation of the present peak with consequent erosion of the unconsolidated chalks. As the peak became more of an obstacle to the movement of ocean waters, erosion increased, removing all but the most consolidated sediments. Although the peak is of different shape and composition, it is subject to erosional processes similar to those which denude the flanks of volcanic seamounts. Small patches of rippled foraminiferal sand which resemble those seen on Lorraine Seamount were observed between the tilted and beveled chalk beds at the crest of the peak.

On the broad level shallow 1100-m summit of the Cocos Ridge to the southeast of the peak, the upper stratification of the unconsolidated sediment is beveled and truncated suggesting a very recent change from deposition to erosion.

ACTIVE FAULTING

A deep trough lies a few kilometers to the northwest of the uplifted block observed on Dive 31. The floor of the trough is filled with 400 m of sediment. An active fault which cuts this sediment along the south side of the trough was examined (Dive 37). The sketch (Fig. 23) which is greatly compressed in the vertical dimension, shows the observed characteristics of the fault. Angular talus blocks lay on the flat sediment-covered floor of the trough (1900 m) several meters beyond the foot of the talus slope. Traversing to the south for a vertical distance of 225 m we ascended this talus slope

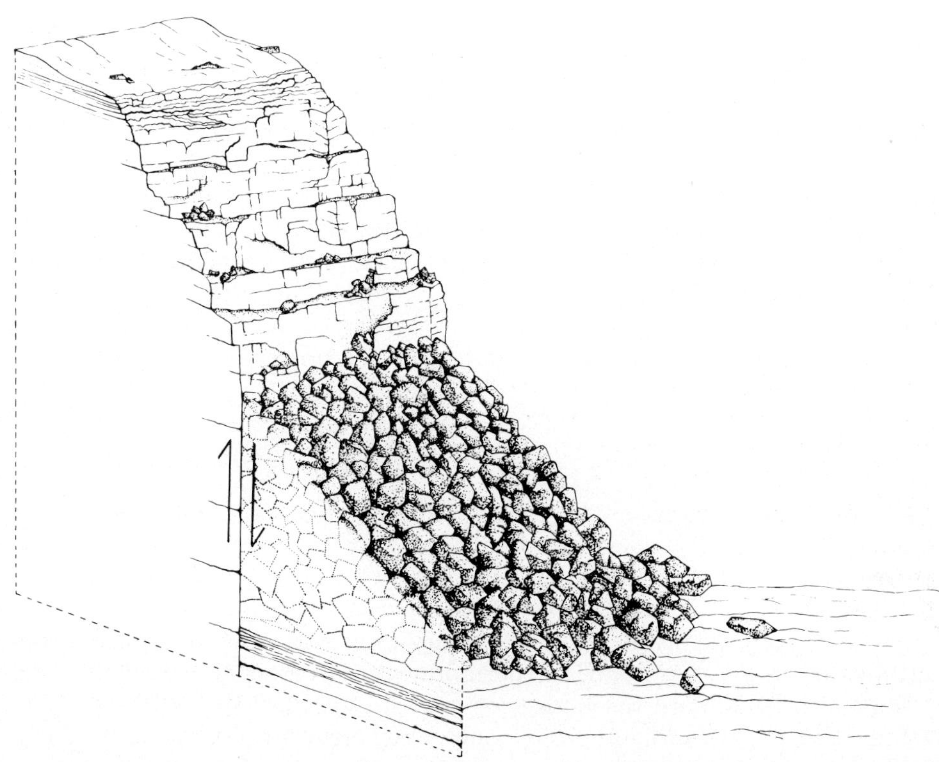

Fig. 23. Fault scarp and talus on the wall of a tectonic depression at the northern end of Cocos Ridge. Based on observations made during a dive (37). The talus and the fault scarp are composed of Miocene cherty chalk. The fault is still active.

composed of cherty chalk blocks resting at their angle of repose. No sediment covered the talus and we could see far down between the blocks into darkness. Above the talus slope (1675 m) only 30 m of the vertical fault surface was exposed. Above the fault scarp the escarpment became a staircase of structural benches. Above 1585 m the sea floor leveled off. Three hundred meters beyond the edge of the upper bench, current scour diminished and the sea floor was covered by recent ooze. Discontinuous and sometimes cantilevered manganese crusts up to about 1 cm thick occur on the flat-lying chalk beds above the lip of the cliff. Nondeposition has been of considerable duration in this location. This fault has apparently been active for a considerable period of time.

The faulting and uplift appear to be related to tectonic activity associated with the impending subduction of the northern part of the Cocos Ridge. As oceanic crust is carried towards a subduction zone it is broadly arched and somewhat elevated. As it comes closer to the outer wall of the trench minor vertical dislocations associated with the arching grow larger and more faults appear. On descent to the trench floor these minor dislocations grow into major faults which cut deeply into the oceanic crust.

Bielicki Seamount rises from the outer wall of the Middle America Trench near or on the northern flank of the Cocos Ridge. It has presumably been within the area of minor dislocations for several million years and has just reached the zone of major faults. Bielicki Seamount is more heavily covered with manganese, a characteristic which sets it apart from other volcanic seamounts of the Cocos Ridge. It also is characterized by many dislocations (Fig. 22) which contrast markedly with the regular depositional lava flanks of the other seamounts. The faulting and fracturing of Bielicki Seamount is entirely in keeping with its tectonic setting on the brink of subduction.

SUBDUCTION

Of the earth's major crustal processes subduction falls within a class of inferred but patently unobservable phenomena (such as granitization and metasomatism) which are known only from end results revealed long after the inferred events by the study of rocks which subsequently experienced a long and complex history. Contemporary subduction can be inferred from earthquake first-motions and, furthermore, some process such as subduction is demanded to balance the budget of crustal growth of the expanding Mid-Oceanic Ridge. Remarkably, the compelling array of arguments cited in support of contemporary subduction has not included visual field observations at proposed sites of contemporary subduction.

The Marianas Trench, a recognized subduction zone, was the site of the first descent by man to the ocean's greatest depths (Piccard and Dietz, 1961). The Puerto Rico and Japan Trenches were later explored with the bathyscaph ARCHIMEDE (Peres, 1965; Bellaiche, 1967). The observers, although noting evidence of contemporary deformation, did not relate their observations to the subduction process.

The observation that Benioff earthquake planes intersect the sea bed somewhere in the trenches implies that differential motion must extend to the sea floor and should have visible manifestations on the sea bed. Although somewhere on or near the landward wall seems a probable site, even the present high accuracy of hypocentral determination is still insufficient to predict the exact location of the sea-bed trace of subduction.

The accumulation in trenches of undeformed stratified sediments has been cited as evidence against subduction, or at least as evidence against presently-active motion related to subduction. The resolution of high-precision echo-sounders and seismic profilers is insufficient to resolve potentially different patterns near precipitous slopes at such great depths.

Subduction now plays an immensely important role in modern tectonic schemes; yet the subject retains a dreamlike character since all details are inferred from indirect measurements or are reconstructions based on a complex superimposed record retained in deformed ancient rocks. It was the lack of a detailed actualistic basis for subduction, particularly in regard to the role and fate of deep-sea sediments, which led us to make four dives to the floor of the Middle America Trench (Fig. 24). Three dives (32, 33 and 35) revealed the consistent pattern described below (Heezen and Rawson, 1977b). On the fourth dive (34) a seamount embedded in the landward wall of the trench was examined (Fig. 25).

Normal faults of the trench axis

A series of northeast-facing scarps lie at the trench axis. Nearly identical characteristics were observed on the five traverses of these cuesta-like features (Fig. 26). The scarps are 10—30 m high and approximately 150 m apart. Along the crest of each ridge broken angular blocks of unconsolidated sediments litter the sea floor. Outcrops of dense gray fine-grained limestone occur a few meters below the crest of each ridge. The outcropping rocks were devoid of the manganese coatings usually seen on deep-sea exposures. The limestone contains a few mineral grains presumably derived from penecontemporaneous volcanic exposures of the oceanic basement. A weak current supports a scattered population of sponges and sea fans which attach themselves to outcropping rocks. The narrow sediment ramps which lap on

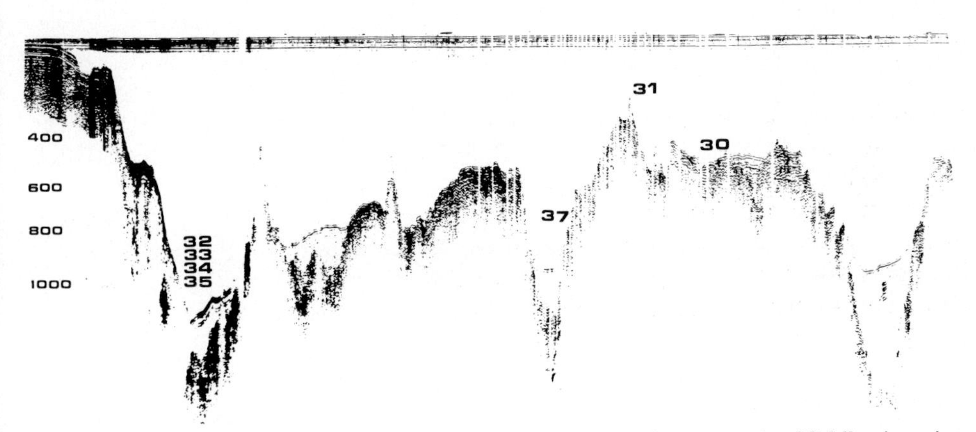

Fig. 24. Seismic reflection profile from Costa Rica south across the Middle America Trench and across the adjacent northern Cocos Ridge. The erosional channel on the elevated crest of Cocos Ridge was investigated during Dive 30. The elevation investigated on Dive 31 exposes tilted chalk uplifted to its present erosional environment following deposition in a deeper, more tranquil environment. A fault cuts through the sediment in the depression to the left of Dive 37, where the sediment-free talus slope of an active fault was observed. Dives in the Middle America Trench located the narrow subduction line. Profile south to north from 7°N84°W to 9°N84°W. Depths in tau (1 tau = 1/400 sec).

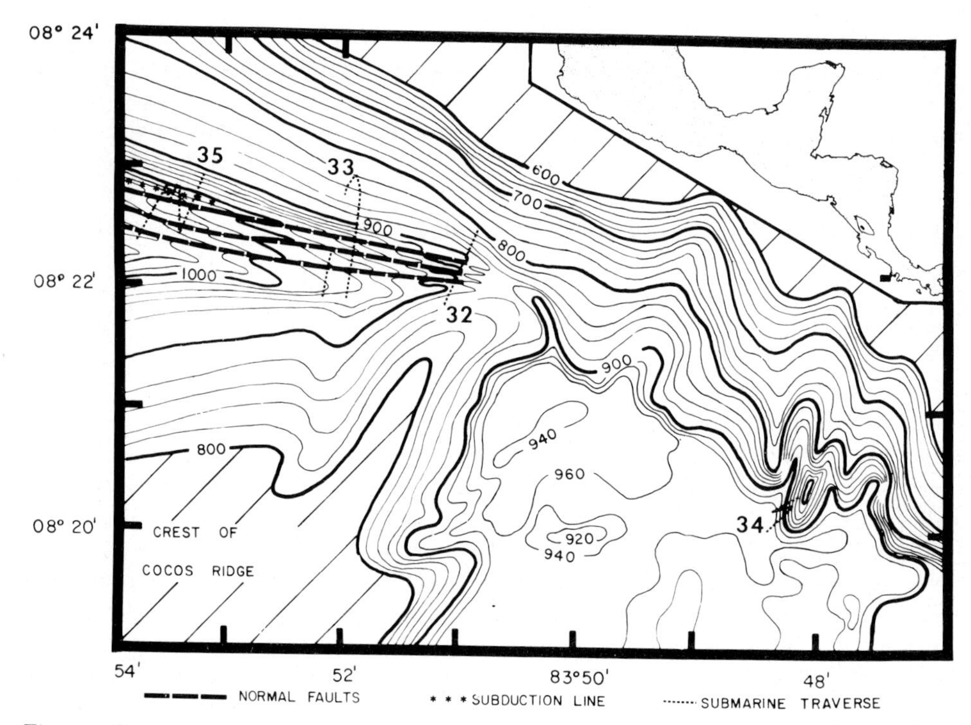

Fig. 25. Southeast Middle America Trench with locations of the dives of DSV TURTLE to the subduction zone. Isobaths in tau (1 tau = 1/400 sec).

the base of each scarp are smooth and undisturbed. It is concluded that the small cuestas represent a series of normal faults of small displacement which downfault the oceanic crust and its pelagic cover (Fig. 26).

Throughout the four trench dives the quantity of large organic particulate matter in the water column was high as compared to the clear waters encountered further seaward on the Cocos Ridge. The strong back-scatter of light from the particulate matter unfortunately resulted in relatively poor-quality photographs.

Trench floor apron

A gently sloping tranquil, ooze-covered bottom extends for 200—500 m from the base of the northeasternmost normal fault of the trench axis to the foot of the landward wall of the trench. With the notable exception of a dramatic deformed zone, which longitudinally bisects the apron, this portion of the trench floor is otherwise devoid of surface expressions of deformation.

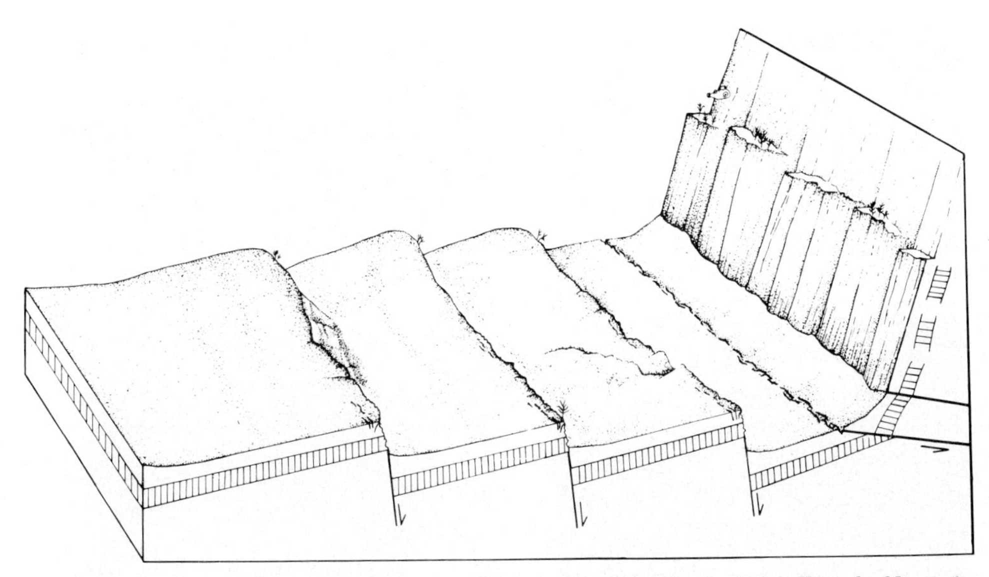

Fig. 26. Subduction zone at the southeast end of the Middle America Trench. Normal faults carry oceanic crust down along trench axis. Smooth trench floor apron is bisected by sea floor trace of subduction. Landward wall to the right. Width of profile approximately 1 km.

Line of active deformation

Deformation has created sharp serrated shapes and angular blocks in the unconsolidated green silty ooze of the trench floor (Fig. 27). Contemporary deformation has thrust the unconsolidated sediment into upturned blocks and has caused the sediment to part, forming narrow bifurcating chasms (Fig. 28). Remarkably, this deformation is limited to an extremely narrow belt less than 30 m in width. Although animals were burrowing new holes into the recently exposed sediment slopes, neither the benthic fauna nor the currents had yet appreciably altered the sharp forms (Fig. 29). Some of the other fissures were slightly rounded and partly filled, thus seeming a little older. But all signs of deformation on the apron, both strictly contemporary and slightly older, were limited to the one narrow zone.

Trench wall

A sharp change in slope marks the boundary between the gently sloping apron of the trench floor and the precipitous trench wall. We observed no deformation at or near the base of the wall during any of the three traverses. Thus, the sharpest break in slope observed in the trench is not associated with any active surface manifestations of deformation. The landward trench

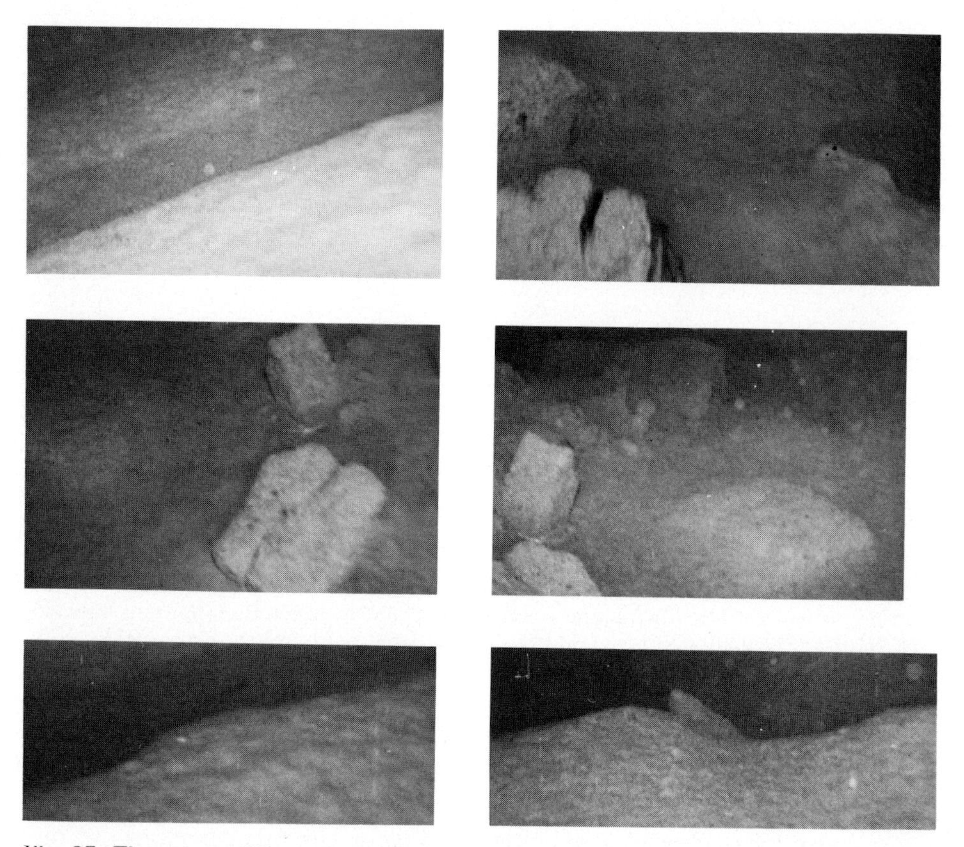

Fig. 27. The unconsolidated sediment is broken into small troughs and ridges at the subduction zone in the Middle America Trench. The blocks shown in these photographs have fallen from the deformed masses of unconsolidated trench sediments. Photographs taken at a depth of ~1875 m from DSV TURTLE on Dive 35.

wall is extremely complex and varied. Much of it is a steep, smooth sediment-covered slope but locally on smaller (~30 m) escarpments sharp edged, V-shaped grooves 0.5—2 m deep and 1—2 m wide cut a geometrical pattern almost precisely up and down slope, often becoming somewhat mantled by sediments before completely disappearing. The V-shaped grooves expose bedding and parting planes in the essentially unconsolidated sediments. Here and there, shells, pebbles and rubble were seen in and near the down-slope ends of the regularly spaced V-shaped notches.

Several narrow, nearly level, benches on the wall revealed outcrops of bedrock and scattered rubble. In marked contrast to the almost barren sediment-covered wall, the filter feeders attached to nearly every boulder or outcrops lend a garden-like aspect to the rocky benches.

Fig. 28. Scene looking northwest across the subduction line in the Middle America Trench. This narrow line less than 20 m wide cuts unconsolidated sediments. Represents a scene 30 m wide from a mosaic of photographs taken on Dive 35.

Seamount at the wall

The axis of the Middle America Trench passes over a shallow saddle of 1600 m less than 1 km east of the site of Dive 32. Southeast of the saddle, sediments ponded between the trench wall and the Cocos Ridge crest have created a small abyssal plain which abuts the trench wall (Fig. 25). As we began Dive 34 at the northern edge of the plain, we anticipated that contem-

Fig. 29. Sharp crests of angular blocks of deformed unconsolidated sediments at the subduction line in the Middle America Trench. Note beer can for scale.

porary deformation might be recognized in the ponded sediments of the plain, but this was not the case. The plain sharply abutted the landward slope without any signs of deformation. At the dive site, the steep slope encountered ran NE—SW at approximately right angles to the regional strike of the trench wall. This slope is composed of undeformed lava pillows and lava cylinders. The scene was identical to depositional lava terrain which we have repeatedly observed on seamounts. This seamount now embedded in the trench wall has not yet undergone significant deformation. It seems to have just reached the subduction zone.

The normal faults of the trench axis carry down the oceanic crust and its pelagic cover. Seismic reflection profiles reveal that normal faults of much larger displacement form the seaward wall of the trench. The zone of deformation within the trench apron marks the sea-floor trace of active subduction. At the present time no movement is occurring at the base of the landward wall. On this wall, benches with outcropping rocks mark structural breaks. The geometrical pattern of V-shaped grooves oriented up and down slope entirely without bifurcation suggest a structural rather than erosional origin although slides and rock falls have occurred along most of these strange V grooves. The trench wall was far too varied in the three traverses to discern a general pattern. Perhaps the most surprising observation was that most of the steep wall is covered by smooth undisturbed ooze.

At the site of Dives 32, 33 and 35, the trench plunges northwest at a gradient of 1 : 15 which would preclude any sediments from ponding along the trench axis. Any turbidites would be transported 500 km to the northwest into the deepening trench. Thus the behavior and pattern of deformation observed here should differ from the deeper parts of the trench in that trench-floor turbidites will be absent from the apron and landward wall sequence. Within the area we observed no features which could be attributed to turbidite erosion or deposition. The moderate currents (<15 cm/sec) left no scour or depositional forms in the almost pure pelagic depositional environment. This tranquility greatly enhanced the visual contrast between the undeformed and the deformed sea floor.

The subduction zone where observed lay along a section of trench where the longitudinal gradient precluded turbidite deposition. Nevertheless, transport of sediment through submarine canyons and channel systems is an important aspect of the subduction process for these sediments are deformed and either accreted to the trench wall or consumed with the subducting plate. Our submersible studies in the Atlantic have repeatedly confirmed the important role of tidal currents in the erosion of submarine canyons.

Submarine canyons on the trench wall. We took the opportunity to examine several canyons off Mexico, Guatemala and Colombia. I will restrict my remarks here to the description of one dive (23) in a large submarine canyon near Buenaventura, Colombia. This canyon cuts a gorge normal to the con-

Fig. 30. Tortuga Canyon in the continental slope of Colombia. Sediment of the canyon floor is shaped into wide low sediment waves without ripples. Floor is devoid of filter feeders or small benthic animals. On lower wall of canyon rocks are nearly devoid of attached animals but higher on the wall rock outcrops support a lush growth of attached filter feeders. From photographs of floor and south wall of canyon taken on DSV TURTLE Dive 23 in March 1975.

tours of the continental slope to a depth of about 1200 m where it is deflected to the right for a few kilometers by a ridge which parallels the slope, further the canyon turns left and cuts through the ridge at a depth slightly greater than 2000 m. We chose to examine the gorge where the floor lay in about 1000 m. We landed near the foot of the right bank, traversed the canyon floor to the south and proceeded up the left bank to a depth of 300 m. The floor of the canyon was devoid of attached life. Attached filter feeders were also absent from the rock ledges which form the lowermost part of the south wall of the canyon (Fig. 30). The floor of the canyon was also nearly devoid of benthic animals. But fish and large crustaceans were abundant in the water. Visibility was impaired by plankton and particulate matter in the water. About 20 m above the floor the outcrops were covered by filter feeders. The scene gave me the impression that turbidity currents may be rather frequent in the canyon. The absence of filter feeders on the floor and lowermost walls despite the presence of an appreciable up-canyon current, the presence of only large robust mobile fauna (mostly swimmers) in the canyon axis in contrast to the abundant diverse fauna of the canyon walls, and the universal but unusual tendency of the large swimmers to move away from us, led us to the opinion that the frequency of turbidity currents was sufficiently great to significantly affect the fauna of the canyon.

The sediments presumably are carried out of the canyon (to depths beyond the capabilities of present syntactic foam submersibles) to the floor of the Panama Basin in the vicinity of the northern extension of the Peru—Chile Trench.

DISCUSSION

I quite arbitrarily chose the submersible investigations described above to illustrate some of the capabilities (and some of the limitations) of submersibles in the pursuit of a better understanding of deep-sea geological processes. As the mapping of the ocean floor proceeds from reconnaissance to more and more detailed surveys and as our understanding of the gross patterns becomes more and more detailed, there will come a time in the next few years when the meaningful study and utilization of the sea floor will require a much expanded submersible program.

The submersible is at the same time a good and a bad reconnaissance tool; good because the human eye and brain on the spot can make observations which no remote instrument can; bad because without an adequate base map and navigation system it is difficult to organize thoughts and see "the wood for the trees". We did not attempt to apply precise navigation systems in our initial efforts preferring to take that time and those resources to increase our reconnaissance coverage. We now have identified several problems which we believe justify more intensive investigation. The subduction zone clearly justifies additional investigation employing bottom transponder nets and a seismic profiler. The visual investigations of most of the Cocos Ridge channels were instructive but would not have benefited greatly from a controlled grid for bed forms were not present. Obviously additional work on the Puna Ridge preferably employing a bottom navigation network is desirable particularly in the further investigation of pillow walls.

Many features appear to be in equilibrium with the physical environment prevailing at the time of the dive. Others seem to be relics of recent events. Both present intriguing problems for submersible investigation. The most satisfying type are those causes and effects that can be watched. The sight of sand carried in tidal flow undercutting the wall of a submarine canyon, the sight of ripple crests changing symmetry as tidal currents change from up canyon to down canyon, the view of the sharply defined narrow subduction plane intersection with the trench floor, all give a clarity and sharpness of understanding unobtainable from remote observations.

There are frustrations. Not trusting to memory or to instant conclusions, we try to photograph everything in color and in black and white, in cine and in stills, on 8 mm, 16 mm, 35 mm and 70 mm film. Often I wish the deep submersibles were transparent so we would not be limited to the view ports. I often curse the designers when I cannot see or photograph because my head or camera will not fit into the space provided. I am continually concerned with methods to communicate our observations to a wider audience. I am concerned and often unhappy with the acuity of my observations and those of my colleagues. People ask "aren't you afraid of being trapped on the sea floor?" I answer, "yes, but for me results are worth the risk." Some ask "how can you stand to be cooped up in a 6-foot ball with two other people and all that gear?" This really presents no problem at all. With the many cameras to shoot and reload, films to label, video and audio tapes to label and change, notebooks to keep and recorders to mark, you really have little time to contemplate anything further than the vastly exciting experience of trying to comprehend and record the significance of what you are seeing. It is the greatest experience I know; it beats by light years partying and sex. And, of course, there is always a bit of that vanity in us so peculiar to scientific explorers which Charles Darwin so aptly described in the last century:

> "Everyone must know the feeling of triumph and pride which a grand view communicates to the mind. In these little frequented (places) there is also joined to it some vanity, that you are the first (human) who ever admired this view."

REFERENCES

ARCYANA, 1975. Transform fault and rift valley from bathyscaph and diving saucer. Science, 190: 108—116.

Ballard, R.D., Bryan, W.B., Heirtzler, J.R., Keller, G., Moore, J.G. and Van Andel, T.H., 1975. Manned submersible observations in the FAMOUS area: Mid-Atlantic Ridge. Science, 190: 103—108.

Bellaiche, G., 1967. Résultats d'une étude géologique de la fosse du Japon effectuée en bathyscaphe ARCHIMEDE. C.R. Acad. Sci., 265: 1160.

Bender, M.L., Ku, T.L. and Broecker, W.S., 1970. Accumulation rates of manganese in pelagic sediments and nodules. Earth Planet. Sci. Lett., 8: 143—148.

Fornari, D.J., Malahoff, A. and Heezen, B.C., 1977. Volcanic structure of the crest of the Puna Ridge, Hawaii: Geophysical implications of submarine volcanic terrain. Geol. Soc. Am. Bull., in press.

Gerard, R.D., Heezen, B.C., Thorndike, E. and Sullivan, L., 1976. Results of the first deployment of the Bottom Ocean Monitor in the Equatorial Pacific. In: IDOE Manganese Nodule Project, Quarterly Report, June 1976. Scripps Inst. of Oceanography, pp. 5—8.

Hamilton, E.L., 1956. Sunken islands of the Mid-Pacific Mountains. Geol. Soc. Am. Mem., 64: 97 pp.

Heezen, B.C. and Hollister, C., 1971. The Face of the Deep. Oxford Univ. Press, London, 659 pp.

Heezen, B.C. and Rawson, M., 1977a. Visual observations of contemporary current erosion and tectonic deformation on the Cocos Ridge crest. Mar. Geol., 23: 173—196.

Heezen, B.C. and Rawson, M., 1977b. Middle America Trench subduction zone. Science, in press.

Heezen, B.C., Tharp, M. and Ewing, M., 1959. The floors of the oceans, I. The North Atlantic. Geol. Soc. Am. Spec. Pap., 65: 122 pp.

Heezen, B.C., Rawson, M., Lynde, R.P. and Nesteroff, W.D., 1976. In situ submersible observations in the Western Puerto Rico Trench. Geol. Soc. Am. Progr. Abstr., 8.

Heezen, B.C., Fornari, D. and Peng, T.H., in prep., a. Mangrove mats on 350 m submerged terrace, west of Hawaii, 4590 years old.

Heezen, B.C., Lonsdale, P., Ballard, R.D. and Bass, M., in prep., b. Malpelo Ridge.

Lonsdale, P., Normark, W.R. and Newman, W.A., 1972. Sedimentation and erosion on Horizon Guyot. Geol. Soc. Am. Bull., 83: 289—316.

MacDonald, G.A. and Abbott, A.T., 1970. Volcanoes in the Sea: The Geology of the Hawaiian Island. University of Hawaii Press, Honolulu, 441 pp.

Malahoff, A. and McCoy, F., 1967. The geologic structure of the Puna Submarine Ridge. J. Geophys. Res., 72: 541—548.

Malahoff, A., Heezen, B.C., Lockwood, J. and Peterson, J., in prep. Denudation of Holocene dikes in Ka Lae submarine canyon, Hawaii.

Menard, H.W., 1952. Deep ripple marks in the sea. J. Sediment. Petrol., 22: 3—9.

Moore, J.G. and Reed, R.K., 1963. Pillow structures of submarine basalts east of Hawaii. U.S. Geol. Surv. Prof. Pap., 475-B: B153—B157.

Peres, J.M., 1965. Aperçu sur les résultats de deux plongées effectuées dans le ravin de Puerto-Rico par le bathyscaphe ARCHIMEDE. Deep-Sea Res., 12: 883—891.

Piccard, J. and Dietz, R.S., 1961. Seven Miles Down. G.P. Putnam Sons, New York, N.Y.

Tepley, L., 1974. Fire Under the Sea. Documentary film produced by Moonlight Productions, Mountain View, Calif.

Truchan, M. and Aitken, T., 1973. Site surveys on the Coiba and Cocos Ridges in the Panama Basin. In: T.H. van Andel, G.R. Heath et al., Initial Reports of the Deep Sea Drilling Project, XVI. U.S. Government Printing Office, Washington, D.C., pp. 473—495.

Wilde, P., 1966. Quantitative measurements of deep sea channels on the Cocos Ridge, east central Pacific. Deep-Sea Res., 13: 635—640.

Woollard, G.P., 1954. The crustal structure beneath oceanic islands. Proc. R. Soc. London, Ser. A., 222: 361—387.

CHAPTER 9

THE SUBMERSIBLE — A UNIQUE TOOL FOR MARINE GEOLOGY

George H. Keller

INTRODUCTION

By means of the submersible the sea truly becomes man's laboratory where he can immerse himself directly into the medium and bring with him his unique sensory perception which only he and no instruments possess. For decades marine geologists have conducted by means of remote techniques their studies of the ocean basins, their sediments, stratigraphy, and tectonic history. The frustrations of these scientists having to work hundreds or thousands of meters above the sea floor, depending on samplers or sensors lowered randomly from the surface, or relying on shipboard echo-sounding systems incapable of adequately resolving the detailed bottom topography, are known especially to those who have had the opportunity to work with a submersible.

Because of these remote techniques and the general quality of the data, the geographic positions of which are commonly not known to better than ±1 km, the resulting analyses and interpretations are often susceptible to question. Far more serious for certain studies, such as stratigraphy, is the lack of knowledge of a sample's position relative to other samples collected, or of the geological features themselves. Whether samples collected from a surface vessel are representative or anomalous is never known, but they are commonly accepted as being representative. More vividly, surface vessel observations may be likened to attempting to define the surface properties or characteristics of Manhatten Island based on a few samples collected from a helicopter hovering over the city on a foggy day. This may be stretching the analogy slightly, but not drastically.

Frustrations are greatest for those who have been to the sea floor and found that variations both vertically and laterally are often very pronounced over relatively short distances. It is clear that the conventional surface-ship operations may often result in erroneous interpretations and misconceptions. The advent of the submersible has brought the geologist from his floating perch many meters above the bottom to within centimeters of his target. This drastic transition from gaining an impression of the sea floor by viewing it from afar (a surface vessel), to one of literally touching the subject is mind boggling. It is a case of actually getting too close to the subject such that the broader relationships commonly understood are lost. Not being able to adequately define the fine details of the bottom characteristics from the surface has led to this initial difficulty of some to piece together their submersible observations. This might be likened to one who has been studying a forest and then turns to investigate the bark of a single tree with a magnifying glass but never having looked at any of the individual trees.

To most geologists "walking the outcrop" is a basic approach to gaining an orientation to the problem, defining relative positions of strata and features, and the collection of meaningful samples. This approach has not been available to the marine geologist until late after the advent of the submersible. Even with the arrival of the submersible this "standard" terrestrial procedure was not feasible owing to the lack of accurate and detailed maps, as well as positioning inaccuracies in defining the location of the submersible. An ability to "walk the outcrop" in an effective manner has become a reality recently with the development of acoustic transponders (benchmarks). By means of such acoustic arrays placed on the sea floor in a study area, submersible positions can be determined with an accuracy of 5—10 m. Although such positions are not absolute but are relative to the acoustic beacons, the geologist is now able to piece together his observations after the fact in an effective manner.

At long last the geologist can get to where the action is, to see and literally touch the subject of his study. As we will see, there is no substitute for getting the eyeball down to where things are happening.

Mapping of bottom features from surface vessels has advanced significantly in recent years. For example, the Navy's narrow-boom multi-channel system, and deep-tow vehicles have drastically improved our capability for very detailed mapping of local topographic features and defining magnetic characteristics. However, the manned submersible, with its eye of a trained observer, is the only way to effectively obtain the detail needed to accurately resolve many problems of concern to the marine geologist. Bottom photographs are extremely important, but unfortunately the field of view is quite limited and such major factors as range, scale, and relationship to surroundings are lacking. Here again the trained eyes of an observer with their depth perception and ability to quickly scan an area thus defining the overall setting are invaluable in piecing together the numerous individual observations made during submersible operations or later in the interpretation of surface-derived data.

In addition to mapping bottom features in a detail never before accomplished, one of the most significant characteristics of the submersible is its use to selectively sample strata and sediments. Back to the concept of "walking the outcrop" (Fig. 1) by being able to view the overall setting and then sampling specific features, outcrops, or sediment deposits the truly representative nature of the samples is assured. No longer does the geologist have to guess at the meaning and interrelationships of the often varied rock specimens collected in a dredge dragged many meters along the sea floor.

Only by perceiving the overall setting of a study area can the interrelation-

Fig. 1. Typical outcrop on the Mid-Atlantic Ridge.

ships of geological events or bottom processes be properly defined. For those concerned with volcanism and tectonic events, the direct relationship of volcanic forms, mineral composition, fissures and associated tectonic processes can only be effectively pieced together by repeated submersible transects across an area. This is quite evident from the observations of Project FAMOUS (Ballard et al., 1975). For those investigating bottom and near-bottom dynamic processes such as erosion, deposition, sediment stability and turbidity, a knowledge of the local setting is most important, if observed variances in these processes are to be understood. An example which readily demonstrates the need to "see" the somewhat larger picture than that encompassed in a photograph or a sediment sample, is found in relating bottom-current flow to ripple-mark characteristics. During dives in Great Abaco Canyon (southeast Blake Plateau) ripple marks were found to be abundant in the calcareous sands and indicated flow along the canyon's axis. However, ripple marks on mounds or topographic promontories in the canyon proper indicated quite different current directions due to the influence of the features themselves. This is not surprising, but the point is that should only a bottom photograph be available from such a site, the current flow direction for the area in general would be inaccurately interpreted. Here again it comes back to the problem of collecting representative data.

No longer does the scientist have to depend solely on laboratory techniques to study deep-ocean sediment dynamics. With his own eyes he can study the interactions of currents and organisms in regard to the manner in which they disturb or modify the sediments. He too can very effectively document how sediment particles are transported along the bottom and in suspension just above the bottom. Submersibles afford a much greater opportunity to conduct "real time" analyses of dynamic events than can be hoped for by any other means.

A submersible also provides the geologist with a means of placing instruments such as sediment traps, lapse-time cameras, and current meters in positions selected to provide representative data of the parameter to be measured (Keller et al., 1973). Here, the submersible provides the data confidence factor no other system can.

Motion of a surface vessel and its influence on personnel and instrumentation performance is no problem to the submersible user. The absence of wave motion yet the versatility to either stay in place or slowly cruise over an area provides the geologist with a rather comfortable, dry, relatively warm and stable platform. The ability of a geologist to stop, turn, look or sample, as well as time to ponder a situation and then decide on a course of action to better define the question occurs only if he can use a submersible.

SUBMERSIBLE DEVELOPMENT AS A GEOLOGICAL TOOL

Other than for military purposes man's early descents into the deep sea were primarily motivated by his interests in marine life. Although not using a submersible, but a bathysphere tethered to a surface vessel, William Beebe in the early 1930's carried out a number of dives to depths as great as 923 m, in one of the earliest subsurface studies of marine life. Biological interests also prompted the early studies attempted from the bathyscaph developed by Auguste Piccard in the late 1940's (Busby, 1976). One of the earliest recorded dives in a manned vehicle to undertake geological studies was conducted by the USSR in 1927 near Sevastopol to a depth of 46 m using a diving bell. The innovative studies of Vening-Meinesz in the late 1920's of recording the earth's gravitation field from aboard a submarine have been a land-mark experiment in geoscience (Vening-Meinesz, 1932).

Prior to 1960 about 14 papers were published dealing with geological observations made from submarine vehicles (Ballard and Emery, 1970). The majority were reports on gravity data collected aboard submarines.

It was not until the late 1950's, when the bathyscaph TRIESTE was made available to American scientists, that a significant number of dives addressing geological problems were conducted. Following these dives the TRIESTE was purchased by the United States Navy. The record-breaking dive of the TRIESTE to 10,912 m in 1960 appears to have triggered the dawning of submersibles. During the next decade numerous submersibles were built of varying sizes, shapes, and capabilities.

Because of the number of submersibles available by the mid-1960's more and more in-place geological studies were conducted. The high operating costs of the vehicles have unfortunately resulted in their rather limited use by the scientific community as compared to the more conventional surface vessels. As with any new technology, considerable time is required to develop an effective use for such a unique tool as a submersible. For the most part, the technology in this case had advanced beyond the science. Some good science evolved from a number of the dives in the 1960's and early 1970's, but the scientists in general were learning an entirely new manner of observation and amassing a number of dives to basically learn what bottom characteristics were like in a number of areas. Because of this entirely new approach, it took the scientists considerable time to get beyond the "gosh-gee whiz" stage and to begin to effectively document their observations. The geologist was particularly handicapped by his use of these vehicles due to the lack of adequate base maps and navigation control. Only a very few of the most experienced users of submersibles were able to reassemble their observations after the fact, in such a manner that a coherent understanding of the problem could be presented. Echo-sounding techniques are now available to provide suitable maps, and the positioning problems in general are solved. The understanding of the geological problems of the ocean basins has advanced to the degree that submersibles can now be used much more effectively to provide the resolution required to complete the study of many of these problems.

SEA-FLOOR CHARACTERISTICS

The submersible is a unique tool and like any tool is best suited for certain things, but is not a utopia for all. It is designed to carry man and his senses, particularly the eyeball, to a very small spot on the sea floor, where he can sample and visually observe in great detail the problem under consideration. Effective use of this unique tool also requires that extensive studies be conducted on the problem prior to use of the submersible itself. The problem must be researched as thoroughly as possible, so that in the final analysis the submersible serves to provide the ground truth for hypotheses formulated from the earlier less detailed studies. Preliminary studies must provide the submersible user with the basis for specific observations and sample sites. Submersible observations contribute the final degree of resolution to the field operations and must be focused sharply on the problem at hand. As stated earlier, before a geologist can effectively use a submersible, a detailed bathymetric map and an accurate positioning system must be available.

Problems best suited for study from a submersible are those requiring very accurate specimen sampling such as in the examination of stratigraphic variations, compositional alterations, or the in-place documentation of dynamic bottom processes. In all these problems the submersible serves the prime purpose of positioning a trained observer so as to ensure that the samples and observations are representative of the problem under study and not taken from local anomalies.

DYNAMIC BOTTOM PROCESSES

To study sediment dynamics and bottom stability, the submersible provides the ultimate means to analyse these processes in the field. Until recently the investigator of sediment transport and near-bottom dynamics in the deep sea has had to base his reasoning and conclusions on either laboratory experiments, poor fluvial analogies, or the use of random bottom photographs which in turn must also be primarily related to earlier flume studies. The investigator can now be placed directly into the environment he is studying. He can readily determine visually the general energy conditions, as well as the presence of local features and their influence on the bottom processes. An observer can easily ascertain the degree of variability of the transport regime and the reasons for such variation by being able to "look over" the immediate surroundings of the study area. No longer, for example in the study of bedload transport, does the scientist have to estimate the energy levels at the sediment—water interface by the suspect methods of extrapolation from near-bottom current measurements. By use of dyes injected into the water from a submersible at the sediment—water interface and immediately above, very detailed analysis of the current microstructure and flow velocity over the bottom are accomplished to a degree never before attained in the field (Cacchione et al., 1976). Remote sensors are being developed to address this problem, but again without the assurance that these measurements are representative of the bottom and near-bottom dynamics, rather than due to the influence of local features such as a large block or mound, some degree of uncertainty in the results will exist.

Biological fluff

Submersible observations have shown that in certain areas a film and in some cases a thin layer (0.5—1 cm thick) of biological "fluff" blankets the sea floor. This organic-rich fluff, takes on the appearance of large fluffy dust masses similar, for example, to those that might be found under a bed in a room not cleaned recently. Although the influence of this "fluff" on the sediments and the dynamics of their transport is unknown, there is undoubtedly some alteration to the sediment properties. The presence of increased concentrations of organic material in fine-grained sediment has been found to alter their stability characteristics, by affecting the cohesion and rheological properties (Rashid and Brown, 1975). The presence of this biogenic fluff has not been noted before by remote sensors such as cameras; and thus is a factor not previously considered in the study of sediment resuspension and transport.

Turbidity currents

The problem of turbidity currents, their mechanism of formation and flow characteristics is of almost universal interest. Although a process widely documented in the sediments of the deep sea it is one not yet observed on a first-hand basis. The inability of man to generate these flows on the sea floor (Buffington, 1961) or to accurately document a flow as it takes place gives cause for thought about the hypotheses dealing with the mechanisms for distributing the sedimentary sequences referred to as turbidites. The author on a number of occasions has, from a submersible, attempted to cause slope failures and the mass movement of surficial sediments downslope. On slopes as great as 70° these attempts have been to no avail. At best it was found that a small bedload and accompanying cloud of suspended sediment would travel down slope a few tens of meters. It appears that only the mass of material initially set in motion moves down the slope rather than serving as a triggering mechanism to generate even a mini-turbidity flow. Turbidity currents are still a mystery but the use of the submersible to investigate sites of known slope failures such as off Newfoundland or on major deltas as off the Ganges or Mississippi Rivers may provide a dimension to the study of this process not before attained.

Suspended sediments

Most studies made thus far on suspended sediments in the sea have relied on remote sampling procedures to define concentrations and compositional characteristics. This has been an effective means of study during these early phases, but again the investigator is faced with the need to obtain more accurate and truly representative data to advance his studies. The advantage of the submersible with its various water samples (Fig. 2) and outboard pumping systems is obvious. Not only are samples taken selectively, but an appreciation is gained for such things as (1) sporadic concentrations of material, (2) transport as influenced by local features, and (3) the interrelation of sediment and biological matter carried in suspension together. The last aspect is most effectively studied from a submersible. In most areas the organic matter comprises the major portion of the suspended particulates. These biogenic particles are quite varied in size and shape, but are considerably larger

Fig. 2. Typical submersible sample tray showing sediment and water samplers (courtesy J. Heirtzler, Science, October 15, 1976).

than the incorporated sediment particles. The suspension dynamics of particles caught up in the biogenic material follow the behavior of "snow", as opposed to that of individual sediment particles, in suspension. Here again the submersible has afforded the trained observer the most effective means of documenting this relationship of particular matter (Nishizawa et al., 1954).

Slope stability

As in the study of any slope-stability problem on land, it is necessary to understand the characteristics of the local terrain as well as the relationship of the specific area to the more general setting. Most studies dealing with the stability of submarine sediments have been conducted from surface vessels, wherein the above requirement has not been met adequately. The belief that sediments will fail if the slope of the deposit exceeds the angle of repose for a particular sediment texture finds more exceptions than the rule in submarine sediments. A number of submersible observations by the author and others have revealed what appear to be stable deposits on slopes exceeding the angle of repose. Buffington et al. (1967) reported slope angles of 43° in noncalcareous sediments off southern California, but slopes of stable sediment as great as 76° have been recorded in the carbonate deposits of the Tongue of the Ocean by Schlee (1967). The submersible has again provided the means of accurately determining slope angles and visually examining the stability characteristics of some rather steep slopes. The ability of the submersible to define local areas of instability and to document such failures as mud flows by tracing the process from where it originates to where it ends is a unique capability no other means of study affords (LePichon et al., 1975). Relatively few sediment stability analyses have been conducted using a submersible; however, it has been instrumental in carrying out in-place testing of geotechnical properties which will be discussed later.

Sediment variability

There are many portions of the sea floor blanketed by the same type of sediments, but there are a great many areas that exhibit considerable variation in sediment distribution. The problem the marine sedimentologist faces in his studies conducted from a surface vessel is how representative of the actual environment are his samples. In dives to as great as 3600 m the author has noted major sediment variations (from silt to sand to gravel) in a distance of 100 m on many occasions. The influence of local features, generally on a scale smaller than can be detected by surface echo sounders, causes considerable variation in the sediment distribution. Currents provide an excellent sorting mechanism as seen on dives along the Mid-Atlantic Ridge where pteropods are concentrated in isolated windrow-like piles. Because of their relatively low density, pteropods are also frequently found concentrated in narrow depressions along with other low density materials such as grasses, seaweed, and paper. One of the most enlightening experiences to the author in his early dives was the great variability in the sediments across portions of the sea floor. When considering the basis for most studies on deep-sea sediments, wherein relatively few samples are randomly collected from a surface vessel and the extrapolation of data over great distances is commonplace, one has to wonder about the significance of many of these investigations. Here the submersible if nothing else, impresses upon those who have made dives that the sea floor is the scene of complex processes and variation is the rule rather than the exception.

Bioturbation

Until one has witnessed first hand the burrowing, and churning of bottom sediments by organisms it is difficult to appreciate how important the process of bioturbation can be. At shallower water depths closer to shore, where productivity is usually high, the bottom activity of the benthos is a formidable force in the reworking of sediments. In areas of high productivity the sea floor appears to be literally alive with polychaets. In such areas, little imagination is needed to see that the biota are causing considerable alteration to the sediment properties. Even on the abyssal plain organisms are found to play a notable role in sediment alteration (Rowe, 1974).

Another important biological process of concern to the geologist is the tunneling in semi-consolidated sediments by crabs and lobsters. In many areas where ridges or slopes of semi-consolidated sediments are exposed, tunneling activities have literally resulted in a honeycomb of passages. Along portions of the wall of the Hudson submarine canyon it is evident that tunneling by red crabs has undermined the wall to the point of collapse. Similar erosional activities by organisms has been reported in Corsair and Block canyons (Dillon and Zimmerman, 1970). In studies of rock-boring organisms off southern California J. Warme (personal communication, 1973) noted from his DEEP QUEST dives that these organisms are contributing significantly to the erosion of the exposed strata. Although much has been published (Berger and Heath, 1968; Clarke, 1968; Rowe, 1974) a full understanding of bioturbation is not possible without in-place observations to delineate the various types of organisms, their interrelationships, concentrations, and burrowing habits. A major step to such an understanding is the ability to position the investigator at the scene for long enough periods to document the interplay of currents, sediment type and the organisms themselves.

Pore water and heat flow

Migration of pore waters and heat flow through sediments have been subjects of interest to geochemists and geophysicists for some time. The collection of pore water from core samples or by in-place probes has not been as effective as desired. Heat flow measurements on the other hand have been more readily obtained by needle-like probes or thermistors attached to core barrels. The problem again is ascertaining the representative nature of the data. Submersibles have come to play an important role in these studies by being outfitted with suitable sensors that can be pushed into the sediments. In the case of water seepage from the sediments to the overlying water a probe was developed which measured resistivity. The probe registered changes in resistivity from within the upper few centimeters thus allowing a comparison with the overlying sea water (Milliman et al., 1967). In recent years considerable interest has evolved in the study of hydrothermal processes on the sea floor. Because of the small size of the vents, in some cases 2—3 m across, and a limited knowledge of their occurrence, temperature sensors dangled randomly from a surface ship have been inadequate to detect these conditions. However, one exception is on the East Pacific Rise where the process appears to be rather pronounced and has been detected by a remote probe (J. Corliss, personal communication, 1976). The remote probe data, however, only reveal the presence of a hydrothermal process, but do not provide information on the distribution of vents, their orientation, relative position, or the degree of mineralization over the surrounding area. All of these aspects are vital to an understanding of the process. This type of problem is ideally suited to a submersible which can carry not only the trained observer to the site of activity, but it can place temperature sensors and obtain water samples from any location deemed important by the scientist. The rapid dilution of both temperature and water properties in the overlying water requires the need for precise sampling in the vents.

Tectonism and the associated faulting activity in the ocean basins is most appropriately studied on a regional scale to adequately define the interrelationships of the various elements in such a global process. These studies are most effectively carried out from surface vessels. Small-scale faulting which would be indicative of recent tectonic activity, cannot be detected from a ship, but must be inferred from bottom-mounted seismographs, or actually verified from submersible operations. Again, the need for "walking the outcrop" is apparent if faulting is to be accurately defined and the movement measured. This is a difficult task for a submersible because of the very rugged terrain and lack of good exposures; but is made easier if the fault plane is overlain by a relatively thin sediment cover. Fracture zone A on the Mid-Atlantic Ridge at 36°56′N is an example. During Project FAMOUS, offsets noted in the semi-consolidated sediments readily attested to recent tectonism and allowed the scientists to determine with great accuracy the extent of the active zone (ARCYANA, 1975).

MORPHOLOGICAL AND GEOLOGICAL DETERMINATIONS

Bathymetry

Not only does the submersible provide a means whereby the scientist can be placed literally into the dynamic process he wishes to study, but it affords him the opportunity to selectively sample and document geological sequences and morphological conditions as they are today. Bathymetric mapping in the deep sea from surface vessels is at the stage of only being able to resolve major features (20—25 m in size) and is not definitive in areas of steep slopes (greater than 45°). These inadequacies have only been improved upon in certain Navy systems which are not commonly available to the scientific community. A marine geologist lacking a suitable base map is definitely at a disadvantage to his terrestrial counterpart when it comes to the use of a submersible and the concept of "walking the outcrop". The lack of adequate base maps has proven to be one of the major frustrations to the submersible user and has resulted in the less than optimal use of such vehicles in a number of cases. Only recently when data from the Navy's system have been made available such as in Project FAMOUS (Ballard et al., 1975) and the Cayman Trough study (Ballard, 1976) has the submersible observer had an adequate map. The use of deep-towed vehicles and unique bottom photography such as the LIBEC system (Brundage and Cherkis, 1975) prior to submersible dives has greatly enhanced the submersible user's "feel" for the terrain and the interrelationships he might expect to find. Unfortunately this array of information is not commonly available and only in Project FAMOUS was such a wealth of accurate data in the hands of the divers before their field work began. In the case of Project FAMOUS, submersible observations did little to improve on the bathymetric map, but they were instrumental in advancing the understanding of tectonism and compositional changes in crustal material in the area of the Mid-Atlantic Ridge. As might be expected, submersible observations do provide the final degree of resolution to a bathymetric map and have been very important in delineating small features (2—20 m in size) which were unknown from echo-sounding data. These are also very important to an understanding of the present and past processes occurring on the sea floor (Emery and Ross, 1968; Minter et al., 1975).

Sample collection

A unique capability of a submersible of great importance is its ability to collect samples from precise sites relative to one another. Thus, by means of the trained observer within the vehicle directing the sampling operation, it is possible to accurately document by photos and notes the orientation and position of a sample relative to its surroundings. This mode of operation provides the only means to selectively sample both representative and anomalous features which must be considered in their relative perspective if a study is to be of any consequence. A prime example of this unique capability was the work carried out by ALVIN on the Mid-Atlantic Ridge where a suite of samples was collected along predetermined transects normal to the inner rift valley. This was done in order to study compositional changes in crustal material away from the zone of crustal formation. Such an investigation could only be effectively undertaken with the aid of a submersible. The great variability of sediment distribution on the sea floor is another good reason for the need of selective sampling. Bottom photographs taken in the rift valley of the Mid-Atlantic Ridge revealed windrow-like deposits of dark fines (Brundage and Cherkis, 1975) which were believed to be made up of volcanic glass. Samples collected by ALVIN showed these "dark fines" to be pteropods coated with manganese with an average size of 5—7 mm, not as fine-grained as initially believed.

The submersible's ability to sample relatively fragile material such as coral and glass-coated pillow lavas has in conjunction with photographic and visual observations provided an insight into a number of problems, previously clouded by dependence on dredged material. In exposing the interiors of reefs by use of explosives and then sampling from a submersible, significant advances have been made in the study of reef morphology, internal structure and growth characteristics (Ginsburg and James, 1976).

Careful collection of glass and manganese-coated pillows, which heretofore commonly lost much of such coatings when collected by a dredge, has provided new insights into such questions as rates of glass alteration and manganese accumulation (J. Moore, personal communication, 1975). The vivid macro- and micro-morphology of the pillows observed in place (Fig. 3) and then selectively sampled for detailed laboratory studies later, has provided the Project FAMOUS volcanologist with a wealth of material that will require years of study.

Geotechnical properties

In recent years the geotechnical properties of submarine deposits have taken on considerable importance as man increases his need to build and work on the sea floor. Over the past ten years numerous studies have been published defining the distribution and variance of geotechnical properties within the upper few meters of the sea floor (Richards, 1961; Keller, 1967; Bennett et al., 1970; Horn et al., 1974). Until recently, investigators of these properties have had to rely on the collection of relatively disturbed core samples from surface vessels. Here too, the submersible has come to play an important role. Not only are cores selectively taken by a submersible (Fig.

Fig. 3. Pillow lavas on the Mid-Atlantic Ridge photographed from ALVIN.

4), but they prove to be relatively undisturbed which is of paramount importance to these studies. A major limitation, however, is the shallow sampling depth of about 1 m attained from the submersible. More recently in-place geotechnical testing of bulk density and shear strength from a submersible to depths of 2—3 m below the sea floor has become a reality (Figs. 5, 6) (Moore, 1963; Buffington et al., 1967; Inderbitzen and Simpson, 1972; Richards, 1972; Perlow and Richards, 1973). In-place measurements of sediment acoustic properties from submersibles has advanced the field of acoustics markedly by providing the necessary ground truth to formulate concepts on sound attenuation and reflection characteristics of deep-sea sediments (Hamilton, 1963). This work was instrumental in verifying that sound speeds in high-porosity sediments are commonly less than those found in the water immediately above the sediment—water interface.

Nodule deposits

Considering the international interests in manganese nodules, it is remarkable that the submersible has not really played any role in the study of these deposits. Admittedly, the "commercial" deposits are beyond the depth capabilities of most submersibles, but little effort has been made to involve these vehicles in the study of manganese nodules even where depth is not a limiting factor. The author knows of only five submersible dives by United States scientists to investigate nodules and slab deposits of manganese and phos-

Fig. 4. Sediment core sampling from a submersible (courtesy Woods Hole Oceanographic Institution).

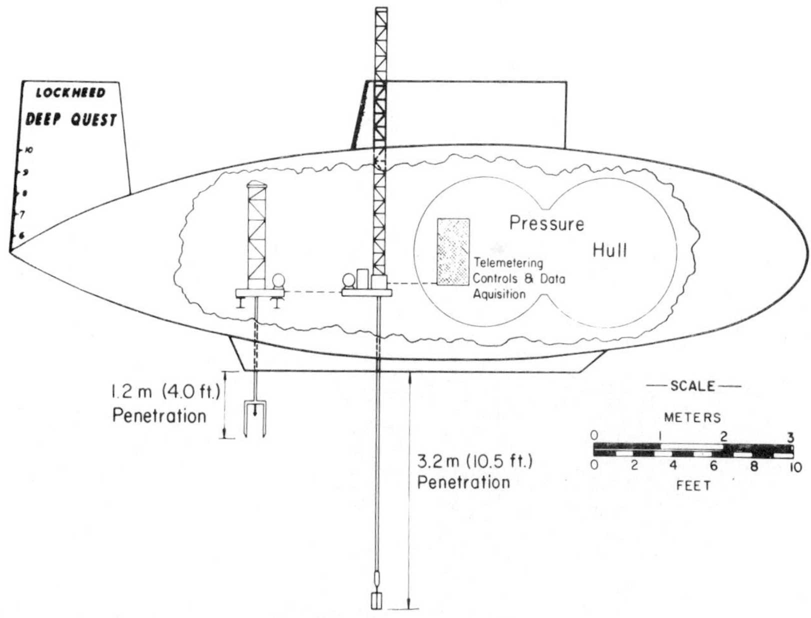

Fig. 5. In-place testing of sediment shear strength (vane) and bulk density (double probe) from DEEP QUEST (courtesy A.F. Richards).

phate. Although limited, these dives on the Blake Plateau provided considerable insight into the occurrence of the different morphological (nodules vs. slabs) and compositional forms (manganese vs. phosphate) (Pratt and McFarlin, 1966; Hawkins, 1968). It became apparent from these dives that generalizations on the occurrence of these deposits could not be made because the immediate environment appeared to be a major factor controlling the specific type of deposit. There is no reason to believe that conditions on the Blake Plateau relate in any way to the Pacific deposits, but an understanding of the dynamics and morphology in the vicinity of such deposits is certainly of considerable importance to those interested in the formation of nodules. Perhaps as studies on the Pacific nodules advance the submersible may come to play a more significant role.

Sub-bottom profiling

High-resolution acoustic profiling studies by submersibles of sub-bottom sedimentary structures and sediment thicknesses to depths of 0—50 m have been limited to examining shallow pockets of sediment in the Gulf of Maine (R. Ballard, personal communication, 1973), short transects across the Hudson submarine canyon and feasibility studies of the use of various low frequencies for such studies (Busby and Merrifield, 1967). More need for these

Fig. 6. In-place testing of sediment bulk density by a nuclear probe from aboard ALVIN (courtesy A.F. Richards).

types of submersible measurements appears to be evolving from commercial interests who are concerned with the burial depth of objects such as pipelines (Buckman, 1975).

SUBMERSIBLE LIMITATIONS

Sea state

As with any unique tool which can do many things better than by any other means, there are a number of limitations to the submersible. Of considerable frustration to many oceanographers is the weather-dependent nature of this vehicle. What may be considered a relatively calm day for shipboard work is often too rough for the launching of a submersible. The low freeboard and poor sea-keeping characteristics of a submersible plus the launching procedures used require relatively low sea states. State 3 (2—2.5-m waves) is usually the upper limit for a safe launch. Frustration also plagues the user of the submersible when during marginal weather conditions he must also consider the chances of worsening conditions and the problems of retrieving the vehicle.

Duration

In the use of deep-diving vehicles with dive durations of eight to ten hours, half of this time may be devoted to the descent and ascent when working to depths as great as 3,000 m. Realizing the effort and cost to get the observer to his study site, one can readily appreciate the anxiety of the scientist when after four or five hours the pilot announces it is time to surface. Such a time constraint obviously puts severe limitations on certain studies of dynamic processes. For the shallow-diving boats, this is not as serious a problem in most cases. Even at these depths, the long endurance (days) of the Navy's NR-1 submersible can provide a unique opportunity to study bottom processes.

Mobility

In a few cases, namely large vehicles such as ARCHIMEDE and TRIESTE II, mobility and speed are severe handicaps. In both these vehicles it is essentially impossible for the observer to get close to an outcrop, because the manned sphere is located beneath a rather large gas floatation chamber. The size of these vehicles prevents their close examination of many features such as would be readily accessible to a submersible such as ALVIN or CYANA. The advantage of these large boats is their great payload capacity of 2700—4500 kg as compared to ALVIN with 450 kg (Busby, 1976). Since the geologist commonly stays within view of the bottom, his speed requirements are minimal, on the order of one knot, but more commonly one quarter to one half knot. Speed limitations only become a problem when the submersible cannot hold its position or advance due to bottom currents.

Visibility

Although visibility is good in the deep sea, up to 60 m in some instances (Busby, 1967), the problem often lies in the field of view provided by the view port available to the observer. Often the forward-looking port is for the use of the pilot and the observer is limited to a view either off to one side or down at some angle. Until one has attempted to decipher the setting of an area through a non-forward looking port, this may not appear to be a serious limitation. In thinking about this for a moment, one can readily realize that viewing an area from only one side port can lend a notable bias to the observations. A simple example will suffice to make my point. With observer view ports to either side, such as in ALVIN, and a submersible traversing parallel to a steep slope the inboard observer sees the outcrop and all its characteristics, whereas the outboard observer may be viewing a distinctly different setting from many meters above the bottom. Having access to a forward looking view port is the only way for the observer to adequately piece together in his mind the relationships he sees as the submersible proceeds across an area.

Vertical escarpments

As pointed out earlier, the selective sampling capability of the submersible is one of its greatest attributes. There are, however, problems for most submersibles when it comes to sampling vertical escarpments. For the boat equipped with a single manipulator, it is almost impossible to sample such a surface because as the hand attempts to collect the sample, it pushes the boat away from the escarpment. The task becomes more feasible but far from easy for the vehicle with two manipulators. Using one manipulator to hold on, the other can attempt to collect the sample. Once having hold of the samples comes the problem of hanging on to it long enough to get it into the sample basket. More times than not the sample slips from the manipulator and ends up with the rest of the talus at the base of the slope. A frustrating experience to say the least.

Reliability

When considering the conditions a submersible must overcome often from high temperatures and motion while aboard its tender, to the temperatures and pressures of the deep, and frequent jolts from striking an outcrop or the bottom, the reliability of these vehicles to remarkable. There are times when failures occur; these often become exaggerated in the mind of the user, because problems with the boat generally means the diving program is halted or even aborted completely. It is not like shipboard programs where parts of the study are almost always carried out despite problems with the ship or its equipment.

FUTURE OUTLOOK

The basic submersible technology, in terms of the vehicle itself, is still slightly ahead of the marine geologist's needs. However, the gap is closing rapidly and the clamour for improved vehicles will be heard within the next three to four years. Now that navigation and data-logging systems are reasonably well developed so the user can reconstruct the scene and his observations with some degree of reality post facto, there are two aspects of a submersible operation that need particular improvement. The two most important of the unique capabilities of a submersible are its ability to allow the observer to *see* the subject he is studying and to *selectively* collect those samples needed for his investigation. It is these same two areas that require improvement for the geologist to be more effective in his use of the submersible.

A number of shallow diving boats have excellent viewing facilities, but the problem is more serious for some of the deep vehicles. As pointed out earlier the observer is not able to see ahead and to the sides, because of the pilot's need for the forward-looking port. The combination of a forward-viewing capability and greater power for longer periods of external lighting would greatly advance the submersible's "eyeball" and photographic observations.

For the geologist, sampling an outcrop is a must, and today this is almost an impossible task if the slope to be sampled is at an angle greater than 45°. Rock drills (Fig. 7) have been used effectively from a submersible, but are not suitable for use on steep slopes. A hydraulic rock hammer has been developed for use from ALVIN, which requires the boat to be firmly on the bottom (R. Ballard, personal communication, 1975), but this does not solve the sampling problem for high angle slopes.

Future needs such as retrieving sediment and rock samples at ambient pressures and temperatures as well as in-place measurements and sampling to greater depths in the substrate will be placed on submersible operations in the years ahead. The increasing need for geophysical and geological data in the higher latitudes, due primarily to offshore mineral development may in future generations lead to entirely different submersible concepts. These would include much larger and longer endurance vehicles capable of conducting surveys below the effect of the severe weather conditions common to these regions.

It is my belief that studies from shallow diving vehicles will continue much as they are today for some years to come. The major advances in geology where the submersible can be expected to play a major role will be in the deep sea. As the geologist completes his regional studies of the ocean basins he begins to focus on the elements comprising the basins themselves. It is a process of ever-increasing the degree of resolution and accuracy of his measurements that will lead him to the final analysis provided by a submersible.

Such submersible studies as those carried out in Project FAMOUS, the Cayman Trough and the Galapagos hydrothermal investigations are the forerunners of the types of geological and geochemical studies that can be expected in the years ahead. The time is approaching rapidly where sufficient background and understanding of various geological processes soon will

Fig. 7. Rock drill for collecting 6 to 8 cm-long cores (courtesy Woods Hole Oceanographic Institution).

be attained so that the unique capabilities of the submersible can be used to effectively solve major problems.

ACKNOWLEDGEMENTS

I am indebted to the ALVIN group at Woods Hole Oceanographic Institution and to my colleagues in Project FAMOUS for providing me with much of the information and photographic material used in this discussion.

REFERENCES

ARCYANA, 1975. Transform fault and rift valley from bathyscaphe and diving saucer. Science, 190: 108—116.

Ballard, R.D., 1976. Windows on earth's interior. Natl. Geogr., 150: 228—249.

Ballard, R.D. and Emery, K.O., 1970. Research Submersibles in Oceanography. Marine Technology Society, Washington, D.C.

Ballard, R.D., Bryant, W.B., Heirtzler, J.R., Keller, G.H., Moore, J.G. and van Andel, T.H., 1975. Manned submersible observations in the FAMOUS area: Mid-Atlantic Ridge. Science, 190: 103—108.

Bennett, R.H., Keller, G.H. and Busby, R.F., 1970. Mass property variability in three closely spaced deep-sea sediment cores. J. Sediment. Petrol., 40: 1038—1043.

Berger, W.H. and Heath, G.R., 1968. Vertical mixing in pelagic sediments. J. Mar. Res., 26: 134—143.

Brundage, W.L. and Cherkis, N.Z., 1975. Preliminary LIBEC/FAMOUS cruise results. Nav. Res. Rep., 7785: 31 pp.

Buckman, D., 1975. Vickers oceanics concentrates on North Sea work. Ocean Ind. Texas, October, pp. 69—71.

Buffington, E.C., 1961. Experimental turbidity currents on the sea floor. Bull. Am. Assoc. Pet. Geol., 45: 1392—1400.

Buffington, E.C., Hamilton, E.L. and Moore, D.G., 1967. Direct measurement of bottom slope, sediment sound attenuation and sediment shear strength from Deepstar 4000. Proc. 4th U.S. Navy Symp. Military Oceanography, 1: 1—10.

Busby, R.F., 1967. Undersea penetration of ambient light and visibility. Science, 158: 1178—1180.

Busby, R.F., 1976. Manned Submersibles. Office of the Oceanographer of the Navy, Alexandria, Va., 764 pp.

Busby, R.F. and Merrifield, R., 1967. Undersea studies with the DSRV ALVIN, Tongue of the Ocean, Bahamas. U.S. Naval Oceanographic Office, Washington, D.C., IR 67-51: 54 pp.

Cacchione, D.A., Rowe, G.T. and Malahoff, A., 1976. Sediment processes controlled by bottom currents and faunal activity in lower Hudson submarine canyon. Soc. Econ. Paleontol. Mineral. Annu. Meet. Progr., 48.

Clarke, R.H., 1968. Burrow frequency in abyssal sediments. Deep-Sea Res., 15: 397—400.

Dillon, W.P. and Zimmerman, H.B., 1970. Erosion by biological activity in two New England submarine canyons. J. Sediment. Petrol., 40: 542—547.

Emery, K.O. and Ross, D.A., 1968. Topography and sediments of a small area off the continental slope south of Martha's Vineyard. Deep-Sea Res., 15: 415—422.

Ginsburg, R.N. and James N.P., 1976. Submarine botryoidal aragonite in Holocene reef limestones, Belize. Geology, 4: 431—436.

Hamilton, E.L., 1963. Sediment sound velocity measurements made in situ from bathyscaphe TRIESTE. J. Geophys. Res., 68: 5991—5998.

Hawkins, L.K., 1968. Visual observations of manganese deposits on the Blake Plateau. U.S. Naval Oceanographic Office, Washington, D.C., IR 68-99: 19 pp.

Horn, D.R., Delach, M.N. and Horn, B.M., 1974. Physical properties of sedimentary provinces, North Pacific and North Atlantic Ocean. In: A.L. Inderbitzen (Editor), Deep-Sea Sediments: Physical and Mechanical Properties. Plenum Press, New York, N.Y., pp. 417—441.

Inderbitzen, A.L. and Simpson, F., 1972. A study of the strength characteristics of marine sediments utilizing a submersible: Underwater soil sampling, testing and construction control. Am. Soc. Test. Mater., Spec. Tech. Publ., 501: 204—215.

Keller, G.H., 1967. Shear strength and other physical properties of sediments from some ocean basins. In: Proc. ASCE Symp. Civil Engineering in the Oceans, pp. 391—419.

Keller, G.H., Lambert, D., Rowe, G.T. and Staresnic, N., 1973. Bottom currents in the Hudson Canyon. Science, 180: 181—183.

LePichon, X., Hekinian, R., Francheteau, J. and Corré, D., 1975. Submersible study of lower continental slope—abyssal plain contact. Deep-Sea Res., 22: 667—670.

Milliman, J.D., Manheim, F.T., Pratt, R.M. and Zarudzki, E.F.K., 1967. ALVIN dives on the continental margin off the southeastern United States. Woods Hole Ocean. Inst. Ref. No. 67-80: 64 pp.

Minter, L.L., Keller, G.H. and Pyle, T.E., 1975. Morphology and sedimentary processes in and around Tortugas and Agassiz sea valleys, southern Straits of Florida. Mar. Geol., 18: 47—69.
Moore, D.G., 1963. Geological observations from the bathyscaphe TRIESTE near the edge of the continental shelf off San Diego, California. Geol. Soc. Am. Bull., 74: 1057—1062.
Nishizawa, S., Fukuda, S. and Inoue, N., 1954. Photographic study of suspended matter and plankton in the sea. Bull. Fac. Fish., Hokkaido Univ., 5: 36—40.
Perlow, M. and Richards, A.F., 1973. Geotechnical variability measured in place from a small submersible. Mar. Tech. Soc., 7: 27—32.
Pratt, R.M. and McFarlin, P.F., 1966. Manganese pavements on the Blake Plateau. Science, 151: 1080—1082.
Rashid, M.A. and Brown, J.D., 1975. Influence of marine organic compounds on the engineering properties of a remoulded sediment. Eng. Geol., 9: 141—154.
Richards, A.F., 1961. Investigation of deep-sea sediment cores, I. Shear strength, bearing capacity, and consolidation. U.S. Navy Hydrographic Office, Washington, D.C., Tech. Rep., 63: 70 pp.
Richards, A.F., 1972. Instrumentation of two submersibles for in situ geotechnical measurements in cohesive sea floor soils. Preprint 2nd Int. Ocean. Dev. Conf. Tokyo, 2: 1329—1346.
Rowe, G.T., 1974. The effects of the benthic fauna on the physical properties of deep-sea sediments. In: A.L. Interbitzen (Editor), Deep-Sea Sediments: Physical and Mechanical Properties. Plenum Press, New York, N.Y., pp. 381—400.
Schlee, J., 1967. Geology from a deep-diving submersible. Geotimes, 12: 10—13.
Vening-Meinesz, F.A., 1932. Gravity expeditions at sea 1923—1930. Netherlands Geodetic Comm., 1: 109 pp.

CHAPTER 10

U.K. EXPERIENCE OF THE USES OF SUBMERSIBLES IN THE GEOLOGICAL SURVEY OF CONTINENTAL SHELVES

R.A. Eden, R. McQuillin and D.A. Ardus

INTRODUCTION

The Institute of Geological Sciences (IGS), with which the authors are affiliated, is a component body of the Natural Environment Research Council (NERC). It is the government organisation responsible for official surveys and compilations of the geology of U.K. land and sea areas and is concerned with all aspects of survey activity at sea. IGS interfaces on a wide front with commercial and academic workers. Its funds are relatively limited compared with those of commercial groups, but it has the strength of providing an ongoing survey not necessarily restricted to precise economic objectives.

The present assessment is an attempt by three IGS authors to put the U.K. geological work with submersibles in perspective. It is heavily slanted in the direction of IGS experience, since much of the relatively recent work by commercial operators is confidential. There have been a number of joint cruises involving IGS with a sister group in the Institute of Oceanographic Sciences (formerly National Institute of Oceanography), and J. Wilson of IOS describes aspects of these in another paper in the present publication (Chapter 7).

COST EFFECTIVENESS OF GEOLOGICAL WORK WITH SUBMERSIBLES

The question of cost effectiveness needs to be faced at the outset, not only because submersibles are comparatively slow and/or costly when compared to other survey tools, but also because the authors differ somewhat in the emphasis of their approach. On continental shelves, where water depths are relatively shallow, a wide range of tools is available for studying geology as it is exposed in the seabed and at shallow depths beneath it. Cost effectiveness of submersibles in this context relates to the following factors:

(1) Is it cheaper overall to achieve the essential objectives of a particular survey by the use of a submersible or by other methods, such as traversing or sampling with a surface vessel?

(2) Is it cheaper to use one type of submersible than another given that both are capable of achieving the required objective?

(3) Is the objective of the survey of sufficient scientific or economic

importance in relation to other objectives to which funds could be diverted?

In practice it is often difficult in general surveys to quantify some of the variables in precise financial terms, partly because survey objectives are concerned with accumulation of data on multiple parameters, not all of which are obtainable by any one method. Moreover, the various methods are not always immediately available in direct competition with each other; for example from 1969 to 1974, of the submersible options only the Vickers Oceanics PISCES system was effectively available in U.K. waters. Another important but unquantifiable factor is the effectiveness of the operator's team. Consideration of cost effectiveness is clearly easier when it is possible to define an objective precisely. If this comprises the collection of engineering data essential to the emplacement of an expensive structure on the seabed, it is probable that actual costs must take second place to the need to use the best available instrumentation for the job.

Variation in microtopography, for example, can only be studied in detail from a submersible, since only poor resolution of the smaller variations is possible from the surface. If data on microtopography are required for a well-defined area such as a platform site, then the essential need is for a submersible system with a maximum capability for precise navigation and recording of microtopography, by whatever method permits the seabed contours to be transcribed accurately on to a map. This is likely to be by the use of sonar or stereo photography or both. Assuming equal capability for several submersible systems, and their equal availability, evaluation of cost effectiveness will depend only on speed of coverage of the ground in relation to costs of operation, bearing in mind the possible effects of weather variations during the work. The same may be said of other engineering studies, such as the survey of a pipeline route. In this case the identification of obstacles to burial of the pipe necessitates a visual system with maximum flexibility, and both bottom profiler and side-scan sonar with good resolution.

The evaluation of cost effectiveness in the type of general survey carried out by IGS is by no means so easy. The area of the U.K. Continental Shelf is approximately 650,000 km^2 of which by 1976 less than half has been surveyed on reconnaissance traverses with an approximate four-mile spacing. It is necessary to compare value for money, on the one hand of pushing on with this reconnaissance traversing, and on the other hand with the benefits of diverting some effort to different types of seabed study as a control on the interpretation of the traversing. It is self evident that geophysical traversing is of restricted value without sampling or drilling to check on the nature of the variations being recorded, and in the IGS survey program this requirement is catered to by follow-up sampling and drilling programs. It is in connection with these follow-up programs that submersibles have been used. The general principle for selection of sites has been quite simple; they have been chosen in critical areas from which observations can be extrapolated as widely as possible by use of other methods. The critical factor under examination at a particular site may relate to solid geology, geomorphology, or sedimentary environment, and in subsequent pages examples of what has been attempted at different sites are described.

In terms of cost effectiveness it is first necessary to calculate that if $\$X$ is available for survey of a particular area, then is it worth-while to spend $Y\%$ of $\$X$ on identifying the solid rocks at a particular location and/or checking in detail on the disposition of the land forms and sediments? It is then necessary to evaluate the possibility and cost of doing the job with a submersible or in some other way. The use of a drill ship to obtain solid-rock samples is a very expensive operation, comparable in daily cost to use of a sophisticated submersible system. If this is the alternative option and seabed exposures are known to occur, it is likely to be cheaper to proceed by using a small corer in the case of extensive seabed outcrops and by the use of a submersible when outcrops are widely scattered or difficult to locate. An alternative to the use of a submersible for some applications, is the use of instrumentation suspended below a surface vessel. This can include television equipment, stereo cameras, and seabed drills. This approach is undoubtedly cheaper in terms of cost per hour than the use of a submersible, but it needs to be borne in mind that suspended packages of this type are in effect "poor men's" submersibles, and the operator is severely penalised by lack of proper control.

The evolution of IGS techniques has passed through the following broad phases:

(1) Geophysical traversing alone.

(2) Geophysical traversing plus sampling, coring and conventional drilling from a surface vessel.

(3) As above (2) plus the use of suspended sensor packages.

(4) As above (3) plus the use of submersibles at critical localities.

IGS suspended sensor packages have in fact evolved into the unmanned drilling submersible CONSUB I by a conscious process of development brought about by specific problems. These have been caused largely by the lack of control induced by wave movement and water currents, which were met in the course of attempting to use drills and cameras hanging below a surface vessel.

HISTORY OF INVESTIGATION ON THE U.K. CONTINENTAL SHELF

The earliest effective use of a submersible in the geological survey of the U.K. Shelf occurred in 1969, when Vickers Oceanics made the submersible PISCES II and its mother ship VICKERS VENTURER available following shake-down trials in Loch Ness. The first charter was by the Natural Environment Research Council, to evaluate the research applications of the vessel

(Natural Environment Research Council, 1970). One of the organisations which participated in the evaluation was the Institute of Geological Sciences, and the geological dives undertaken on this first cruise, in Lower Loch Fyne off western Scotland (Fig. 1) were subsequently described in Eden et al. (1971). Following the successful NERC cruise of 1969, IGS again chartered PISCES II in 1970 (Fig. 1) and on this occasion twenty dives were carried out in a period of ten days off western Scotland. This was an ambitious program aimed at studying selected aspects of most of the accessible geology of the area. The scientific staff comprised a number of IGS geologists together with representatives from IOS and from several British Universities. One objective of the 1970 work was to familiarise as many marine geologists as possible with the seabed environment, and partly for this reason a crew of two geologists plus a pilot was carried on most dives. This practice had also been followed in 1969, but was abandoned after 1970 in the interests of efficiency of working in the very confined cabin space. The results of the 1970 cruise were described in Eden et al. (1971).

In 1971 IGS again chartered PISCES II for dives off the western U.K. coast with the objective of extending and consolidating the earlier work; on this occasion thirteen dives were carried out in seven days. The National Institute of Oceanography (now Institute of Oceanographic Sciences) also chartered PISCES II for periods of one week in 1970 (8 dives) and 1971 (12 dives) for sediment and faunal studies in the western English Channel under the direction of Drs. Flemming and Wilson. The 1970 and 1971 cruises by IGS were regarded as of the nature of extended evaluations of the method for general geological purposes, and the overall conclusion was that for IGS survey it was of value but expensive. It was concluded that the use of manned submersibles had a place for occasional special investigations on which it was important to get a man "on the spot", so that he could thoroughly examine a particular area with a high degree of flexibility. It became however, the opinion that an unmanned system would be cheaper and should serve most of the relevant IGS needs. It was hoped that it could be carried aboard a general geological or drilling vessel and deployed on an occasional basis as required. Much of the reasoning hinged on the consideration that it should be cheaper to obtain seabed rock samples in this way, than to use a drillship. For this reason IGS engaged in development in the first place of suspended packages of cameras, television and rock drills (Eden and Ardus, 1972) and in 1973 a contract was let to the British Aircraft Corporation to carry out a design study for a small unmanned drilling submersible. This study led to the construction of the unmanned submersible CONSUB I by BAC, under joint funding by IGS, and the then Department of Trade and Industry. First sea trials of CONSUB I were in 1974 off Shetland and Norway, using the NERC vessel CHALLENGER.

The 1974 cruise by RRS CHALLENGER was a joint venture between IGS, the Continental Shelf Unit of the Royal Norwegian Council for Scien-

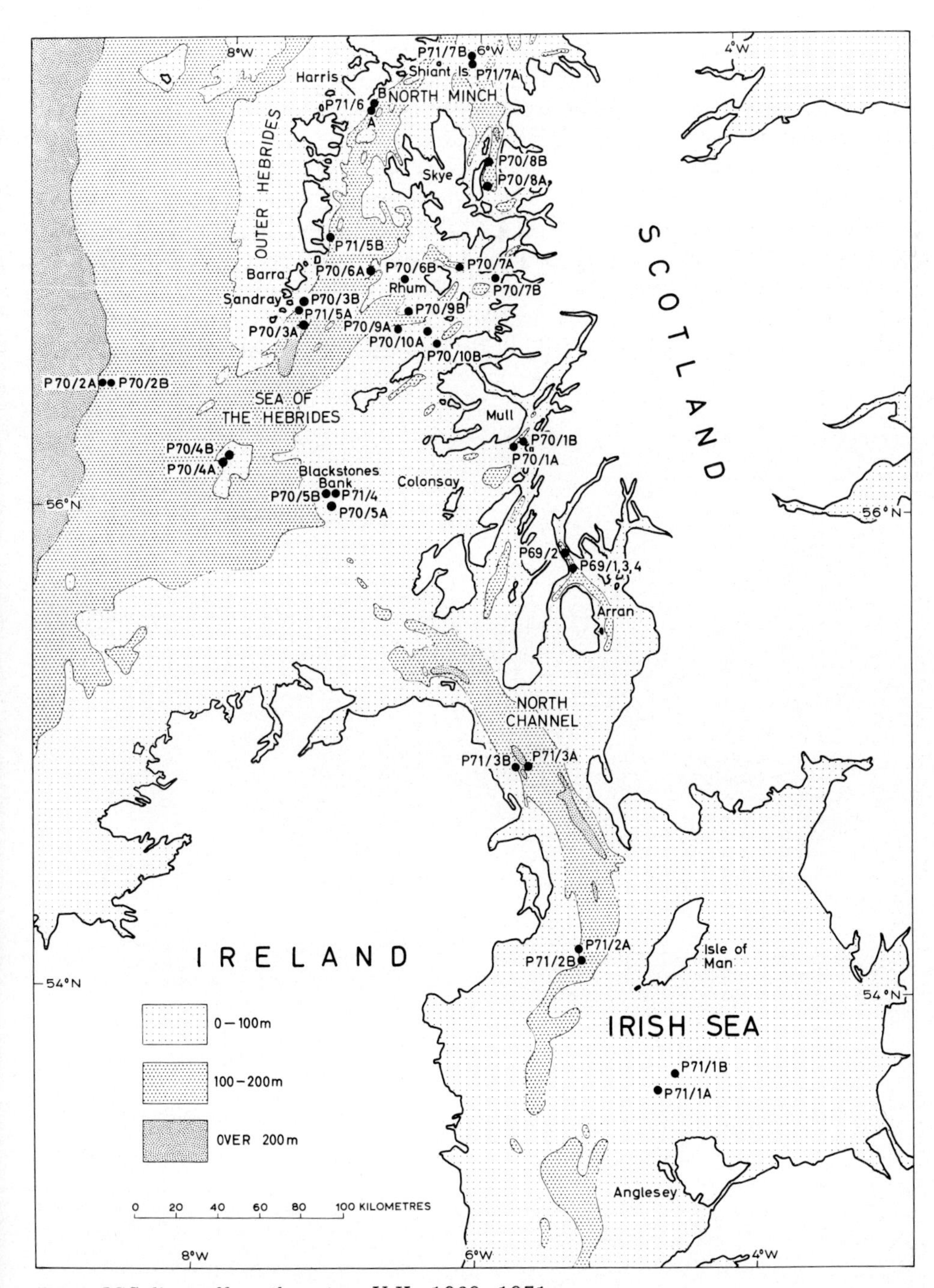

Fig. 1. IGS dives off northwestern U.K., 1969–1971.

tific and Industrial Research with their unmanned vessel SNURRE, and Heriot Watt University (Edinburgh) with the unmanned submersible ANGUS. The object of the cruise was to obtain operational experience of the three unmanned vehicles and to permit the responsible scientists to compare techniques. Although some geological work was done in 1974 all three systems were then very much in the development stage, but work has since continued on a number of other cruises. CONSUB I is now being operated commercially in the North Sea by BAC and an improved version, CONSUB II, is in production. The extent to which CONSUB has lived up to initial IGS hopes is discussed later in this account. Whilst this unmanned system was under development, IGS and IOS jointly mounted the most ambitious of the

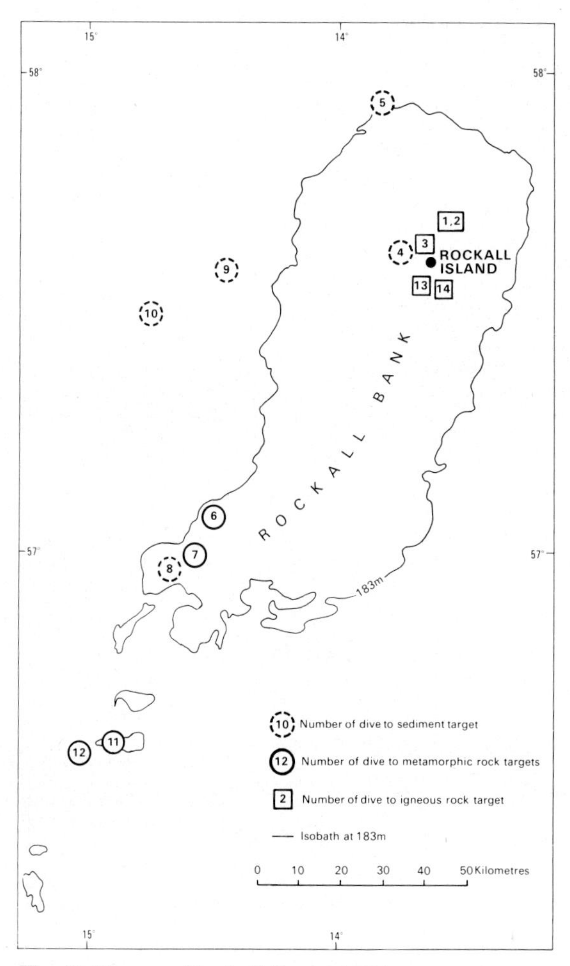

Fig. 2. Dives on Rockall Bank, 1973.

manned submersible cruises which has so far been undertaken for geological research on the U.K. Continental Shelf.

In June 1973, PISCES III and its mother ship VICKERS VOYAGER were chartered to take a mixed IGS/IOS group to Rockall Bank (Fig. 2) for an investigation partly aimed at obtaining a representative suite of rock samples from outcrops known to occur in several areas, and partly at studying the biogenic sediments now accumulating on the bank. This was an outstandingly successful venture resulting, in the course of fourteen dives in seven days, in the acquisition of 924 seabed photographs, 35 rock samples and 28 samples of sediment. The rock samples came from areas previously little known apart from a small number of localities painstakingly tackled by divers, a suspended rock drill and a rock dredge. Since 1973 the requirement for submersibles for geological work on the U.K. Shelf has been transformed by the needs of offshore construction engineers for pipeline and site inspections in the North Sea as described in Chapter 12. Their needs have included preliminary surveys to locate routes free of rock outcrops and gas emission craters (pockmarks), trench inspections to check on the presence of large boulders, and microtopographical survey of sites for gravity structures. Instrumentation for these purposes has increased in sophistication, and the number of submersibles available has increased greatly. As well as small manned vehicles, they now include a number of diver lock-out vessels. In 1976 unmanned submersibles scheduled to operate include CONSUB I and the Canadian vehicle TROV. In addition the six-man submersible AUGUSTE PICCARD, operated by Horton Maritime Explorations Ltd. is scheduled to be available for continuous seabed traversing using only a small support ship in attendance. In parallel with improved geological instrumentation there has been a rapid improvement in precision-navigation facilities (see Chapter 7) which are essential for engineering studies and highly desirable for all geological applications.

GEOPHYSICAL SITE SURVEYS AND INTEGRATION WITH DATA FROM SUBMERSIBLES

The ability to operate in submersibles effectively will depend firstly on the quality of geophysical data acquired not only in detailed site surveys, but also in regional surveys within the area of submersible investigation, and second, on the position-fixing accuracy attained throughout the whole program.

Site surveys

The site-survey methods most commonly used are those of bathymetric surveys using echo-sounders, seabed surveys using side-scanning sonar, and sub-sea bottom investigations using continuous seismic-profiling equipment. If operated as an integrated investigation, these methods will provide a high-resolution, three-dimensional picture of geological structure at and immediately beneath the seabed, with depth of penetration to levels well beyond those significant in planning subsequent submersible investigations. The principles of equipment operation will not be discussed here (for further information see McQuillin and Ardus, 1976), but some examples of survey results will be discussed to illustrate the type of pre-dive information which can be obtained.

An important class of target is that of hard-rock exposure on the seabed. Once located, this can be subject to visual and morphological inspection before sampling with small coring or other apparatus. On the continental shelves of northwestern Europe the occurrence at seabed of any extensive tract of pre-Quaternary rock exposure is relatively rare. More usually such exposures are small in area and are associated with topographic features, ridges or channels, or they occur within limited zones of tidal scour. Over the rest of the shelf, a layer of recent and Quaternary sediments up to a few hundred metres thick blankets the underlying strata placing them beyond the reach of conventional submersible tools. Thus, for investigation of pre-Quaternary geology, it is a necessary function of the pre-dive survey that seabed exposure is reliably identified and accurately located. Echo-sounding surveys can be used to locate topographic features which can be interpreted as rock exposure. Such interpretations are only reliable if supported by evidence from other geophysical or sampling data. A more reliable indication of rock exposure is obtained from side-scanning sonar records (sonographs). In some cases it is possible to confidently predict the occurrence of pre-Quaternary outcrop using sonograph patterns alone; but in general it is necessary to confirm such an interpretation with concurrently surveyed seismic-profile data.

As an example of some of the problems to be met in selecting dive sites, data collected by IGS in an area west of Orkney off northern Scotland will be used. A hull-mounted single side-scanning sonar was used operated at 38 kHz and the seismic-profile data were obtained using a sparker profiling system operated in the range 2—5 kJ. Data were collected during a reconnaissance survey of a large area on a grid-line spacing of 5—10 km. The area is of complex geology with Mesozoic rocks faulted into grabens and half-grabens in a complex Palaeozoic and Precambrian basement. A possible object of submersible use could be that of locating, studying and sampling both basement and Mesozoic outcrop. Mesozoic rocks are identified by well-defined stratifications on seismic profiles. Basement targets can be located without too much difficulty. A complex sonar pattern is typical of that produced by an irregular, eroded basement rock surface, with sediments filling depressions between craggy ridges of hard igneous, metamorphic or highly compacted sedimentary rocks. This particular area of intermittent exposure is sufficiently large that low positioning accuracies might be tolerated during submersible operations. In Fig. 3, irregular topography is associated with thin pre-Quaternary cover over either Mesozoic or Palaeozoic sedimentary rocks. From the sonar patterns exposures of pre-Quaternary rocks cannot be identified with any degree of certainty, though a combination of sparker and sonar records indicates a limited area where there is a high probability of exposure. Submersible navigation and position fixing should allow easy location of an approximately 50 × 50 m site.

In areas where geophysical records are not definitive the manned sub-

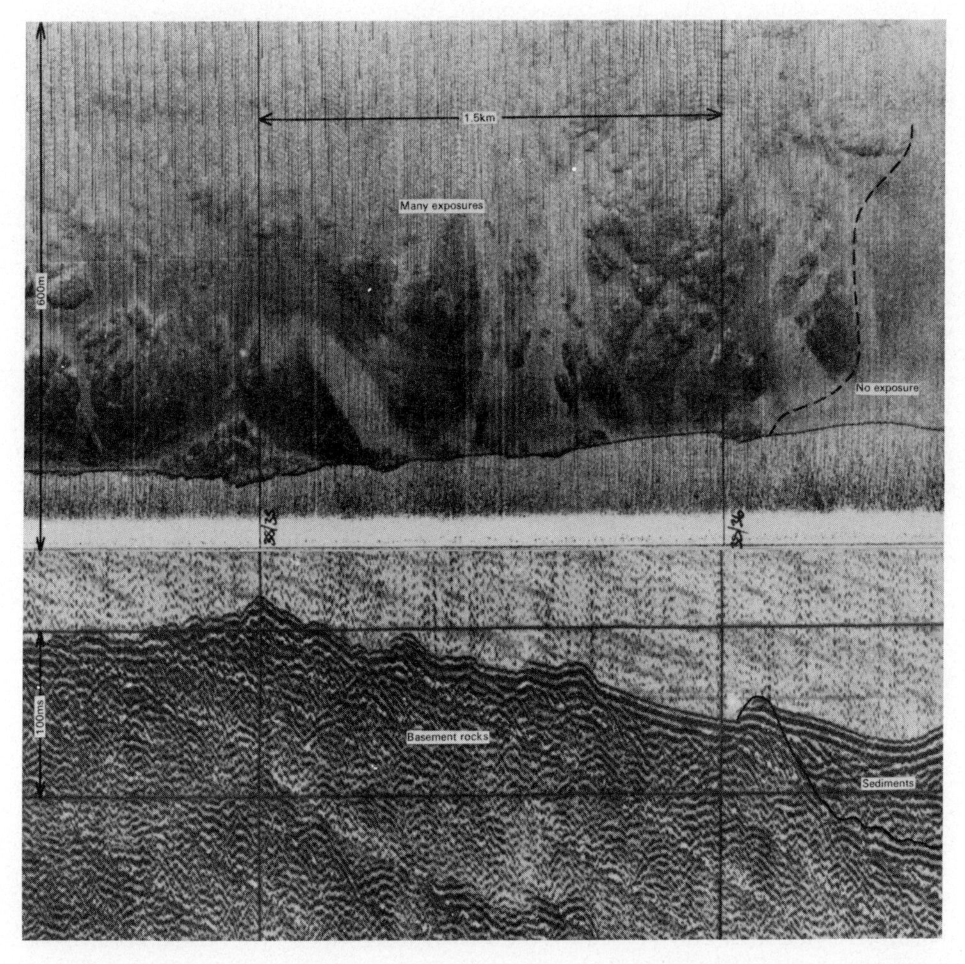

Fig. 3. Sonograph and seismic profile record over irregular sea bottom formed from thin glacial deposits overlying poorly bedded sedimentary sequence with possible limited exposures of pre-Quaternary rocks. Site: approximately 50 km NW of Orkney.

mersible can collect geological data and samples of greater value and reliability than could be obtained using remotely operated equipment from a surface ship. The geologist can make a detailed visual appraisal of geology as seen on the ground. He can also study relationships between the various rock units inspected, as well as collect information which will help assist interpretation of sonographs obtained not only on this traverse but also in other adjacent continental shelf areas. He can judge whether any samples obtained are truly representative of a particular rock unit and ensure that samples are from rock in situ and not from large erratic boulders.

The geophysical records shown in Fig. 3 were all obtained during a reconnaissance survey programme from a ship making gravity and magnetic, as well as seismic, sonar and bathymetric measurements. Because the principal aim of such a survey is not that of site surveying for submersible dives, the data suffers from some deficiencies, mainly in terms of lack of resolution. The sonar used is of relatively low frequency to obtain adequate range in deeper waters and is attached to the side of the ship so that operations are relatively safe and simple on a ship involved simultaneously in a wide range of tasks. For detailed site surveys, a deep-tow, fish-mounted sonar operating at a much higher frequency would give higher resolution and is more appropriate for these applications. A record obtained with an instrument of this type is shown in Fig. 4. On the record a detailed picture is seen of an area of Carboniferous outcrop off the east coast of Scotland. With such a picture to work from, and good (say ± 10 m) position fixing it should be possible for a geologist in a submersible to sample at key localities and make a detailed interpretation of even a complex area of sea bed geology.

Similarly, conventional sparker data of the type seen in Fig. 3 are not sufficiently well resolved to determine with confidence the presence or absence of a thin sediment cover (<1 m). But a pinger-type sub-bottom profiler can provide the required higher resolution and pinger records are often obtained during reconnaissance surveys. Nevertheless, the cost of submersible operations is usually sufficiently high to justify detailed local pre-dive surveys over pre-selected sites, both to pinpoint the best localities to be visited, as well as to provide extra geophysical data for interpretation alongside data collected by the submersible. Such pre-dive surveys can be conducted overnight by a submersible mother ship with submersible operations confined to daylight hours.

So far, our main concern has been that of the search for pre-Quaternary rock exposures. Submersibles have a value also in the study of seabed sedimentary environments. For example, sonographs are widely used to map the distribution of different types of seabed sediments. But in some areas distinctive patterns can be seen which obviously relate to variations in sediment type or topographic form, but which cannot be adequately explained by comparison with sample data obtained using sediment grab, or vibracorers operated from surface ships.

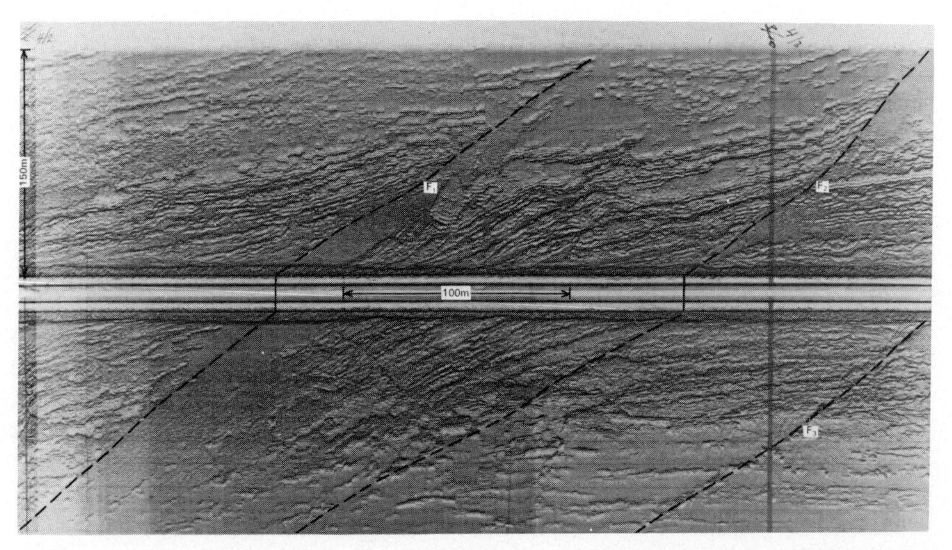

Fig. 4. Dual side-scan sonar record over Palaeozoic rocks cropping out on seabed showing three sub parallel fault zones. Site: approximately 2 km offshore Berwickshire coast, Scotland. Sonar: Klein Hydroscan.

Integration of data into geological interpretations

Geophysical methods usually play a prominent role in dive-site selection. The data collected during a dive has intrinsic value, providing precise information on geology along a short profile. To extend the application of this information into surrounding areas it needs to be fed into the process of geological interpretation of regional geophysical surveys. Frequently in an area use of geophysical data, in particular seismic data, produces a well-defined structural map, but it lacks any reliable stratigraphical identification of the units involved. Drilling and sampling can provide geological control essential to the production of a meaningful geological map. Submersibles can be used for the same purpose, and have the advantage of being able to search for critical exposures and, with modern positioning systems, provide direct correlation between a rock layer sampled and a seismic reflector which can be mapped. However, for such direct correlation to be possible, both the original regional geophysical survey and the submersible survey must be accurately positioned.

Most shelf regional geophysical surveys are conducted to a positioning accuracy with reference to a map coordinate system of better than ±50 m using either radio positioning or integrated satellite navigation systems. Submersible surveys can be located with reference to seabed acoustic responders to an accuracy of better than ±10 m. The responders can be located with reference to a mother ship and through it to a map coordinate system with a comparable accuracy to that of the previous geophysical surveys. Thus, in normal circumstances, a positional uncertainty between regional geophysical survey and submersible survey would be approximately ±60 m. This is often tolerable, but the situation can be improved if a pre-dive geophysical survey is made from the mother ship positioned with reference to seabed responders. It should then be possible to correlate geophysical records and submersible positions to an accuracy equivalent to that of the acoustic positioning system, probably better than ±10 m.

Two examples will help to illustrate the problem, first from a site which has been sampled using a submersible, and second from one where sample data would be useful. The first site was on Dive Location P71/6A (Fig. 1) in the North Minch off northwest Scotland and a submersible investigation is described in Eden et al. (1971). The dive was made to investigate an area of westerly dipping strata lying between two branches of a major northeasterly trending fault system: the Minch Fault. To the west of the western branch the rocks are known to be Lewisian gneisses, to the east of the eastern branch a thick sequence occurs of gently dipping Permian and Mesozoic strata. The age of the fault-bounded belt was unknown prior to the submersible investigation. The submersible was able to investigate a number of rock knolls in the vicinity of this profile which were for the most part covered with boulders. On one of these, exposures of bedrock were discovered beneath the boulders and it was possible to obtain a core using a rock drill mounted on the submersible. On laboratory investigation, the core was found to contain miospores suggesting a Late Palaeozoic age, possibly Dinantian or Namurian. The importance of this identification is not limited to that of giving an indication of the age of the rocks within the narrow fault bounded zone. It also indicates the possibility of Carboniferous rocks beneath the Permian and Mesozoic infill of sedimentary basins in this part of the U.K. Continental Shelf. On this occasion the submersible was able to obtain data which would have been most difficult to acquire by any alternative method other than by mounting a costly drill-ship operation.

The second site is in the Irish Sea where a sparker section shows a faulted boundary between a folded sequence presumed to be Carboniferous, and a gently dipping sequence presumed to be Permian or Mesozoic. It is difficult to assess from this record whether or not bedrock is exposed. A pinger profile would have enabled a better evaluation of the likely success of a submersible traverse. However, assuming that intermittent exposure does occur a submersible profile along the seabed for about 2 km would (1) establish the age of both units, (2) check the best alternative means of obtaining samples in this area, and (3) give valuable stratigraphic control to an interpretation of the grid of sparker lines already surveyed in this part of the Irish Sea.

The relationship between submersible studies and regional surveys of Continental Shelf geology is summarised in Fig. 5. The diagram shows the main

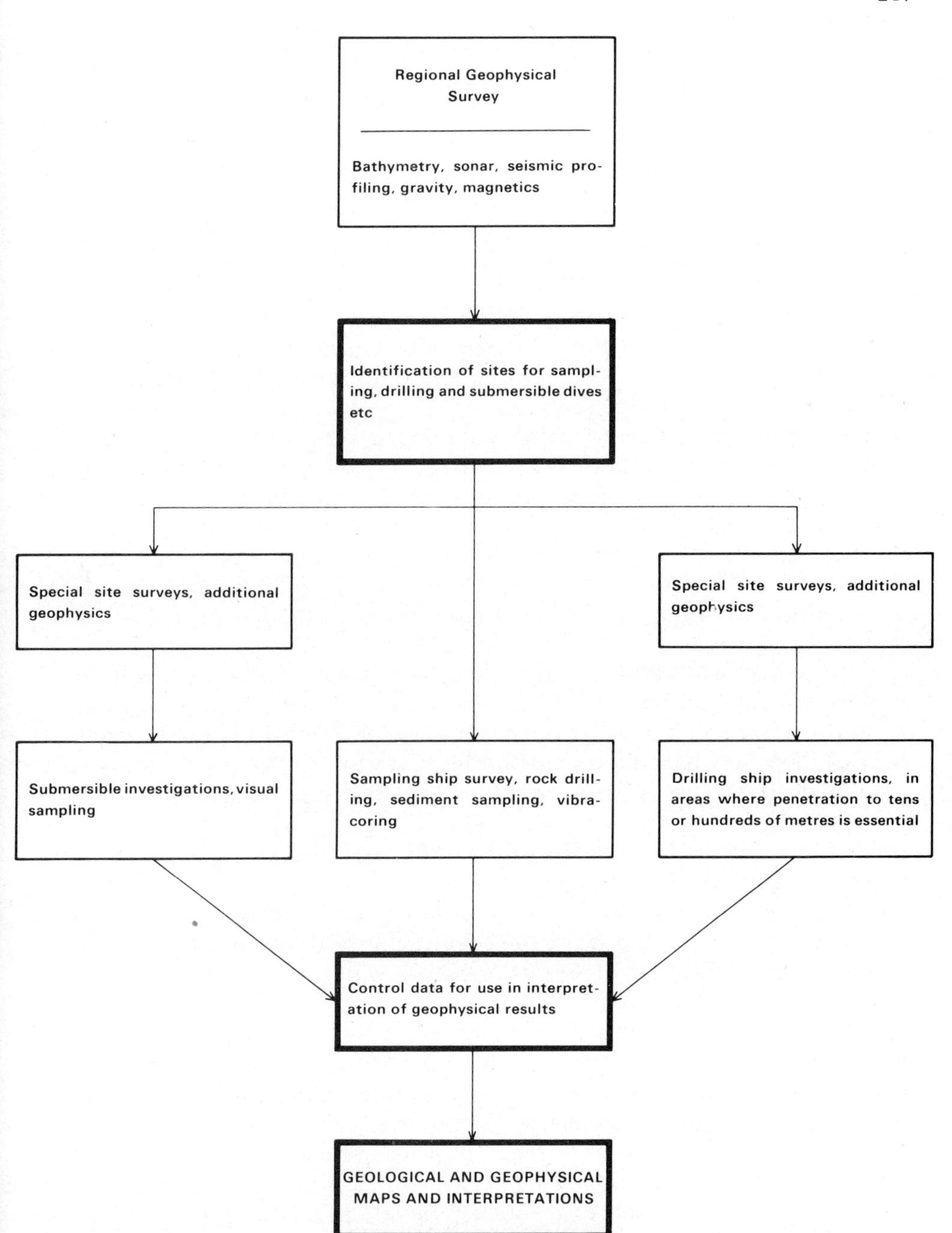

Fig. 5. Diagram showing relationship between submersible studies and other geophysical and geological exploration activities on the Continental Shelf.

activities involved in such work as they affect the planning of dives and subsequent use of data.

OPERATIONAL CONSIDERATIONS IN WORK WITH SUBMERSIBLES

Safety of men and vehicles

In all operations in which IGS has been concerned it has been necessary to continually monitor a changing weather situation, and the highest degree of independence of weather has been desirable to keep loss of time to a minimum. For operations in U.K. waters an ability to endure poor weather in safety clearly carries a high premium.

Because of the stern A-frame handling system developed by Vickers Oceanics the PISCES class of submersibles proved capable of safe launch and recovery, right from the early days of operation in 1969, in a wide range of sea states. During the 1969 cruise, wave heights of up to 2.5 m were noted during uneventful recoveries aboard the 33-m mother ship VICKERS VIKING, and in 1973 the 83-m mother ship VICKERS VENTURER was able to carry out fourteen dives in seven days in exposed conditions on Rockall Bank.

The weather limitations of the Vickers Oceanics method of launch and recovery of small manned submersibles is a subject on which a great deal of experience has accrued, but there is an ill-defined cut-off beyond which the method becomes hazardous. This problem relates to the following factors:

(1) The submersible has to be manually attached to its recovery cables by a swimmer.

(2) Risk of collision between submersible and mother ship increases with deteriorating weather.

(3) Snatch effects on the cable occur due to different motions of the submersible and mother ship while the submersible is attached but in the water.

(4) Pendulum motion of the submersible while it is suspended increases as the sea movement increases.

These possible dangers have to be evaluated by the operator's party leader in deciding whether or not to launch, and when to abort the dive if weather deteriorates as it proceeds. The problem can be reduced but not eliminated by improved launch and recovery procedures such as those which Vickers Oceanics have introduced since 1969. It is not proposed to discuss these here except to say that they largely relate to working with the motion of the sea in every way possible, thereby reducing the rate of relative movement of the submersible and its mother ship.

Integral large systems such as submarines with the endurance of the 28.5-m AUGUSTE PICCARD or the 15-m BEN FRANKLIN do not require to be recovered and are able to operate continuously below the air—sea interface with a high degree of immunity to surface conditions. For cost effectiveness, however, they require to be employed on continuous survey tasks, such as traversing along a selected route, rather than on the spot examination of widely spaced critical localities for which IGS has used PISCES.

Similar to small manned vehicles, unmanned systems have to be recovered through the air—sea interface, where they encounter the same sort of problems. From the point of view of safety in poor weather they have, however, three important advantages:

(1) Risk to life is not involved, even though the machine itself is obviously not to be lightly placed in jeopardy.

(2) They are already attached by a control cable and a swimmer is therefore not required.

(3) As they do not contain heavy personnel spheres and life-support systems they are relatively small and light, and lower momentums are therefore involved in any snatch or pendulum effects which may occur.

Accuracy of navigation

Accuracy of navigation of a submersible depends on:

(1) Fixing an initial reference point, normally a mother ship or acoustic responder.

(2) Fixing the relative position of the submersible to the initial fix.

No scientist can be content without the best possible fixing of the position of his observations, and engineering data for commercial purposes is of little or no value unless it can be related directly to a chosen site. The initial fix, can, however, only be as accurate as available marine methods permit. Most early IGS work with submersibles was by the use of Main Chain Decca, plus radar where possible, in fixing the position of the mother ship. More precise methods are available in the North Sea. For the 1973 Rockall Cruise Loran C was used, together with radar transponders installed on buoys which were referenced by the use of Loran C.

The simplest method of navigation of the submersible relative to the mother ship is by dead reckoning from heading, speed and elapsed time, checked against position at launch and recovery. This is crude but effective, and it is the method used during the IGS/NERC cruises of 1969, 1970 and 1971. In addition, on the 1971 cruise the submersible carried a surface float on a taut line, around which the mother ship circled. The ship's position was continuously recorded by a standard Decca Plotter, and a line joining the centres of the circles therefore represented the course of the submersible. The use of a taut-line float becomes increasingly difficult below about 150 m of water depth, because the float tends to submerge due to drag effects, but the method is practicable over wide areas of the Continental Shelf.

Acoustic transponders laid by the mother ship on the seabed and picked up on the sector-scanning sonar aboard the submersible were also used from the early days with PISCES II. Increasingly sophisticated computerised methods of integrating the signals from such transponders into a track are now becoming available.

A serious problem was encountered in attempting to locate the submersible by the integration of acoustic signals received on a short base line laid out along the length of the mother ship. This arose because of the confusion induced by the reception of signals not only from the submersible itself but also from reflections on rock outcrops. In open areas this problem does not exist, but many geological dives are amongst rugged terrain. On Rockall Bank in 1973, for example, acoustic fixes obtained by the use of a short base line system were of indifferent value. It was because of this known problem that prior to the 1973 cruise IOS developed a system which depended on measuring the E/M variations of the moving water mass as perceived by a probe mounted on the submersible. Results are measured in X—Y terms and later computer processed to produce a track. This work, which is discussed by Wilson in Chapter 7, is still in the development stage but has produced useful results.

Payload and utilisation of vehicle

Similar to working from a surface vessel it is possible to measure and record data relating to a large number of parameters from a submersible, and because of the high operational costs of the vessel it is very desirable to make best possible use of time available. In this respect a small submersible is more limited than one such as the AUGUSTE PICCARD, which has abundant space for instrumentation and can carry the personnel required to operate it. A submersible of this class can mount a complete suite of geophysical tools, comparable to those of a surface vessel. But much of the instrumentation can be expected to be more effective because of the improved stability, isolation from wave noise and ability to control distance to the target. Instrumentation for the AUGUSTE PICCARD is now being developed.

For a small submersible, manned or unmanned, tools can be considered in the following categories:

Sonar systems: echo sounders * and sub-bottom profilers; sector scanners * and side scanners.

Visual systems: direct observation * and recording; television * and video-recording *; photography and stereo photography.

In situ measuring systems: geotechnical properties; acoustic properties; geochemical properties; geophysical properties.

Sampling systems: sediment scoops; rock drills and manipulators.

In the list above the items marked by an asterisk can be expected to be available as part of the standard system of most submersibles, and for the Vickers PISCES class all were available from the time of the first cruise in 1969. Standard sub-bottom profilers and side-scan equipment can be fitted within the limitations of power and space available. Both give much improved results compared to those operated from a surface vessel, partly because of the stability of the platform and partly because of proximity to the target. It may, however, be necessary to tape incoming signals if the recorder is too large to enter the hatch, and in this case on-the-spot quality control is not possible. This is a development problem only.

Direct observation by the human eye is a valuable facility of which fullest use should be made. Continuous taped comment should be made on every aspect of what is to be seen, and the operator should be encouraged to keep up a constant stream of observations. These should include (1) size and shape of rock outcrops and rock fragments, (2) strike and dip of bedding and fractures, (3) sediment fabric and bed forms, (4) organic markings on sediments, and (5) rate of change in the nature of any of the parameters recorded. It is advisable to tape comments on two recorders simultaneously, in case of failure of one. If there is any doubt, an operator should use the facility of the submersible to stop and look around or over objects to fully appraise them. This can be done rather more effectively in a manned than in an unmanned vehicle, because of the availability of several portholes and the stereo vision of the human eye. In most respects, however, the technique of making visual observations and recordings from the two types of vehicle is surprisingly similar. IGS experience with marine cameras on manned submersibles has been that because of the possibility of failure it is advisable to carry at least two good-quality cameras outside the vehicle plus one which can be rigidly mounted inside so as to take photographs through the porthole. Light levels in U.K. waters are normally such that long exposures have to be made if an external flash unit is not used. However, with a rigidly mounted camera excellent-quality colour photographs can be taken through a porthole using exposures of up to a second. At least one outside television camera and linked photographic camera should be mounted on a pan and tilt mechanism where it can examine and record the seabed closely. As with audio records the operator should be encouraged to take the maximum number of photographs. He should normally aim to return with all frames on his film used, if necessary using them up in the final part of the traverse.

A most effective sediment sampling system for a manned submersible is by the use of the hemp bag scoops devised by IOS, described in Chapter 7. Six or eight of these were standard external fitments on the later IGS dives with PISCES. In 1970 external sediment core tubes modelled on those developed for ALVIN (Winget, 1969), were used with some success, but it was necessary to cause the submersible to go heavy if there was normal resistance to inserting the tube into the seabed.

Rock sampling from a manned submersible carrying a good manipulator is most easily carried out by seeking an in-situ piece of rock which is backed by a weak joint plane, so that it can be detached by the manipulator and placed in an external basket after being described and photographed as in

Fig. 6. Fig. 7 illustrates an outcrop of jointed Precambrian basement suitable for sampling in this way.

IGS has, however, also made extensive use of small rock drills. One used on PISCES II in 1970 when the external manipulator failed, was derived from a suspended drill system (Eden et al., 1970). Fig. 8 illustrates the type of core obtained with a small drill. A small Hyco drill (Figs. 9 and 10) with exchangeable barrels was carried by PISCES III on the 1973 Rockall cruise and proved entirely practical, but it was not used exclusively because of the excellent capability of the PISCES III manipulator. The availability of a drill is, however, important when joint-bounded fragments of solid rock cannot be abstracted as on some glaciated pavements (Fig. 11). The moral, here as in most marine work, is to carry adequate back-up equipment. It is most inadvisable to rely on only one, or even two, means of taking external samples. Manipulators and external drills are highly vulnerable to collisions with rock faces, which are apt to occur fairly frequently in irregular rocky terraine.

Fig. 6. Telechiric arm of PISCES III with joint-bounded sample of in-situ dolerite, viewed through porthole of submersible. Rockall Bank, 1973, Dive 13. IGS photograph UW/P73/13H/1.

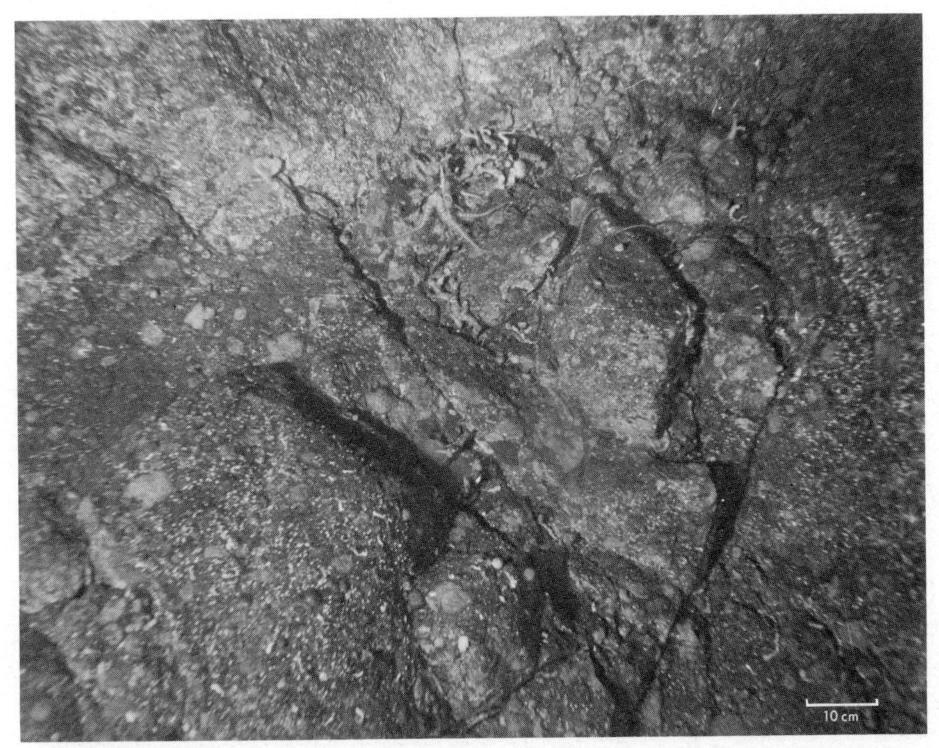

Fig. 7. Jointed Lewisian granite, Stanton Banks. Dive P70/4A (Fig. 1). IGS photograph UW/P70/4A/3.

CONSUB I has no manipulator but carries a hydraulic rock drill mounted on its pan and tilt head. This drill can be advanced to a rock face from a relatively sheltered position within the space frame of the vehicle, and when drilling is complete it can be withdrawn by the same mechanism. Needless to say a drill which is allowed to penetrate sufficiently far to become jammed into its own hole can effectively anchor the submersible to the seabed! For this reason weak links or eject mechanisms are essential. IGS per-

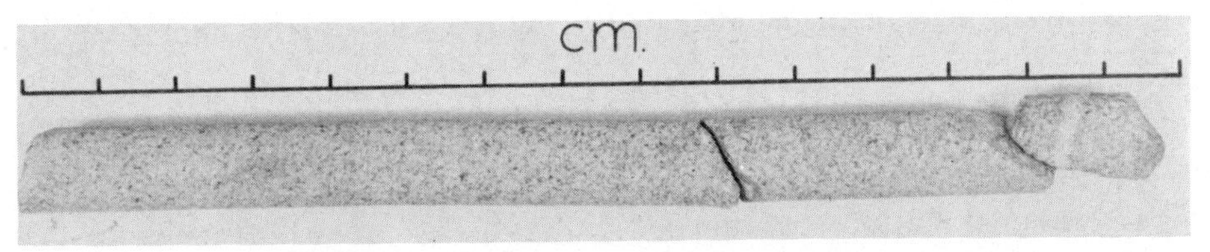

Fig. 8. Core samples obtained with BAC drill mounted on CONSUB. Dive site 8, sample NV 30. Top is to the left.

Fig. 9. PISCES III as mobilised aboard MV VICKERS VOYAGER for Rockall Bank cruise 1973. Note the Hyco drill and three replacement drill barrels mounted in a rack, and the telechiric manipulator above the somewhat battered sample collecting basket. To the right of the basket is a pan and tilt mechanism carrying TV and photographic cameras and lights. Another photographic camera is mounted on the bar above the portholes. The small mushroom shaped object above the bow is the E/M transducer used for navigation. IGS photograph UW/P73/11H2/12.

sonnel have had one or two anxious moments in this respect both with PISCES and CONSUB, but it has not yet been necessary to jettison equipment to free the vehicle.

In the interests of cost efficiency it is desirable to keep any submersible system in operation during as long a proportion of charter time as is practicable without impairment of efficiency. Aside from steaming time this will depend mainly on:

(1) Weather limitations

(2) Availability of operators and clients' personnel to work some sort of shift system

(3) Operational matters such as position fixing and pre-dive surveys.

The greater degree of independence of weather which can be expected with a large submersible and with unmanned systems have already been men-

Fig. 10. Jointed dolerite covered in epifauna, as viewed through the centre porthole of PISCES III. Note the Hyco drill and rack with four replacement barrels. Rockall Bank. IGS photograph UW/P73/13H/10.

tioned. It is worth adding at this stage that servicing of a manned submersible relates mainly to re-charging batteries and attending to life-support systems. This gives a significant advantage to unmanned vehicles operating on surface power, which might be expected to work round the clock using a shift system for the crew. It is therefore our belief that there is an important future for unmanned vehicles, the commercial development of which has had to await the availability of more sophisticated control systems and operational procedures than are required for manned submersibles.

Survey procedures

Once a site has been selected the first step is to ensure that it is fully documented before the cruise begins. IGS procedure has been similar for both drill ship and dive sites in that a file of essential data has been set up at an early stage, into which copies of all relevant bathymetric, geological and geophysical records can be accumulated, together with a statement of objec-

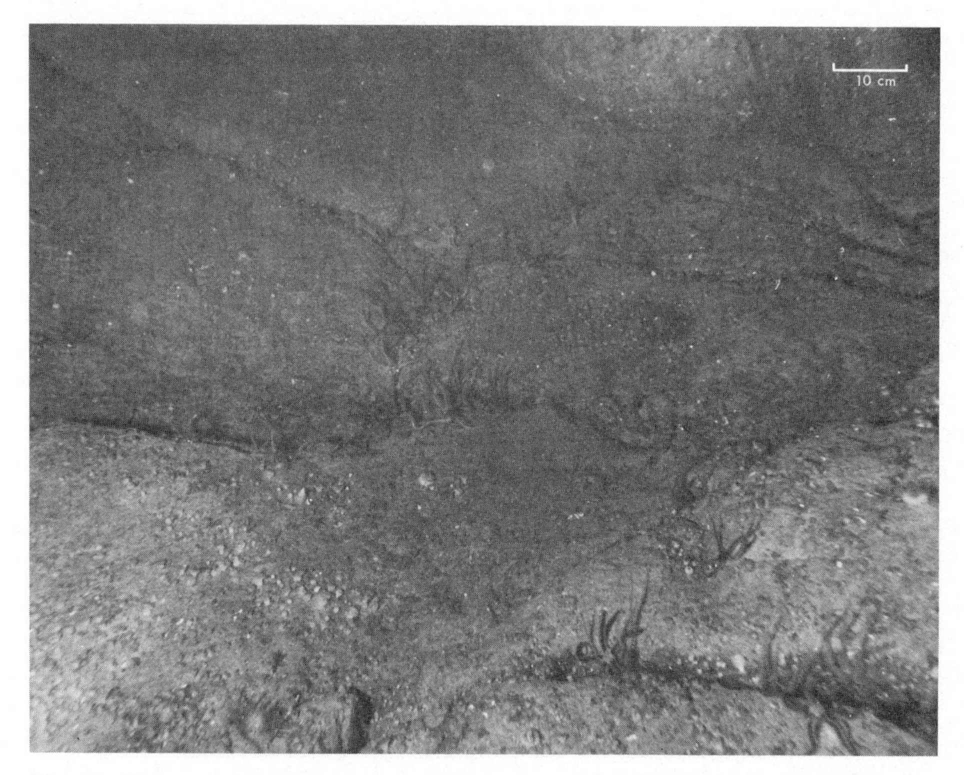

Fig. 11. Glaciated pavement of Lewisian rocks on Stanton Banks. Dive P70/4A (Fig. 1). IGS photograph UW/P70/4A/10.

tives. This file accompanies the party leader to sea, but a copy is left at base so that if necessary problems can be discussed by radio telephone. Practice has been to identify more sites than can be dealt with on any one cruise. As far as possible these are sites with varying degrees of exposure to different wind directions, so that in poor weather a choice can be made. The sites are assigned priorities which are borne in mind by the senior scientist in making his assessment of the next day's work in conjunction with the operator's party leader and the ship's captain. It is because of the availability of a selection of sites and the indented nature of the coast of the northwestern U.K. that 100% utilisation of available diving time was possible on the PISCES cruises of 1970 and 1971. The weather was by no means uniformly good during these cruises, and their success depended on judicious use of shelter and issuing into more exposed location as and when the weather permitted.

A second-stage contingency plan has also been prepared for all IGS cruises with submersibles, involving back-up instrumentation for use if the weather is such that diving is not possible or if an irretrievable systems breakdown occurs. Back-up instrumentation has ranged from light-weight equipment such as Shipek Grabs and side-scan sonar to a suspended seabed drill and a magnetometer. Once on site the first requirement is for a site survey with echo sounder and/or side-scan sonar to check the main features of the target area. The detail of this survey will depend on how much is already known of the site and on availability of time. At least a triple echo-sounder traverse along close-spaced parallel lines across the general target area is desirable, in order to pick up the strike of topographical features.

Recent IGS practice with PISCES submersibles has been to group scientists as a number of teams of two, several dives being assigned to each team. On their assigned dives one partner dives in the submersible and one is responsible for coordination of the scientific side of the relevant surface work, including the pre-dive survey; on the next dive the roles are reversed. The surface coordinator is responsible for passing to the dive scientist copies of records and track plots obtained on the pre-dive survey, so that the two can converse by acoustic telephone as the dive proceeds, with the same documentation available to both. Fig. 12 (Rockall Bank 1973 dive 3) illustrates the type of site plan which, together with copies of geophysical logs, is held by both scientists. For the 1973 Rockall dives the pre-dive track plots on board both the mother ship and the submersible carried a latitude and longitude grid, hence fixes on the positions of the submersible obtained by the mother ship could be communicated to the crew of the submersible to assist movement towards the planned target. Any other grid could of course have been equally suitable.

As the surface coordinator is carrying out and copying the results of the pre-dive survey, for which a simple copier is extremely useful because of the limited time available, his partner is making final dive preparations. He is required to check his tape recorders, including spare batteries and cassettes, and with the pilot to check items such as video recorder, cameras and sampling gear. If he has an internal camera he checks this and its means of rigid attachment to a porthole. All the smaller items including note book and pencils are carried aboard in a plastic bag, and the dive-site data and plans will be passed to him attached to a clipboard. He wears light footwear and normally takes a spare pullover. All loose items are carefully stowed ready to hand, to avoid damage to delicate gear in the confined cabin space. The dive scientist commences both video and audio recordings with a statement of the dive number and the date, place and time of launch. On the 1973 Rockall dive an elapsed time digital clock was started as soon as the submersible was on the seabed, and all subsequent times were referred to this, with occasional cross references to real time.

Because of the cost of operations, and the probability that any particular dive will not be repeated, it is important that all possible records of the dive be obtained and permanently filed for future reference. These include all operational data relating to the movements of the mother ship and the submersible, video and audio tapes and photographs. For the 1973 Rockall

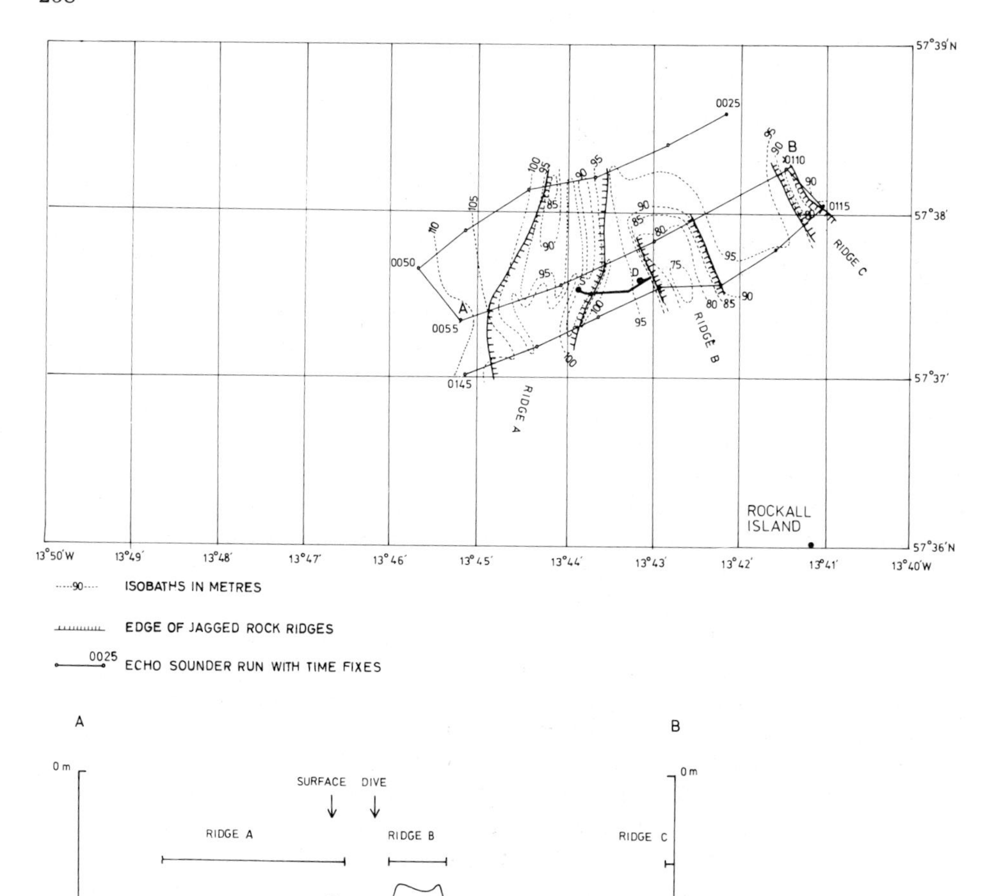

Fig. 12. Site plan for Rockall Bank dive P73/3 (Fig. 2), with Abstract from echo-sounder record.

dive, Hunter of IOS developed and fitted a six track pen recorder linked to the submersible's systems and to the elapsed time clock (see Chapter 7). This machine recorded the following parameters in graphical form: (1) depth (from pressure gauge), (2) heading (from gyro compass), (3) elapsed time in hours, minutes and seconds, (4)—(5) X—Y readings from the E/M log (see Chapter 7). The sixth channel was used by the dive scientist to mark events, such as collection of samples, by means of a push button. The graph output is kept with the dive data and can be annotated as required during or after the dive. The need for a constant stream of audio observations from the dive scientist has already been stressed. Once the dive has been completed an immediate de-briefing is required to be held so that samples can be identified and labelled, any lessons learnt assimilated and future plans decided. It is also desirable after the de-briefing to make an immediate transcript of the audio tapes and finally to work the dive data up into a running assessment, with diagrams suitable for publication. In practice this procedure is only likely to be feasible on board ship if a draftsman/secretary is assigned to assist full time, and it is our recommendation that this should be done.

The majority of these procedures are simple common sense and it may be thought that they can be picked up fairly easily as work proceeds. It is our experience, however, that most scientists find it difficult to achieve proper efficiency without practice, and therefore some advance training and discussion on dry land is a distinct advantage. Records of earlier dives can be used for such training and in particular dive scientists can be persuaded to try to keep up a continuous running commentary by practice with video recordings.

SUMMARY OF MAIN IGS PROJECTS WITH MANNED SUBMERSIBLES

Dives in Lower Loch Fyne 1969 (PISCES II)

Only four geological dives were carried out during this first evaluation of PISCES by scientists of the Natural Environment Research Council. The dives were arranged at short notice and were not sited with geological targets in mind, nor was any special instrumentation available, the only external tool being a torpedo recovery clamp. The four dives did, however, permit the construction of a visual concept of the geomorphology of this glacially over-deepened trough (Fig. 13), which is flanked by impressive submerged cliffs and floored by an undulating valley-fill of which the top comprises bioturbated mud locally with manganese modules. Because of the finite beam width of an echo sounder it is difficult from the surface to distinguish between a vertical cliff and a steep slope. This series of dives proved that both are present in Loch Fyne, as are gentle overhangs, and that most of the rock surfaces are smoothed or plucked by glacial abrasion. It was noted that fine-grained sediment is adhering to slopes with an inclination of up to 70%.

The top of the cliffs constitutes the top of a U-shaped valley. It lies well below present sea level and significantly further below local sea level during glaciation. It is therefore concluded that the hollow was not eroded by a valley glacier but by a fast-moving ice stream within an ice sheet. To the best of our knowledge this is the clearest demonstration of such an effect anywhere in the U.K. In the central mud plain five facies were distinguished and it was established that tidal scouring is preventing mud deposition in the area

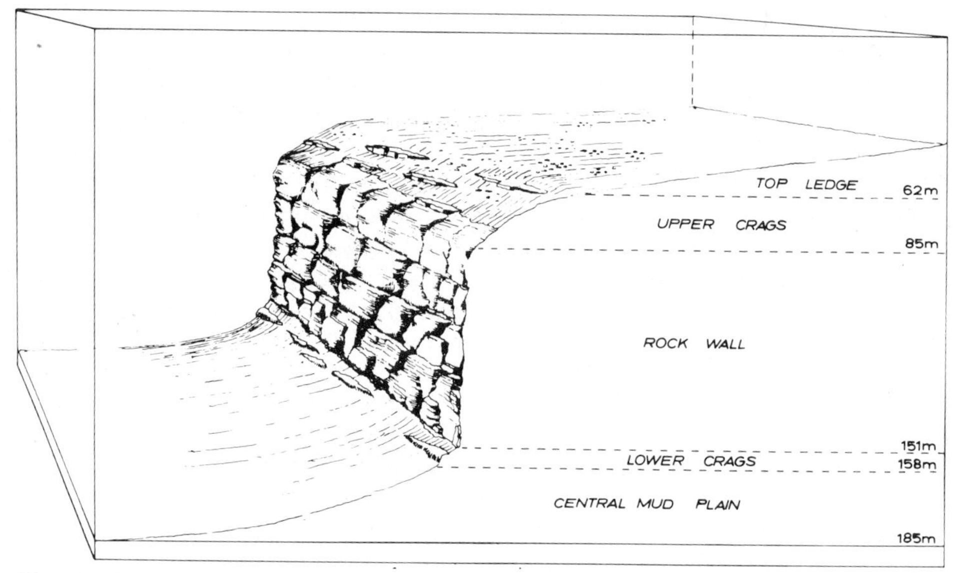

Fig. 13. Artist's impression of the rock wall on the north-east side of Lower Loch Fyne. True to scale drawing compiled by G.A. Goodlet from photograph and descriptions.

examined. It was noted that burrowing organisms are bringing manganese nodules to the surface and thereby delaying their burial.

Dives off western Scotland 1970 (PISCES II)

Twenty dives were undertaken in ten days, time on the seabed varying from just under two hours to four and a half hours. Two scientists were aboard PISCES on most dives, one as dive scientist and the other as observer. Rock sampling was planned to be carried out mainly by use of the submersible's manipulator, and this was achieved successfully early in the cruise, although in some glacially smoothed terraine no loose in-situ material could be located. Two collisions with rock faces eventually put the manipulator out of commission and it was later replaced by a temporary battery-operated drill held in a torpedo clamp. This drill was used successfully at the end of the cruise, but in the middle part of the cruise no sampling could be done. It proved possible, however, to recognise several outcrops of Torridonian Sandstone on the basis of their morphology alone. Sediment sampling was planned by the use of the manipulator to handle six short plastic core tubes lodged in an external rack. The system was based on that developed for ALVIN and described by Winget (1969). One core was obtained on the first dive, but thereafter no soft cohesive sediments suitable for coring were encountered until after the manipulator had been put out of action. Before this occurred, difficulty with the tubes had been experienced, however, in that adequate penetration of the seabed was difficult to obtain with the submersible in its normal near-neutral buoyancy condition.

The twenty dives were carried out on a wide range of types of seabed. The first day was devoted to two short shake-down dives in the Firth of Lorne. Subsequently the weather forecast being suitable for exposed area work, the mother ship proceeded overnight to a location at the top of the continental slope 80 km WSW of the southern end of the Outer Hebrides. Thereafter operations moved back into more sheltered sites in the Sea of the Hebrides and the Inner Sound of Raasay. The general geology of the area had been ascertained in some detail before this cruise (Binns, 1974). Ridges of Precambrian basement rock lie between basins containing Mesozoic and Tertiary strata, the latter capped by thick basaltic lavas. The area has been glacially scoured and scattered with glacial detritus so that most visible rock outcrops are of hard basement material, Tertiary lavas or intrusive igneous rock. A clear picture of the geomorphology of the whole area, supported by a large number of photographs, was obtained during the dives. The edge of

Fig. 14. Sand and gravel at the top of the continental slope. Dive P70/2A (Fig. 1). Depth 230 m. Note the sand shadows indicating transport from south to north. IGS photograph UW/P70/2A/38.

the continental slope is underlain by Quaternary morainic material, probably thick, and between depths of 160—230 m winnowed and washed by the action of a sea level substantially lower than the present. A partly consolidated spread of poorly sorted but well-rounded cobbles and other lithic material is now being crossed by sand apparently travelling north, on the evidence of ripple marks and sand shadows behind boulders (Fig. 14).

Further inshore deep hollows filled with bioturbated mud as in Lower Loch Fyne are interspersed between glacially smoothed rock ridges (Fig. 11). The shallower ones are swept bare by wave action and the deeper ones encumbered with angular glacial erratics. Between the ridges and the hollows lies a belt where glacial erratics are partly buried in mud (Fig. 15), sometimes just the tops of the larger ones being visible. There are indications of wave-eroded gulleys down to 60 m on Blackstones Bank (Dives 5A, B, Fig. 1) and 90 m on Stanton Banks (Dives 4A, B). Below these depths and around these two banks are great spreads of a stabilised cobble pavement comparable to that at the top of the continental slope. The gulleys contain spreads of shell sand disposed as fresh sand waves, commonly symmetrical and aligned at

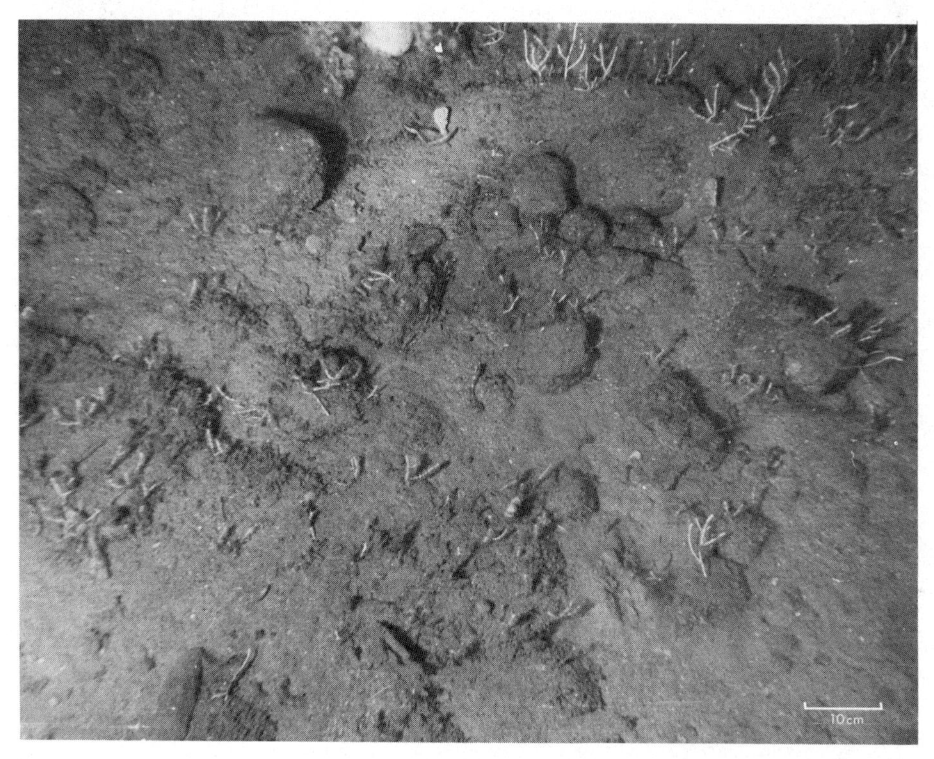

Fig. 15. Morainic material in process of submergence in mud. Ridge in trench east of the Outer Hebrides. Dive P70/3B (Fig. 1). Depth 100 m. IGS photograph UW/P70/3B/9.

right angles to the gulley walls. A dive in the Inner Sound of Raasay was designed to examine a trench reaching 323 m and comprising the deepest part of the U.K. Continental Shelf, which proved to be floored by soft bioturbated mud. The steep slope westwards up from this hollow was covered with similar mud adhering to slopes of 60° or more. In this situation there was some fear of the possibility of accidentally creating a mudslide and the climb up the slope was abandoned. It is to be noted, however, that although soft, mud in the deep hollows is semi-consolidated, and cracks rather than flows when the weight of the submersible is placed on it (Fig. 16).

Dives in the Irish Sea and off western Scotland 1971 (PISCES II)

Thirteen dives were undertaken in seven days, operating in a wider area than in 1970, but repeating work at some of the earlier sites where rock sampling had not been achieved due to damage to the PISCES manipulator in the previous year. The manipulator was used as on the earlier cruise, but

Fig. 16. *Nephrops* plain with sea pen and *Calocaris* burrow, North Minch, dive P71/6B, depth 131 m. Note the fashion in which the mud has cracked under the weight of the drill of PISCES (left). IGS photograph UW/P71/6BD/2.

better protection had been devised against accidental damage. The manipulator collected a total of 20 samples of loose cobbles and in-situ rock. It was also used to handle hessian bag sediment scoops of the type devised by IOS and described in Chapter 7, obtaining 14 sediment scoop samples. In addition, IGS fitted a small rock drill to the torpedo clamp, similar to that used in emergency conditions in 1970. This was a one-shot drill, intended for use when sampling could not be achieved with the manipulator. It obtained small cores from 7 of the 8 locations at which it was used (Fig. 8). In choosing dive sites preference was given to locations of suspected or known rock outcrops of uncertain nature. In an area scattered with glacial erratics, gravity coring and rock dredging are of very limited value for sampling solid rocks and it proved to be so in 1971 that the use of PISCES achieved identification of solid rock at eight sites, doing so at a significantly lower cost than by the alternative procedure of using a drilling ship. Some of the sites were, moreover, at depths in which the available drilling ship, MV WHITEHORN, could not have anchored.

The general geological environment was for the most part similar to that of the previous year, and procedures were much the same except that one scientist only was carried in the interests of greater working efficiency. Three sites (71/6A, B, Fig. 1) were investigated at which geophysical data suggested the possibility of outcrops of bedded strata expected to be of Mesozoic or Palaeozoic age, but at which resolution of the records was inadequate to be certain that there was not a thin sediment cover. All three sites proved to be heavily scattered with glacial erratics and in one (71/6B) no exposures were found. At one of the others (71/7A) two small whalebacks of Mesozoic limestone protruded slightly from amongst the erratics and were sampled. At another (71/6A) a smooth glaciated pavement of solid rock was located below boulder erratics and sampled with the drill. The emplacement of this sample is illustrated diagrammatically in Fig. 17 and its importance is discussed previously above on p. 246. This is a classic example of an important finding which in 1971 could not have been achieved by any other method than the use of a manned submersible.

Dives on Rockall Bank 1973 (PISCES III)

This cruise was carried out jointly by IGS and IOS, using the 83-m mother ship VICKERS VOYAGER. Rockall Bank is the shallow eastern part of the Rockall Plateau, itself a foundered microcontinent about the size of Ireland left isolated in the North Atlantic by the process of continental drift. The bank is separated from the U.K. by the Rockall Trough, up to some 3000 m deep in the latitude of the north Irish coast, and its only visible expression is a group of three rocks, of which the largest is the 19-m high Rockall Island. Rockall Trough contains a number of seamounts of which the Anton Dohrn Seamount rises sharply from a depth of some 2300 m to a plateau level at

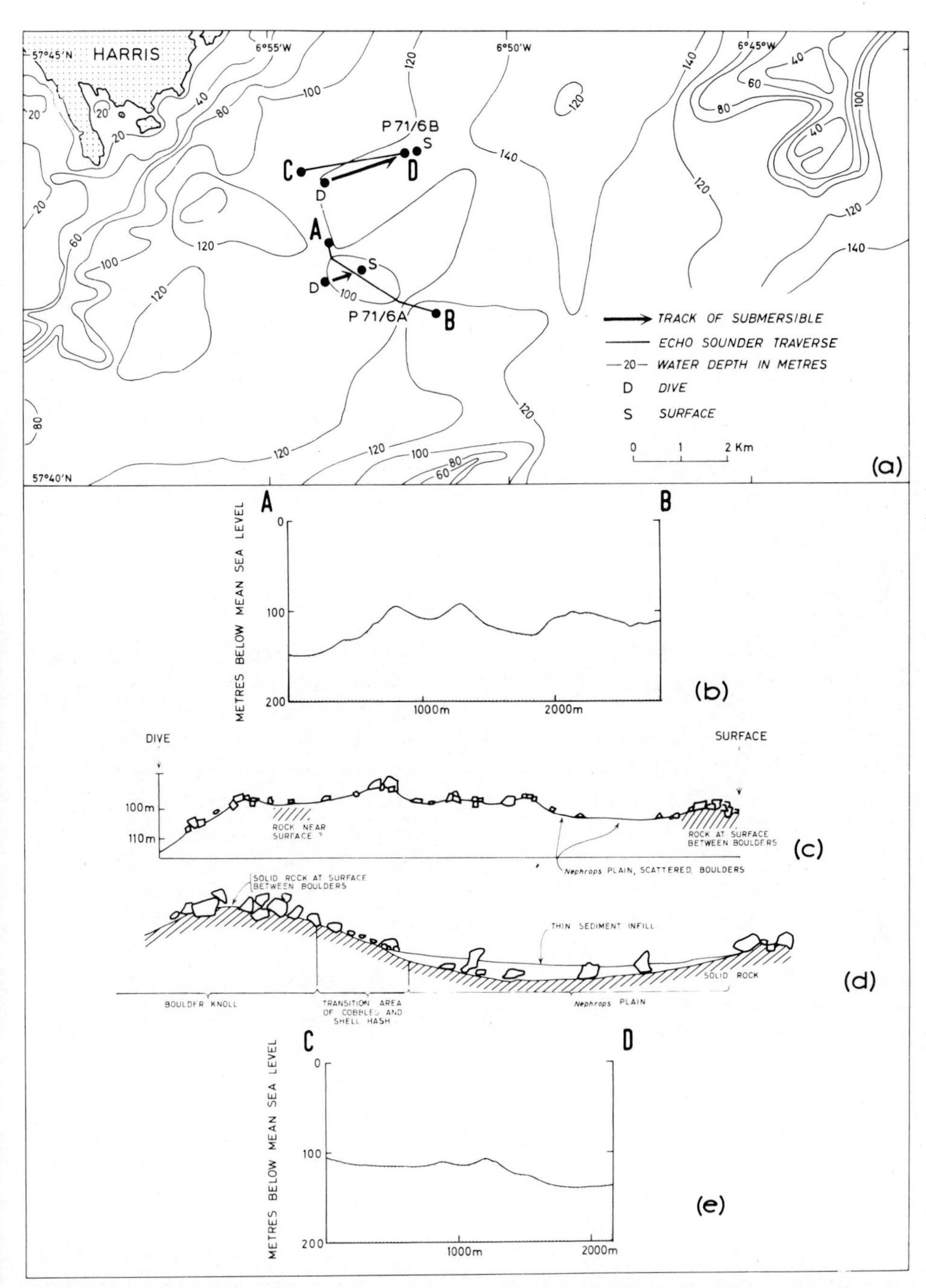

Fig. 17. North Minch dives, 1971. a. Location of dives. b. Echo-sounder traverse near site of dive P71/6A. c. Schematic section along line of dive P71/6A. d. Schematic sketch of relationship of *Nephrops* plain to boulder mounds, dive P71/6A. e. Echo-sounder traverse near site of dive P71/6B.

ca. 750 m. Rockall Plateau has been studied for a number of years by IOS, and latterly jointly by IOS and IGS. The 1973 dive objectives included:

(1) Study and sample a number of rock outcrops and probable outcrops identified in earlier cruises.

(2) Study and sample the sediments, largely biogenic, which are at present being generated in this isolated area of relatively shallow water.

(3) Study iceberg plough marks, a submarine landslide and coral reefs.

Wilson discusses in Chapter 7 the studies of sediments and reefs, and his findings are not repeated here. A total of fourteen dives (Fig. 2) were carried out in seven days; nine were primarily intended to study solid rocks and five to study sediments. Working depths ranged from 100 to 460 m. A total of 924 photographs were taken, and 35 rock and 28 sediment scoop samples were collected. Most rock samples were collected by the use of the PISCES manipulator, although a Hyco drill was carried and was used successfully on a number of occasions. It had been uncertain whether or not glacially smoothed surfaces similar to those off western Scotland would occur, presenting problems to sampling by use of the manipulator alone. In the event the manipulator proved effective, the availability of the Hyco drill and a back-up IGS drill from the 1971 work were regarded as an essential precaution.

As in 1971 the PISCES crew was restricted to two. Geological diving duties were the responsibility of three specialists, each of whom was assigned to one dive in three; this proved to be as much as each could conveniently manage. There was a non-diving back-up team of eight geologists and technicians to help with sample examination and servicing cameras, the IOS six-channel pen recorder and the IGS reserve drill. It was agreed to start work on an igneous complex in the shallow northern area, where Rockall Island provided a useful reference point, to work south into deeper-water areas where reliance would have to be placed wholly on navigation by Loran C, and finally to consider two relatively deep dives to depths of some 600 m. These two deeper-water dives would have been to depths substantially in excess of those previously experienced by the operators' personnel aboard and there was concern as to the availability of retrieval systems should the submersible be in difficulties. Because of these factors and deteriorating weather the decision was taken to abandon the two deep dives and divert to additional planned sites around Rockall Island. It is of note that two months later, whilst on contract to other clients, PISCES III sank in somewhat shallower water off southern Ireland and was successfully retrieved, although with the greatest difficulty.

The solid-rock provings made in the course of this work are indicated on Fig. 2. They show a Precambrian basement to the south and a Mesozoic/Tertiary igneous complex to the north. There is no doubt that this sampling would have been exceedingly difficult to achieve by any other method available in 1973, since previous experience in the continuous long swell in this exposed area had demonstrated the problems associated with the use of suspended drills. The overall picture of Rockall Bank which has emerged is of a great plateau of white carbonate sand from which rise several extensive patches of rocky outcrop forming ridges and pinnacles. There are spreads of cobbles, some associated with a low sea level at a depth approaching 200 m. There are no indications of glaciation, and it may be inferred that Rockall was a large and substantial ice-free island at times during the Quaternary. The modern carbonate sand is in places disposed in well-developed sand waves which end abruptly against rock outcrops. There are flourishing colonies of cold-water corals around the 200-m isobath which are discussed by Wilson in Chapter 7. This cruise was the culmination of IGS work with PISCES and technique had at this stage been brought to a satisfactory level of efficiency and reliability. Costs were, however, beginning to increase alarmingly, partly because of the growing commercial requirement for submersibles to survey sites of hydrocarbon installations in the North Sea.

SUMMARY OF IGS WORK WITH REMOTELY OPERATED VEHICLES

Following the development of the unmanned vehicle CONSUB I (Fig. 18) for IGS by the British Aircraft Corporation Ltd. the IGS involvement in submersible operations has been largely confined to the use of remotely operated vehicles. This design and construction project, jointly funded by IGS and the Department of Industry (DOI) with technical support from the Marine Technology Support Unit of the Atomic Energy Research Establishment (AERE) Harwell, was aimed to produce a submersible facility available to IGS on a continuous basis. It was initially intended to be deployed as part of the back-up equipment carried by the IGS-chartered drilling vessel MV WHITEHORN for use during bad weather downtime. In addition, it was to be suitable for use from ships of opportunity. The IGS drilling operations have now extended into deeper water by the use of a dynamically positioned ship. MV WIMPEY SEALAB, limited in time by its higher cost to intermittent operations, whereas the previous vessel was operational for 300 days per year on a continuous basis. The larger vessel, without the constraint of anchoring, is better able to combat adverse weather and consequently little time remains for the deployment of ancillary systems as the proportion of drilling time, the priority activity, increases.

The CONSUB system has therefore been used by IGS from the RRS CHALLENGER, which is equipped with an aft-mounted hydraulic A-frame, and omni-directional bow thruster. In addition British Aircraft Corporation Ltd. have entered into a contract with IGS and DOI to use the vehicle for commercial applications outside the IGS period of priority use. The requirement for this arrangement exhibits the vehicle's competence to conduct tasks beyond those of geological data aquisition.

Fig. 18. The IGS remotely operated submersible CONSUB I showing the payload of TV camera, stereo camera system and rock drill mounted on the pan and tilt assembly with the static TV camera mounted on the starboard side of the chassis frame. The photograph was taken before the fitting of remote instrumentation and an integral navigation system.

Dives in North Sea in October/November 1974

The first sea operation undertaken by the CONSUB system was a joint prototype development exercise and a geological investigation. Sites were selected because of their geological validity but the equipment performance was strictly monitored and different techniques employed in launch, cable handling, on-bottom procedure and recovery to optimise the system. To enhance the value of the cruise it was planned in three legs. On the first, CONSUB was used alone on sites in the Firth of Forth and around the Orkney Islands. On the second, SNURRE, the vehicle being developed by the Royal Norwegian Council for Scientific and Industrial Research, was also carried and work was undertaken around the Shetland Islands, while on the third the Heriot Watt University (Edinburgh) vehicle ANGUS was added and operations proceeded off the Norwegian coast, around the Shetland Islands and again in the Firth of Forth. The exercise proved valuable both in confirming that a remotely operated vehicle can achieve many of the geological objectives required in submersible use and for an exchange of ideas between the three teams.

Dives in the Sea of the Hebrides and Irish Sea in April 1975

The format of this cruise followed that of the previous one but the CONSUB vehicle was the only submersible carried. RRS CHALLENGER was equipped with two point forward wireline moorings to provide station keeping in deeper water, but as on the earlier cruise the submersible was restricted to navigating by dead reckoning based on direct observation of the seabed plus depth and compass heading data provided by the vehicle.

Dives on the East Shetland Platform and in the Orkney area, Minch, Sea of the Hebrides, Malin Sea and Clyde in May 1976

This cruise undertook an investigation of rock outcrops using the CONSUB system and the IGS one-metre rock drill and for the first time ship to vehicle navigation control was available. This allowed operations to be undertaken with the vehicle and ship both free and with the ship maintaining position relative to the underway vehicle by a station keeping exercise. Geological traverses beyond the immediate limit of a point tethered umbilical were thereby achieved. The longest deployment was more than eight hours, during which all planned geological requirements were achieved apart from the recovery of a rock core, which was successfully drilled but lost from the single-barrel coring system in use.

TYPES OF SUBMERSIBLE VEHICLES AND THEIR RELEVANCE TO GEOLOGICAL SURVEY AND SITE INVESTIGATION

The manned vehicle provides a means whereby the geologist can directly observe the seabed and record his impressions verbally, by VTR and by still photography. In addition, dependent upon the ancillary tools carried, he is able to (1) run precisely controlled sector or side-scan sonar or acoustic profiling, (2) collect samples and cores, or (3) undertake geotechnical measurement in a controlled manner, having first ascertained the relevance and reliability of the data or material to be collected. However, he works generally under conditions of limited space, restricted visibility and, despite being physically independent of the ship, is strictly controlled with a view to a safe recovery. The primary restriction on this type of vehicle is the requirement to carry personnel and in consequence a very sophisticated life-support system with duplication and back-up apparatus to cover component malfunction. The first penalty to be felt is that of cost, as the life-support necessary in the vehicle increases its size and maintenance requirements, demands adequate safety provision during launch and recovery and thereby ties it to a specialised or specially fitted mother ship rather than a ship of opportunity. Bottom time is limited by the need to replenish life-sustaining equipment and to recharge power supplies, while throughout the operation sufficient contingency time must be allowed against each of these parameters to cope with unexpected circumstances. In addition, the endurance of the personnel to maintain efficient observations has a distinct limit, which is influenced by such factors as the nature of the work and the posture that is sometimes necessary to adopt within the confined space available. The weather obviously controls activities, and it is not only the wind and sea state which limit operations but also visibility, as the surface recovery operation, in which swimmers may be involved, is largely a visual exercise.

The greatest benefit of the manned vehicle is its independence, with no ties to the surface vessel. However, the absence of this link demands that the vehicle be self sufficient with regard to power, and therefore range, speed, manoeuverability and vertical control are usually limited. The latter is often achieved by variable buoyancy and alterations can be time consuming, while activity demanding a downthrust on the seabed, e.g. drilling, may only be achieved by adjustment to an adequate negative condition. In contrast to the manned submersible the remotely operated vehicle may be viewed more as a mobile work package of much more extended duration, and configurations adopted have therefore been very much more varied. Despite this, one severe constraint exists and is common to all. This is the umbilical link between the surface vessel and the vehicle, which is subject to current drag. The major expenditure of power by the cable controlled submersible is concerned with cable towing, and the interplay of forces on the umbilical determines the performance achieved.

Various modes of deployment may be selected for the umbilical which can be listed as follows:

(1) Operations using a negatively buoyant cable to a clump block on the seabed beyond which a buoyant section transmits power, instructions and information to and from the vehicle. The initial design specification for the IGS CONSUB I defined a performance requirement of 50 m operating radius around a clump block in a 2-knot current to depths of 600 m, although in all operations to date CONSUB has in fact been used with a continuously buoyant cable. The clump block may take the form of a simple weight which may be disposable or it can be a docking module perhaps with automatic reeling facilities for the buoyant clump-to-vehicle tether. The use of a clump block creates a further storage and handling requirement, minimised if the two units lock together for launch and retrieval. The clump block in this mode copes with all the element of drag deriving from the main section of umbilical and the vehicle simply combats its own drag, relatively low at speeds below 2 knots, and that of the link. The most demanding condition

to be achieved is that of operating at the limit of the achievable range across current from the clump block. A cable scope minimum of the order of 3 is necessary at this limit position.

(2) Operations using a negatively buoyant umbilical to a mid-water clump block, again either recoverable or disposable, with the cable beyond buoyant as with the clump block on the seabed. The weight of the cable and clump block minimise the effect of the current on this section and remain largely controlled by the ship, deviating from a point beneath the lifting point on board only through drag due to current or ship motion. The clump block in this condition is subject to the motion of the ship, unless heave compensation has been incorporated on board, and this may be translated into longitudinal motion in the clump-to-vehicle tether. Thus, the deep ocean unmanned vehicle developed by the U.S. Naval Undersea Centre, San Diego, described in Chapter 4, incorporates a mid-water block with automatic reeling facilities for the outer section of cable and a heave-compensated crane system. However, this style of deployment is not necessary in the shallower continental shelf depths.

(3) A variation of (2) is to use a negatively buoyant umbilical with a section near to the vehicle buoyed. The length buoyed depends on the required operating radius and the water depth, while the non-buoyant section is limited to a length which will not reach the seabed despite the extremes of ship motion in the prevailing swell. This creates no additional handling difficulty, given that a cable-handling facility and suitable fairlead are fitted to the ship, but with the CONSUB system it is only relevant in water depths in excess of 150—200 m.

(4) Operations using a positively buoyant cable. Buoyancy is achieved by attaching buoys to the umbilical, but these impose a further drag and therefore limit vehicle performance. By using Kevlar in place of steel as a strain member a significant weight saving is achieved without an increase in cable diameter and if this is used in conjunction with a high degree of multiplexing, to reduce the number of conductors required, a minimal cable weight and diameter can be achieved. This is the format used for CONSUB II by British Aircraft Corporation Ltd. A positively buoyed cable demands a further energy output by the vehicle to drag the cable down from the sea surface in addition to combating the lateral drag forces.

In continental shelf depths when a large operating radius is required, to use a heavy cable arrangement as in (3) together with a scope of 3 on the external cable section may cause that section of umbilical to approach the surface with the risk of tangling with itself or with the ship propellors. In contrast, using a buoyed cable provides a surface indication of the lie of the cable which can assist in the station-keeping exercise, although as this can sometimes be misleading the vehicle position is better monitored by shipboard or seabed-transponder navigation. A heavy umbilical passes down and remains safely clear of the vessel, but in the buoyant mode the loops of cable hanging between buoys can be a danger to the ship, and the buoys themselves (6 m apart as used by IGS) can snag together if tension is removed from the umbilical. If positive buoyancy could be achieved by means of a continuous flotation jacket on the cable this would be very desirable. It would provide a means of protecting the cable from mechanical damage, remove the loops suspended between buoys which lie at a vulnerable depth relative to the ship's propellor and also to the vehicle when it is on the surface. It should also eliminate problems with the buoys, both their tendency to snagging and the difficulty of their shipboard handling. However, the efficiency of such flotation materials does not match that of buoys and therefore a severe drag penalty would be imposed while the cable stiffness, and therefore its bend diameter, would increase with consequent handling and storage penalties.

It should be noted that the behaviour of water flow around a Kort-nozzle propellor at very low ship forward speeds can create a water-cell with circulation aft through the nozzle and forward to port and starboard. Buoyed loops of umbilical are liable to follow this flow forward and round toward the propellor shaft if care is not exercised. Similarly the lift of the ship transom in heavy swell conditions causes an inboard movement of water, but this can be of more concern when effecting a submersible recovery aboard a stern handling vessel. Whichever cable-handling arrangement is selected an operating area can be predicted for every vehicle for a given current condition and a given depth; this area is known as the footprint of the submersible. Given a static deployment platform this footprint takes the form of a circle around the ship's clump block in zero current conditions, which becomes an oval, displaced and distorted downstream with increasing current conditions. Fig. 19 compiled by the Marine Technology Support Unit, Harwell, illustrates the footprints of some European machines.

In a geological role, the submersible is often required to undertake a defined traverse during the course of which stops and deviations are made to make detailed examinations and to take samples, cores or measurements. This is analogous to the role of the submersible in pipeline inspection and, in each case, the vehicle needs to be deployed from a non-anchored ship capable of maintaining station relative to the vehicle while making good the required course at the pace dictated by the submersible. A restraint, similar to the vehicle footprint, is therefore placed on the ship, which must operate within a restricted area beyond which tension will be put into the umbilical and the vehicle pulled off station. To achieve success in running an underway traverse it is necessary to have a clear indication of the position of the ship and the submersible, ideally related to geographic coordinates and the position of known seabed features, and to achieve a high degree of cooperation between ship and vehicle control. In CONSUB operations aboard RRS CHALLENGER the omni-directional White Gill bow thrust unit and E/M log have been of great importance in achieving success in underway traversing. The

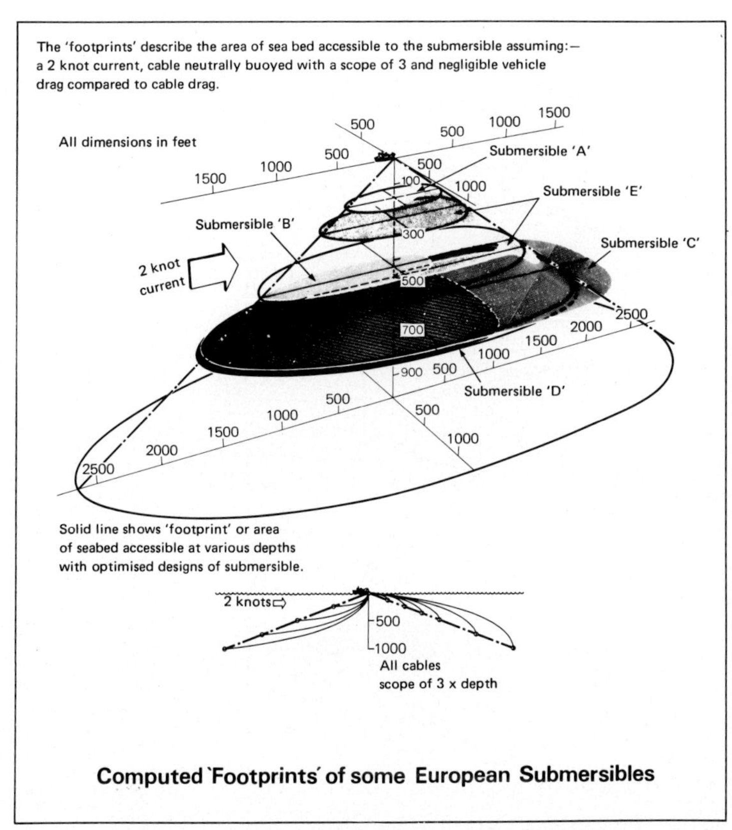

Fig. 19. Computed footprints of some European submersibles. These footprints predict the relative performance of cable-controlled submersibles in a 2-knot current which is uniform through the water column. If the speed of the current is reduced the area and depth of the footprint all increase accordingly, but retain their relative disposition and proportions. For example, in a 1-knot current the footprint dimensions are four times greater and their depths four times deeper than shown. Diagram published by courtesy of the Marine Technology Support Unit, AERE, Harwell.

latter provides a measure of the ship's motion forward and transversely through the water body.

Various different thruster and payload arrangements have been selected for unmanned submersibles and Fig. 20 compiled by the Marine Technology Support Unit, showing a comparison between a single pair of tiltable thrusters and a paired horizontal and vertical array, illustrates the nature of the compromise involved in the choice. Here the single pair of thrusters achieve their performance while requiring lower power transmission and therefore a reduced cable area with consequently less drag. However, compared to the separate horizontal and vertical control, pilot awareness and ability are excessively exercised to achieve a desired course. The configuration of the vehicles

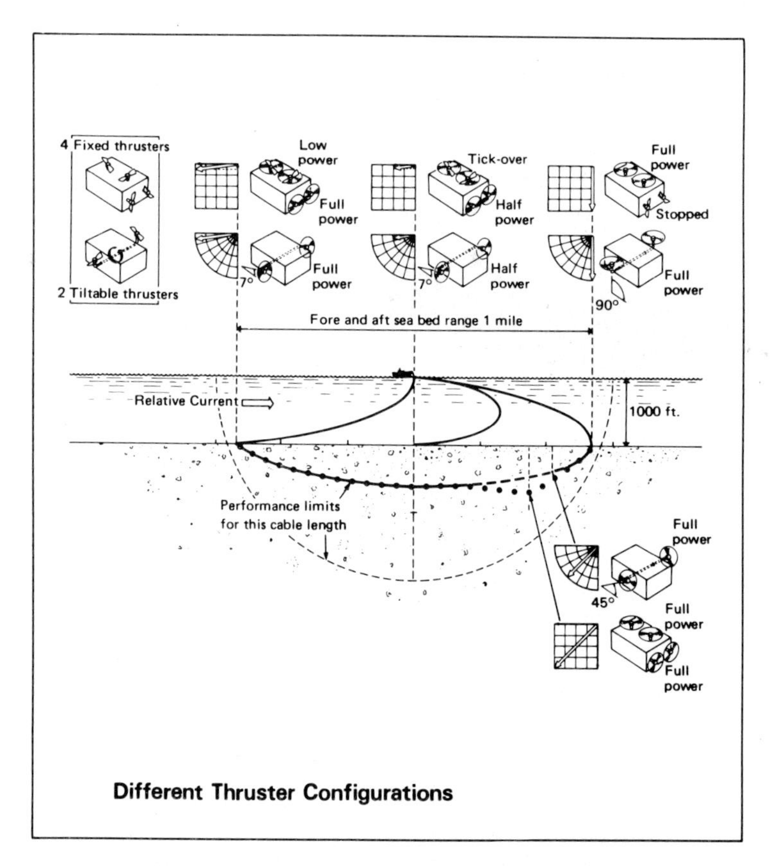

Fig. 20. Different thruster configurations. This diagram shows two of the many possible thruster configurations. Note that similar performance to the four-thruster arrangement is achieved by the pair of tiltable thrusters, with half the installed motor power. Diagram published by courtesy of the Marine Technology Research Unit, AERE, Harwell.

on the first IGS unmanned submersible cruise provided an interesting contrast. ANGUS carries two horizontal thrusters and uses buoyancy adjustment for vertical control. SNURRE is fitted with three tiltable thrusters at 120° to one another in a horizontal plane together with one central vertical thruster. CONSUB I has two pairs of 0.5-m diameter 5 hp Kort-nozzle propellors, one pair horizontal, the other vertical. The latter are adequately powerful to combat drilling thrust and hold the vehicle steady on the sea floor.

CONSUB I has been found to have excellent flight characteristic, being competent to maintain course "hands off", while being very manoueverable. The vertical thrusters provide a delicate ability to hold a course above the sea bed and to land with minimal disturbance. CONSUB II being constructed by

BAC Ltd. has two horizontal thrusters, one vertical and one transverse to permit sideways movement, which is achieved by the IGS CONSUB I by reversing and turning. The IGS CONSUB I is 2.7 m long, 1.8 m wide and 1.7 m high and has a space-frame construction of aluminium, hollow, compartmentalised tubing. The vehicle has a positive buoyancy in the order of 20 kg and normally carries a payload consisting of two TV systems, one static, for use by the pilot, and the other fitted on a pan and tilt assembly together with a stereo still-photography system and a rock drill. Colour TV has been substituted in place of the rock drill on some dives during the BAC Ltd. use of the vehicle for engineering inspection in the North Sea. The rock drill is limited at present to the recovery of a single core on each dive, though this is a tool which ideally requires a multiple capability. Similarly the presence of a telechiric arm for sample collection is desirable. Examples demonstrating the successful use of CONSUB I for inspections including precise navigation along electric cables and detailed study of epifauna are shown in Figs. 21 and 22, respectively. It is evident from studying these figures that the details in these examples are comparable with those that might have been obtained with a manned submersible.

The geologist, in the relative safety and comfort of the ship's operations room (which can be a rapidly mobilised control-system container), derives the benefit of continuous observation of the seabed, he can combine this with the opportunity of discussing with colleagues any problems encountered as they happen, while still having the ability to review them after the dive by means of VTR. The most important features of a submersible in a geological operation, apart from simply conducting a visual observation, are twofold. First, knowing the vehicle position both relative to the ship for control and in absolute terms for the value of the data gathered, and second the ability to collect sample material or data. In most instances the unmanned vehicle is competent to achieve results equivalent to the manned vehicles on the continental shelf, while operating costs are significantly lower.

Alternative and additional systems which may be considered are:

(1) Diver lock-out vehicles. These, as with deep diving technology in general, do not have a ready geological application. Geological work required, namely, (1) direct observation, (2) microtopographic measurement, (3) sampling of seabed deposits and outcrops, (4) geotechnical measurement, (5) sonar traversing and acoustic profiling, (6) type locality definition, and (7) personnel training is not easily undertaken by a non-geologist diver, and there are few geologists competent to participate in the higher-risk activity at greater depths than 40 m. The limited time available to the diver at depth is more appropriately allocated to tasks which cannot be achieved by any alternative means.

(2) The small submarine, the 28.5-m AUGUSTE PICCARD being developed by Horton Maritime Explorations Ltd., which will operate without a mother ship and will be self supporting except for a fishing vessel used to deploy

Fig. 21. Photograph from CONSUB I demonstrating one required feature of an unmanned vehicle — ability to follow a precise course. Inspection of a 15-cm electric cable crossing a glaciated pavement of Precambrian rocks off Helle, Norway. IGS photograph UW/C74/3R/19.

seabed transponders and to act as a safety support. This will have a six man complement, 90 man day life support, 1600 km surface range, 160 km sub-surface range and a maximum speed of 6 knots. Battery charging and surface running are conducted on diesel power. It will be of interest to compare the cost of this method with the conventional manned submersible with its mother ship.

(3) The atmospheric diving suit JIM developed by DHB Construction Ltd. is not likely to have great geological application despite its excellent ability to observe and undertake numerous manipulative tasks. Its role is essentially

Fig. 22. Photograph from CONSUB I demonstrating another required feature of an unmanned vehicle — ability to carry out close recording of detail. Epifauna on jointed Devonian sandstones 6 km south-southwest of Lerwick, Shetland. IGS photograph UW/C74/1L/15.

local, as its traverse ability is limited to walking about, which is not an easy task in areas of irregular terrain, boulder fields or rock outcrops.

(4) One area where the submersible may play a developing role is in drilling to depths in the order of 200 m into the seabed for geological survey purposes to identify critical horizons or in a site investigation role. The Tours 430 is being developed by Ingenieurkantor Lübeck for this purpose.

REFERENCES

Binns, P.E., McQuillin, R. and Kenolty, N., 1974. The geology of the Sea of the Hebrides. Inst. Geol. Sci. Rep. No. 73/14: 43 pp.

Eden, R.A. and Ardus, D.A., 1972. Sampling solid rock outcrops at Continental Shelf depths; the problem and a discussion of methods. Oceanol. Int., 72: 392—395.

Eden, R.A., Flinn, D. and Harrison, E., 1970. Experimental work with the Harrison Hard Rock Corer, a light weight underwater drill. Proc. Geol. Soc. London, 1662: 73—74.

Eden, R.A., Ardus, D.A., Binns, P.E., McQuillin, R. and Wilson, J.B., 1971. Geological investigation with a manned submersible off the west coast of Scotland 1969—1970. Inst. Geol. Sci. Rep. No. 71/15: 49 pp.

Eden, R.A., Deegan, C.E., Rhys, G.H., Wright, J.E. and Dobson, M.R., 1973. Geological investigations with a manned submersible in the Irish Sea and off western Scotland 1971. Inst. Geol. Sci. Rep. No. 73/2: 49 pp.

McQuillin, R. and Ardus, D.A., 1976. Exploring the Geology of Shelf Seas. Graham Trotman Dudley, London.

Natural Environment Research Council, 1970. A report on the research applications of the submersible PISCES II. NERC Publ. Ser. C.2: 1—20.

Winget, C.L., 1969. Hand Tools and Mechanical Accessories for a Deep Submersible. Woods Hole Oceanographic Institute, Woods Hole, Mass.

CHAPTER 11

SUBMERSIBLES: GEOLOGICAL TOOLS IN THE STUDY OF SUBMARINE CANYONS

Harold D. Palmer

INTRODUCTION

In their extensive review of submarine canyons and other sea valleys, Shepard and Dill (1966) attribute the first discussion of valley-like depressions on the sea floor to Dana (1863, p. 441). This work, a text on Geology, included a description of the Hudson shelf valley in the New York Bight. On the basis of further work by Lindenkohl (1885), and additional studies by Dana (1890), this valley became the focus of speculation regarding possible origins for such features. With the first thorough contoured bathymetric charts prepared by Davidson (1887, 1897), the presence of extensive canyon systems on the west coast of the United States became known. Subsequent charting of the continental margins of the world revealed canyons and sea valleys on most shelves and slopes.

The controversy regarding the origin of these significant features still continues, and the reader is referred to summaries provided by Shepard and Dill (op. cit.), Whitaker (1976), and a collection of new data soon to be published (Stanley and Kelling, 1977). In these references, it is evident that the manned research submersibles have played a key role in the study of submarine agents and processes which act (or have acted) to form and maintain these features. The past decades have witnessed the penetration of canyon heads by diving geologists employing SCUBA techniques (Dill, 1964a,b; Chamberlain, 1960, 1964). But these studies were limited by decompression requirements and the slow development of techniques for working underwater with prototype equipment. For years the lower portions of these intriguing features remained beyond the direct view of the diving scientist.

Studies of submarine canyons from surface ships are made difficult by the nature of sea-floor geometry and the sampling techniques available to oceanographers. Because the beam pattern of most echo sounders is quite broad, it is impossible to obtain an accurate profile of the canyon slopes and floor. Thus, features of low or moderate relief on the floor and walls may not be resolved, and only an approximation of the profile surface may be obtained. By slowing the ship, or drifting while sounding, some improvement may be realized, but increasing depth always degrades the echo. Similarly, lowering of devices such as corers, grabs or cameras into the axial portions of the canyon risks the loss of these instruments because an accurate measure of the true depth and bottom configuration is not possible. For these reasons, the deployment of the observer and limited sampling capability within a research submersible became an obvious solution to the reconnaissance of canyons. With the initial dives (Dill, 1961; Shepard et al., 1964) geologists saw for the first time the true nature of canyon floors and slopes below the range of SCUBA techniques. The results were startling, and the remainder of this chapter will highlight several findings which represent some of the discoveries made by diving geologists.

CANYON HEADS

West Coast, North America

The proximity of the Scripps Institution of Oceanography to a major canyon system — the La Jolla—Scripps Canyon complex — has permitted extensive studies in canyon processes by geologists of that institution and neighboring marine research centers. The most extensive series of investigations took place in February, 1964, when 8 dives were made to depths as great as 300 m with the small diving saucer, or SOUCOUPE, belonging to Cousteau. Observations made by the five scientists involved are reported by Shepard et al. (1964). Perhaps the most significant findings of this survey lay in the determination of active sediment transport and erosion at depths to at least 190 m, and the observation of steps as high as 8 m along the axis of the canyon. In some areas this canyon becomes so narrow that the SOUCOUPE (diameter 3 m) could not maneuver to the floor of the canyon. All observers noted the accumulation of talus aprons along many wall margins, and Shepard, in a later paper (1966) describes the presence of talus blocks of sandstone and shale bedrock which protrude through the axial sediment fill of the canyon floor.

Shepard (1967) reported finding a concrete block at a depth of 170 m which Dill had placed in the shallow (30 m) head of Scripps Canyon one year earlier. Other indications of down-canyon sand transport included the lack of growth on the walls near the sediment in forming the canyon floor; and the observation of an anchor chain at 190 m stretched taut in a down-canyon direction after having caught on a rocky pinnacle on the wall. The contact between rock wall and axial sediment was often seen to be polished, and to truncate borings of pholads and other animals which excavate a habitat in the rock walls, similar to conditions found in the upper portion of the canyon.

Current meters emplaced in the headward portions of the Scripps and La Jolla Canyons were transported from their original locations to a site several kilometers down canyon. Search and recovery efforts with one of the NEKTON submersibles were undertaken, and the meters were found tangled with kelp (a shallow-water alga) well below 100 m. The results of these studies (see Shepard et al., 1964, 1969 for southern California canyons) suggest that internal waves and tidal influence are responsible for currents in canyons which may reach velocities sufficient to entrain sediment and transport it both up and down the axes of submarine canyons.

Through extensive studies using SOUCOUPE and other small submersibles, we have detailed information regarding submarine canyon agents and processes in canyons adjacent to the tip of Baja, Mexico. Shepard and Dill (op. cit.) present numerous photographs and maps of the Cabo San Lucas and Los Frailes Canyon systems, and again highlight the importance, at least locally, of canyons in the transport of sand and silt from the nearshore zone to deep basins adjacent to the peninsular mainland. Here, hanging tributaries, undercuts, and significant scour and polishing of submarine outcrops along the walls of the canyons testify to the effectiveness of these features as conduits for transporting coastal sediments to deeper offshore sediment sinks.

East Coast, North America

Unlike most of the submarine canyons off the western coast of North America, the east coast equivalents have heads which lie tens or sometimes hundreds of kilometers offshore. On the east coast, we must deal with a different structural situation in which broad wedges of Mesozoic and Tertiary sediments extend eastward to a shelf break which lies 40 to more than 200 km from the coastline. Thus, investigations of the canyon heads require operations far from shore, and the result has been fewer studies than in west coast counterparts.

Because of the gentle nature of the shelf break, and as a result of the relatively soft (sedimentary) rock exposed on the upper continental slope, the heads of east coast canyons are often broad bowl-like depressions. Nevertheless, active sediment transport is occurring in the canyon heads even at great distances from the shoreline. Stanley (1974) has described the pebbly mud forming the axial fill in the headward portions of Wilmington Canyon. In this single dive with the DEEPSTAR 2000, the author and Stanley observed mud slopes inclined to 16° along the canyon's wall. At a depth of about 420 m, the slope gradient decreased abruptly and the muddy bottom gave way to a pebbly mud, with some larger cobbles and rock fragments. The axial fill appeared to be less than 100 m wide, and a sharp increase in the shell content within the pebbly muds was obvious.

The mechanism for sediment transport in this and other east coast canyons cannot be related to coastal processes as are the west coast canyons. Thus, we must seek processes active on the outer shelf and upper slope which can produce the types of deposits observed in Wilmington Canyon. Stanley (op. cit.) suggests debris flow (as defined by Middleton and Hampton, 1973, p. 20) as a possibility. However, new evidence lending importance to internal waves as active agents in sediment transport in canyons is accumulating (Shepard et al., 1964). It appears that this mechanism, plus storm-wave agitation, may provide sufficient energy to generate and maintain sediment transport at the shelf break and upper slope.

The Manned Undersea Science and Technology (MUST) Office of the National Oceanic and Atmospheric Administration (NOAA), has, for several years, sponsored a series of dives on the shelf and upper slope off New England and the Middle Atlantic states. These dives have often involved research in the heads of submarine canyons, and although some of the objectives were biological observations and sampling, geological information was also obtained. During studies on deep-water lobsters, aspects of habitat reconnaissance yielded geological information in Veatch and Norfolk Canyons (Anonymous, 1974). Tilefish weighing as much as 45 kg were observed in and around excavations in the clay walls of Veatch Canyon. Dense populations of lobster, shrimp, crab and demersal fish were discovered in "pueblo-like communities" carved into soft walls at depths between 130 and 180 m. These excavations weaken the wall rock and promote failure through slumping, spalling and mass wasting.

Under the MUST Office program, Keller completed studies of currents and sediment patterns in the axis of Hudson Canyon. In a series of 6 dives with ALVIN during the 1972 season, he performed coring operations and collected suspended sediment samples as well as detailed bottom photography. (See Chapter 9 for details.) In this, the largest of the east coast canyons (although Great Abaco Canyon at the southeastern end of the Blake Plateau is the deepest) evidence of erosion was found only in the headward portions at and near the shelf break. Depositional features in the form of ridges of semi-consolidated silts lie at various angles to the axis and probably serve to restrict flow along the bottom. Current measurements from ALVIN and meters deployed in the axis showed relatively low velocities which maintained a net down-canyon flow regime (Keller et al., 1973).

Gulf Coast, North America

Relatively few diving programs have been carried out in the Gulf of Mexico. The NEKTON boats have performed a series of dives under MUST sponsorship, but few canyon investigations have been undertaken. The most extensive series of dives were those employing DEEPSTAR 4000 in the "Gulfview" program (Gaul and Clarke, 1968). During May and June, 1967, 11 dives were performed; 5 for biological studies and 6 for geological research. Dive sites were selected on the basis of bathymetric expression, and three included the head and axial portions of DeSoto Canyon. Others were in the broad depression termed the Mississippi Trough southeast of the Mississippi delta.

The DeSoto Canyon dives provided both biologists and geologists with the opportunity to evaluate biological modification of soft substrate in the Canyon axis. These dives, to depths from 630 to 750 m, revealed a rich assortment of bottom fauna whose activities in burrowing, grazing and excavating the sea floor left numerous tracks, trails and other microtopographic features ("lebensspuren"). Similar to the New England canyons, and as will be pointed out in the discussion of central valleys, biological activity may locally be the most efficient erosional agent within canyons. In the DeSoto Canyon area, depressions larger than the submersible were observed in the stiff sediments of the sloping walls. No explanation for these features was offered by the diving scientists. Linear furrows were also observed, again of unknown origin. Based on a lack of bottom currents, the fine-grained nature

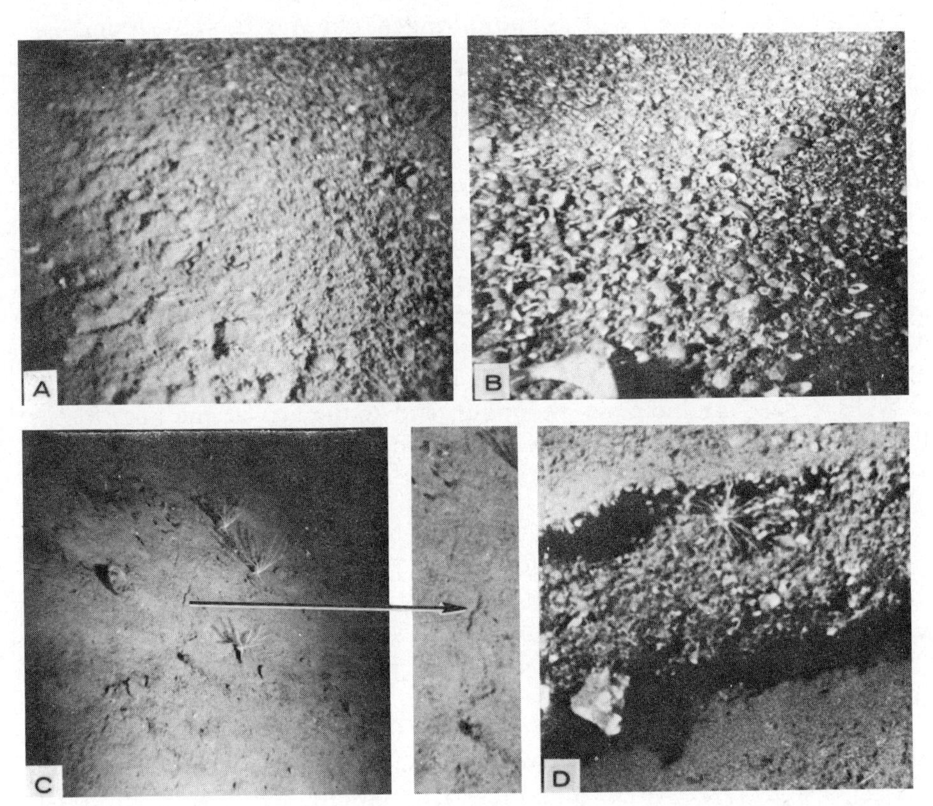

Fig. 1. Examples of various facies and outcrop features characteristic of the heads of submarine canyons. A. Gravel (at right) lying as a veneer on bioturbated muds in the axis of Cap Creus Canyon. B. Co-mingled gravel and shell deposits in the axis of Lacaze-Duthiers Canyon. C. Crack along an incipient slump plane is exposed on the sloping mud walls of Cap Creus Canyon. D. An undercut in the rock ledge exposed in the head of Lacaze-Duthiers Canyon reveals abrasive action of sand as it passes down the canyon axis. All photos from the SAUCER SP. 300 vehicle; depth range from 150—300 m. (Photos courtesy D.J. Stanley.)

of the canyon sediments and lack of evidence for transport of materials, it appears that DeSoto Canyon is inactive at the present time. The major agent affecting the canyon is biological disturbance of the substrate.

Mediterranean examples

Because of the numerous canyons which head close to the northern shoreline of the Mediterranean Sea, and through the activities of Cousteau and others who support diving operations, many observations have been made in canyon heads and valleys in this area. Dill, in three dives made with the SOUCOUPE in the canyon off Villefranche, noted steep sediment-covered slopes displaying slump scars leading toward a narrow V-shaped bottom. Artillery shells presumably dumped at this site during World War II, were found embedded in these sediments, and were tipped toward the canyon axis, a situation Dill attributed to slow gravity creep. Peres et al. (1957), dove in the same area with the French bathyscaph ARCHIMEDE and observed extensive mud walls within this canyon.

In an extensive series of dives with the SAUCER SP 300 vehicle (Dangeard, 1962; Reyss and Soyer, 1965), French scientists examined in detail the heads of the Lacaze-Duthiers and Cap Creus Canyons in the western sector of the Gulf of Lions. Their observations, summarized by Got and Stanley (1974), show a remarkable variation in the sedimentary facies present in the canyon heads. Examples of some observations are shown in Fig. 1. Evidence for recent sediment movement is present as small clefts in the mud walls suggesting incipient slumping, and undercuts at the base of the walls proper. Axial sediments were composed of mixtures of gravels, muds and pelecypod-rich relict shell debris from outcrops at the upper edge of the canyon walls. Bioturbation of the axial sediments is extensive as observed elsewhere in all canyons.

CENTRAL VALLEYS

West Coast, North America

Probably no other canyon in the world has received the attention given the Scripps—La Jolla Canyon complex off southern California. Thorough treatments of geological aspects of this feature are given in Shepard and Dill (op. cit.) and will not be treated here. However, two features recently described by the author (Palmer, 1976) will serve to illustrate the effectiveness of biological activity in the erosion of canyon walls and the utility of submersibles in exploring portions of canyons where access by other means is difficult or impossible.

Diving operations with the DEEPSTAR 2000 conducted off the southern

California coast included geological research in the central valley of the La Jolla Canyon. Two dives were made to inspect the canyon walls where detailed bathymetric maps (Buffington, 1964) had shown the existence of steep slopes. During the first dive, the submersible encountered the east wall of the canyon at a depth of 316 m. At this location the walls consist of siltstones, and extensive excavations by small galatheid crabs are in the process of destroying the wall rock by promoting slumping, spalling and mass wasting (Figs. 2 and 3). These examples offer dramatic proof of the dominance of biological agents in many present-day submarine environments. But as the second dive showed, this has not always occurred in La Jolla Canyon.

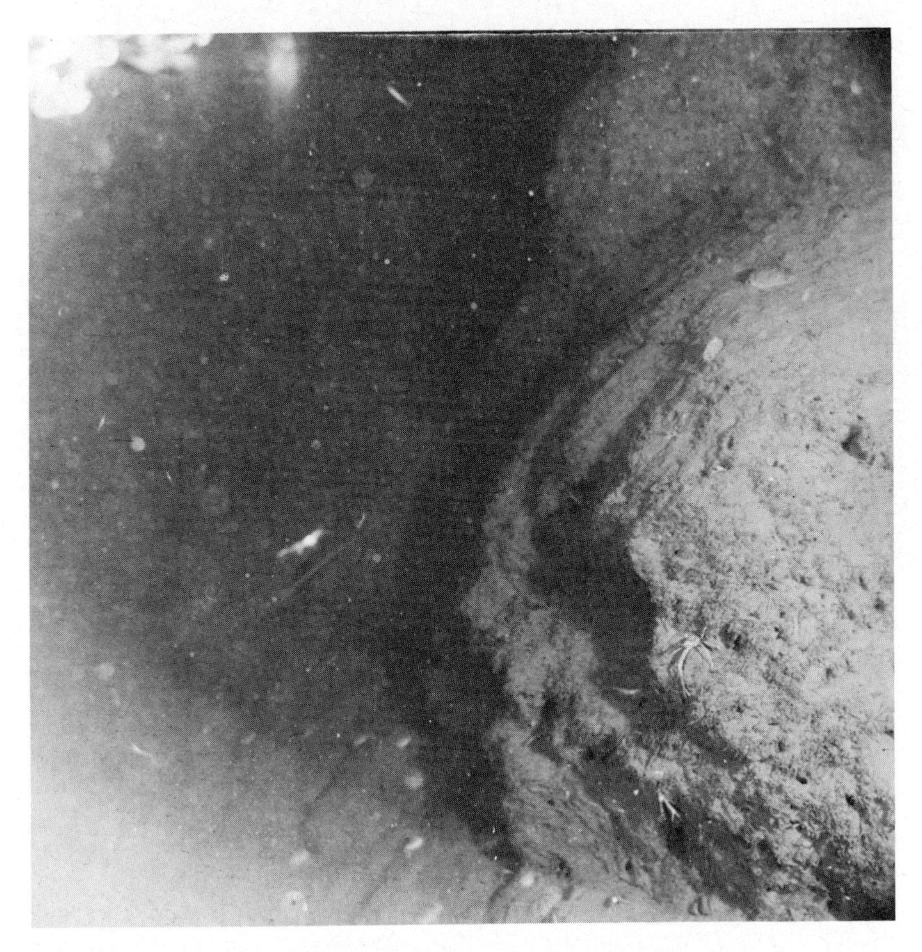

Fig. 2. This outcrop of siltstone exposed in the wall of La Jolla Canyon shows the effects of bioerosion. The talus apron (lower field of view) displays slump structures induced by grazing urchins (white objects at the bottom and on the terrace at top right). Galatheid crabs and their burrows are conspicuous at right. The field of view in the foreground is about 2 m. Depth, 316 m; photo from DEEPSTAR 2000.

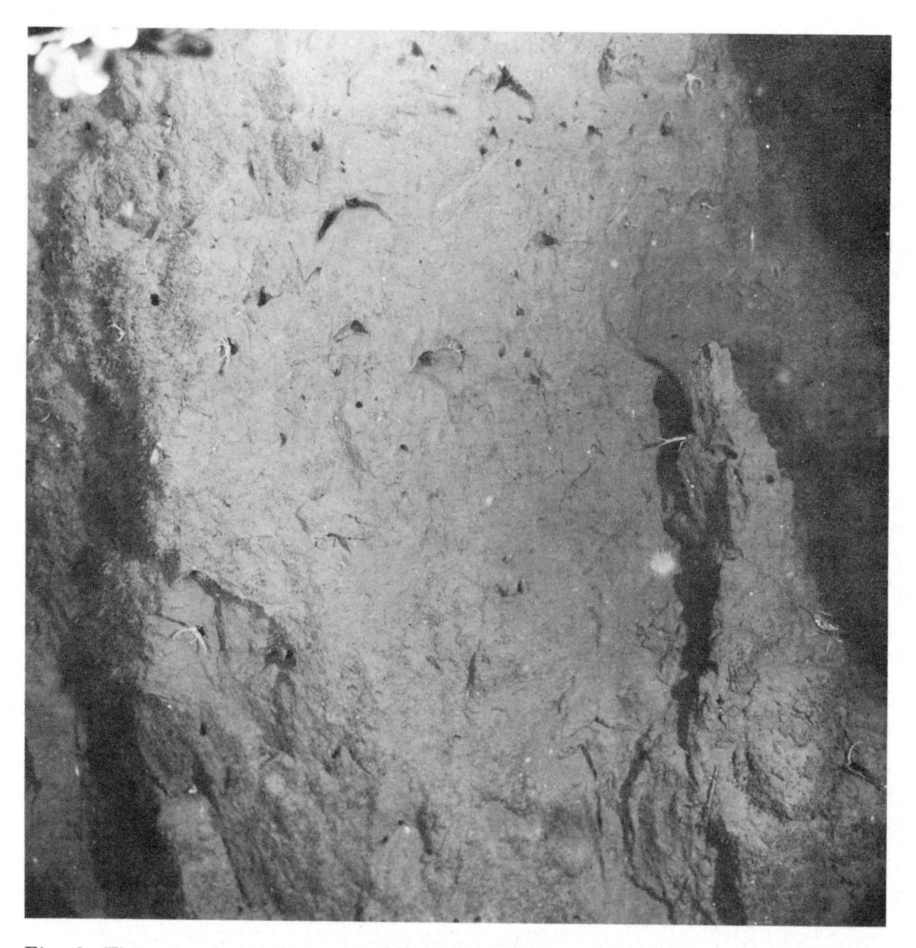

Fig. 3. This view, some 10 m to the right of Fig. 2 shows fresh slump scars in the siltstone wall and spalling effects resulting from the activity of burrowing crabs. The burrow density is about 7/m^2. Scale and depth as in Fig. 2; photo from DEEPSTAR 2000.

The second dive site was located in an area where superimposed isobaths on Buffington's map implied a vertical wall. The axis of the canyon at this point (460 m) was a gently sloping sandy mud surface with abundant benthos. The traverse up the axis passed over similar materials, but at a range of some 18 m from the canyon wall, the upward scanning sonar indicated an obstruction over DEEPSTAR. Continued exploration in this area revealed the presence of an extensive undercut in the canyon wall, and reconstruction of the various sonar tracks permitted preparation of a scaled sketch of the condition at this site (Fig. 4). It lies at the base of a steep wall on the outer bend of a sharp curve in the canyon, and it is believed to represent a relict erosional feature associated with Pleistocene lowered sea levels. Accelerated

Fig. 4. Sketch to scale of the undercut encountered in La Jolla Canyon. This feature on the outside bend of the central valley at a depth of 445 m, is attributed to accelerated submarine erosion of the canyon walls which accompanied lower stands of sea level.

sediment transport through this canyon during lower stands of sea level is suggested by evidence on the La Jolla Fan, and this erosional feature is believed to have originated through submarine erosion by flow and creep of high volumes of sediment.

The impressive canyons systems at the southern tip of Baja California, Mexico have received the attention of many scientists interested in canyon processes, for this area consists of granite bedrock as opposed to most canyons which are cut in softer sedimentary strata. Much of the submersible work has been accomplished by Dill, and a detailed account will be found in Shepard and Dill (op. cit.), with additional geologic data in Shepard (1964). The spectacular sandfalls filmed by SCUBA divers by Limbaugh and Dill occur in these canyons, and striking examples of other submarine sedimentary processes in canyons are to be found beyond the range of SCUBA. In dives with SOUCOUPE, Dill has documented fluctuations in the sediment fill of the axis through discolorations and truncated growth of sessile forms at a depth of 250 m in Los Frailes Canyon (Fig. 5), and at greater depth he observed the introduction of pebbly talus from tributaries to this canyon

Fig. 5. The discoloration in the granite outcrop is caused by the polishing effect of moving sand which has cleaned off the attached growth (dark area). An undercut is evident at the base of this outcrop, and the height of the color break implies a recent loss of 1 m of sediment at this site. Such features support the contention that in some areas canyons act as conduits for sand and gravel moving from the shelf to adjacent basins. Depth is 250 m in Los Frailes Canyon. (Photo from the DIVING SAUCER, courtesy R.F. Dill.)

(Fig. 6). Clearly, here we have examples of the dynamic nature of canyon sedimentation without appealing to coastal runoff as a major factor in sediment budgets. Southern Baja California is quite dry, and the sedimentary processes active in these canyons are wholly marine in nature.

East Coast, North America

The availability of the submersible ALVIN to projects active in east coast studies has led to numerous dives within the central channels of the many canyons which lie beyond the shelf break. Examples of ALVIN involvement in the Hudson Canyon program of the MUST Office of NOAA have been mentioned, but it should be noted that dives in that area extended to depths as great as 1826 m in the canyon. More recent work reported by Cacchione et al. (1976) again with ALVIN, has pointed out marked contrasts in the sediment facies between depths from 2800 to 3400 m, and those described by Keller from work above 2000 m. This recent work (August, 1975) revealed a fine to medium grained sandy substrate essentially devoid of fine-grained accumulations of muds which typify the shallower portions of Hudson Canyon. Apparently, current speeds are sufficient to remove, or preclude

Fig. 6. A slide deposit of freshly broken angular rubble is shown entering the axis of Los Frailes Canyon, Baja California. This material, which has passed down a tributary to the main canyon, consists of a different rock type than the granite walls of the central valley. The absence of scour pits and any sediment cover on the fragments implies a recent depositional event. (Photo by R.F. Dill from DIVING SAUCER, depth about 230 m.)

the deposition of, muds which may reach these depths. However, only one example of current-induced effects (ripples) was observed by Cacchione and his colleagues.

ALVIN has been active in other areas, as reported by Trumbull and McCamis (1967). This dive, this first descent into an east coast canyon, was made in October, 1966. The region surveyed lay between 1460 and 1310 m, and included the flat floor and eastern sidewall of Oceanographer Canyon off Georges Bank. These observations support the conclusion of mass transport of large blocks of rock by axial creep or debris flow. Surprisingly few organisms were seen, but borings and excavations attributed to benthos were noted. Within the axial portion, protruding blocks of consolidated material were sampled and determined to be of Pleistocene age, suggesting that it was an exhumed older fill.

Further to the east, the ALVIN participated in dives the following year in Corsair Canyon. The results of this dive are described by Ross (1968), who noted abundant life in the 1604 and 1216 m interval inspected. Three specific outcrops on the canyon wall were noted, and in places these materials had failed and fragments were found at the base of the wall. Undercuts beneath some outcrops imply strong current activity, and portions of the axial floor exhibited ripples and scour marks around cobbles. A paucity of sessile forms suggests that strong currents prevail in this portion of Corsair Canyon.

Additional east coast canyon dives were made in 1967 with the DEEP-STAR 4000. Dillon and Zimmerman (1970) report on dives within Block and Corsair Canyons where the objectives were to assess differences in sedimentation and morphology which might have resulted from their relatively different positions with respect to the Wisconsin ice sheet. However, the main contribution from these dives consisted of a detailed accounting of biological erosion of outcrops along canyon walls. A layer of mobile surficial sediment perhaps 0.5 m thick was noted in the axis of Block Canyon, but was absent in Corsair Canyon. Currents during the dives were weak, but eroded and subdued ripple marks in the axes suggest that occasional currents strong enough to move sediment must pass through the canyon.

European areas

The earliest work with submersibles in submarine canyons was undertaken in the Mediterranean Sea by the French operating TRIESTE, the FNRS III, and ARCHIMEDE. These were (and are) large and relatively clumsy vehicles, and thus they were used in canyon areas where high maneuverability was not a prerequisite to safe-diving practices. In one case, the FNRS III struck the side slope of a canyon near Toulon and generated a small "turbidity current" which moved down the canyon slope into the axis (Peres, 1959). A similar experience with the same submersible was reported by Dangeard (1960), who noted that sediments rolling at the base of frontal surge formed striations in the bottom sediments, and chutes and groves in the adjacent canyon walls.

SUBMARINE FANS

West Coast, North America

The provinces of fans consist of a proximal zone near the mouth of the canyon, an intermediate area of channels and more gentle slopes, and finally a distal fan where anastomosing shallow channels distribute the finest materials brought across the fan through fan-valley channels. Again, we must turn to examples revealed to us from the La Jolla Canyon in considering the agents and processes active at the far end of the canyon system.

An excellent summary of such deposits is provided by Stanley and Unrug (1972) but the definitive work employing in-situ observations coupled with samples and geophysical surveys is furnished by Shepard et al. (1969). In this latter work, many of Dill's photographs from the TRIESTE I, II and DEEPSTAR 4000 illustrate processes active in the deeper regions impacted by the presence of submarine canyons. These are depositional environments — the areas in which materials swept down the canyon systems come to rest, either temporarily or for extended periods.

Beyond a depth of about 550 m where the rock walls of the canyon give way to softer depositional materials forming the proximal fan, the main fan valley curves westward and south into the broad San Diego Trough. Observations in the interval between 550 and about 1080 m revealed a spectrum of depositional and erosional features which confirms the complex interactions of processes which both form and destroy the fan deposits. The degree of relief which confronted the diving scientists was surprising; vertical "scarps" more than a meter high were observed in many areas. Displaced blocks of consolidated materials originating from these truncated outcrops were seen scattered over much of the channel floor and cracks, crevices and partially displaced blocks in these outcrop areas proved that the slump/spall process was currently active. Again, as in the upper sediments of this canyon, biological activity was clearly the dominant aspect in small-scale erosion of channel walls. However, some larger features on outside bends in the fan-valley channel appear to have a "fluvial" characteristic, i.e. undercuts and precipitous banks on the outer margins of curved portions of the valley. Moore (1965), in a dive with the TRIESTE II into the fan valley at a depth of about 880 m, noted overhanging clay ledges and slumped blocks of channel wall sediments which had slid into the channel axis; in some cases a distance of perhaps 40 m from the site of their disengagement. Moore observed numerous small scars and "chutes" leading down-slope from outcrop areas, and concluded that active erosion and displacement of blocks of clay materials was taking place in the fan valley.

Evidence for strong currents at these depths is also provided by Dill's observation of large sand waves with wave lengths of 15—30 m and heights of up to a meter. These features were noted at a depth of 1097 m where the sandy channel deposits merge with the relatively flat floor of San Diego Trough. These observations, coupled with sediment data from numerous cores recovered from the channel and its enclosing fan, suggest that for some time the fan system has been a dynamic area of alternating erosion and accretion. The youngest axial materials — fine-grained muds — appear to be but a thin veneer over coarser materials which flooded the fan area during lower stands of sea level.

Because of the relatively deep location of most fan-valleys, fewer studies of these features with manned submersibles have been possible. Deep diving requires larger submersibles and support vessels, and thus the smaller vehicles which are so useful at depths of less than about 300 m cannot participate in such deep operations. Submersible studies in fan-valleys other than the La Jolla Canyon have certainly taken place but they are not included herein.

SUMMARY

Marine scientists have found that the manned submersible provides an unparalleled extension of surface capabilities involving the personal factor, because the scientist/observer can make decisions on the spot as to changes in scope and plan for observations in confined areas such as submarine canyons. Tethered systems are a definite asset in many operations, but the deployment of the scientist in a one-atmosphere environment, complete with a payload of sampling and photographic equipment, cannot be duplicated by any alternative system.

The late Ron Church, a talented pilot and photographer in the DEEPSTAR submersible crew, has been quoted as saying that "a submersible is the best underwater camera housing made." This philosophy — that a recording of observations is the prime mission of the submersible — has great merit. Considering the power limitations restricting most operations, and since most instruments for sampling and measuring from a submersible are prototype devices prone to failures, the concept of submersibles as primarily observation platforms for visual and photographic studies emerges as their logical role. Also, considering the uses made of submersibles described by various authors in this volume, the most economical and, perhaps, the best return on the diving dollar investment, becomes apparent. It occurs when the trained observer is provided with a device to place him in direct visual contact with areas of the sea floor which would generally be unreachable through conventional surface sampling systems.

The question in the use of any oceanographic system is one of resolution. How well or accurately do parameters need to be resolved? The use of acoustic systems from surface ships to determine the configuration of the sea floor might be compared to the use of Mariner orbiters around the planet Mars. These "remote" approaches could determine gross aspects of the surface beneath them. But the fine detail, which contained the critical information on processes affecting these surfaces, awaited vehicles which could approach the surface close enough to resolve features of a scale undetectable by the remote systems. On Mars, Viking provided such information and on our water planet, the manned submersible — in canyons and elsewhere — provides the platform for scrutiny of small-scale features which reveal the nature and magnitude of agents and processes at work upon the sea floor.

Some of the examples described in this chapter support the contention that a major function of the research submersible is to provide a window into regions which cannot otherwise be examined with care and consideration for changing bottom conditions. The deep undercut in the La Jolla central valley is one example of a dive plan which was altered as a result of unexpected circumstances. Other observations by biologists, physical oceanographers and chemists, have resulted in discoveries regarding the disposition of living resources in canyons. In Hudson Canyon, tilefish of 45 kg were observed in proximity to excavations in soft wall rock of the canyon, but the largest tilefish ever caught in this area was less than half this size. Considerations regarding waste disposal on the continental shelves often include discussion of the canyons as "natural conduits" to the deep ocean basins, and therefore ideal depositories for refuse which will be carried away from coasts. But current measurements and observations of sediment transport suggest that there may be significant transport *up* the canyons as well as down.

It is obvious that we are still learning basic facts about submarine canyons. They play a complex role in shelf water circulation, and provide a habitat and nursery for economically valuable species. Therefore, they may not be considered as one-way conduits for waste disposal. In addition, in many other respects they influence physical and biological processes operating on the shelf and slope. Submersibles provide a key tool in evaluating the role of submarine canyons in the overall continental margin environment. Therefore one would hope that programs permitting the exploration of these features would continue, as man's use of the outer shelf and slope increases.

ACKNOWLEDGEMENTS

D.J. Stanley, Smithsonian Institution, and R.F. Dill, West Indies Laboratory, Farleigh Dickinson, kindly provided photographic examples of their research in submarine canyons. Other materials provided by J. Vadus, MUST Office of NOAA, assisted in the preparation of this chapter. Discussions with many colleagues, especially F.P. Shepard and N.F. Marshall, Scripps Institution of Oceanography, were most helpful. Figs. 2, 3, and 4 reproduced with permission, Geological Society of America.

REFERENCES

Anonymous, 1974. Manned Undersea Science and Technology Fiscal Year 1973 Report. NOAA, Rockville, 60 pp.

Buffington, E.C., 1964. Structural control and precision bathymetry of La Jolla submarine canyon, California. Mar. Geol., 1: 44—58.

Cacchione, D.A., Rowe, G.T. and Malahoff, A., 1976. Sediment processes controlled by bottom currents and faunal activity in lower Hudson Submarine Canyon. In: Program, Annual Convention A.A.P.G./S.E.P.M., New Orleans, pp. 47—48 (abstract).

Chamberlain, T.K., 1960. Mechanics of Mass Sediment Transport in Scripps Submarine Canyon, California. Thesis, Scripps Institution of Oceanography, La Jolla, Calif., 200 pp.

Chamberlain, T.K., 1964. Mass transport of sediment in the heads of Scripps submarine canyon, California. In: R.L. Miller (Editor), Papers in Marine Geology, Macmillan, New York, N.Y., pp. 42—64. (Shepard commemorative vol.).

Dana, J.D., 1863. A Manual of Geology. Philadelphia, 798 pp.

Dana, J.D., 1890. Long Island Sound in the Quaternary era, with observations on the submarine Hudson River channel. Am. J. Sci., Ser. 3: 425—437.

Dangeard, L., 1960. Glissements de vase sous-marine et phénomènes de compaction. Observations faites en Bathyscaphe. C. R. Acad. Sci. Paris, 251: 2224—2225.
Dangeard, L., 1962. Observations faites en SOUCOUPE PLOGEANTE au large de Banyuls. Cah. Oceanogr., 16: 19—28.
Davidson, G., 1887. Submarine valleys on the Pacific Coast of the United States. Calif. Acad. Sci. Bull., 2: 265—268.
Davidson, G., 1897. The submerged valleys of the coast of California, U.S.A., and of Lower California, Mexico. Calif. Acad. Sci. Proc., Ser. 3(1): 73—103.
Dill, R.F., 1961. Geological features of La Jolla Canyon as revealed by dive no. 83 of the bathyscaph TRIESTE: U.S. Naval Electronics Lab. Tech. Memo No. TM 516: 27 pp.
Dill, R.F., 1964a. Sedimentation and erosion in Scripps submarine canyon head. In: R.L. Miller (Editor), Papers in Marine Geology (Shepard commemorative vol.). Macmillan New York, N.Y., pp. 23—41.
Dill, R.F., 1964b. Contemporary Submarine Erosion in Scripps Submarine Canyon. Thesis, Scripps Institution of Oceanography, La Jolla, Calif., 269 pp.
Dillon, W.P. and Zimmerman, H.B., 1970. Erosion by biological activity in two New England submarine canyons. J. Sediment. Petrol., 40: 542—547.
Gaul, R.D. and Clarke, W.D. (Editors), 1968. Gulfview Diving Log. Pub. 106, Gulf Univ. Res. Corp., College Station, Texas, 37 pp.
Keller, G.H., Lambert, D., Rowe, G. and Staresnic, N., 1973. Bottom currents in the Hudson Canyon. Science, 180: 181—183.
Lindenkohl, A., 1885. Geology of the sea bottom in the approaches to New York Bay. Am. J. Sci., Ser. 3(29): 475—480.
Middleton, G.V. and Hampton, M.A., 1973. Sediment gravity flows: mechanics of flow and deposition. In: G.V. Middleton and A.M. Bouma (Editors), Turbidites and Deep Water Sedimentation. Pacific Section, S.E.P.M., pp. 1—38.
Moore, D.G., 1965. Erosional channel wall in La Jolla Sea-fan valley seen from Bathyscaphe TRIESTE II. Geol. Soc. Am. Bull., 76: 385—392.
Palmer, H.D., 1976. Erosion of submarine outcrops, La Jolla submarine canyon, California. Geol. Soc. Am. Bull., 87: 427—432.
Peres, J.M., Piccard, J. and Ruivo, M., 1957. Resultats de la Campagne de Recherches du Bathyscaphe F.N.R.S. III. Bull. Inst. Oceanogr. Monaco, 1,092: 1—29.
Peres, J.M., 1959. Observation en bathyscaphe de l'instabilite des vases bathyales Mediteraneennes. Rec. Trav. Stn. Mar. d'Endoume, 29(No. 17): 3.
Reyss, D. and Soyer, J., 1965. Etude de deux vallées sous-marines de la mer catalane. Bull. Inst. Oceanogr. Monaco, 65 (1356): 27 pp.
Ross, D.A., 1968. Current action in a submarine canyon. Nature, 218: 1242—1244.
Shepard, F.P., 1964. Sea-floor valleys of Gulf of California. In: Tj.H. van Andel and G.G. Shor Jr. (Editors), Marine Geology of the Gulf of California. Am. Assoc. Pet. Geol. Mem. 3: 157—192.
Shepard, F.P., 1966. Deep-diving vehicles and geology: Workshop on research use of deep manned vehicles. Woods Hole Oceanographic Institution, 8 pp. (unpublished).
Shepard, F.P., 1967. Submarine canyon origin: Based on deep-diving vehicle and surface ship operations: Rev. Geogr. Phys. Geol. Dyn., 9: 347—356.
Shepard, F.P. and Dill, R.F., 1966. Submarine Canyons and Other Sea Valleys. Rand McNally, Chicago, 381 pp.
Shepard, F.P. and Marshall, N.F., 1969. Currents in La Jolla and Scripps submarine canyons. Science, 165: 177—178.
Shepard, F.P., Curray, J.R., Inman, D.L., Murray, E.A., Winterer, E.L. and Dill, R.F., 1964. Submarine geology by diving saucer. Science, 145: 1042—1046.
Shepard, F.P., Dill, R.F. and von Rad, U., 1969. Physiography and sedimentary processes of La Jolla Submarine Fan and Fan-Valley, California. Am. Assoc. Pet. Geol. Bull., 53: 390—420.

Stanley, D.J., 1974. Pebbly mud transport in the head of Wilmington Canyon. Mar. Geol., 16: M1—M8.
Stanley, D.J. and Kelling, G., 1977. Submarine Canyons and Fan Sedimentation. Dowden, Hutchinson and Ross, Stroudsburg, Pa., in press.
Stanley, D.J. and Unrug, R., 1972. Submarine channel deposits, fluxoturbidites and other indicators of slope and base-of-slope environments in modern and ancient submarine basins. In: J.K. Rigby and W.K. Hamblin (Editors), Recognition of Ancient Sedimentary Environments. S.E.P.M. Spec. Publ., 16: 287—340.
Trumbull, J.V.A. and McCamis, M.J., 1967. Geological exploration in an east coast submarine canyon from a research submersible. Science, 158: 370—372.
Whitaker, J.H. McD. (Editor), 1976. Submarine Canyons and Deep-Sea Fans, Modern and Ancient. Dowden, Hutchinson and Ross, Stroudsburg, Pa., Benchmark Papers in Geology, 24: 460 pp.

CHAPTER 12

USE OF SUBMERSIBLES IN THE CONSTRUCTION OF SUBMARINE PIPELINES

Dolf van den Berg

INTRODUCTION

Construction of submarine pipelines may require assistance of a submersible in one or all of the following stages of a pipeline project:

(1) Pipeline route survey. Where the seabed is of a complex nature, severely restricting the possibilities for a suitable pipeline route, direct visual inspection may be of great importance. This enables verification of geophysical data obtained from surface-orientated survey equipment, as well as the determination of size and location of isolated obstacles.

(2) Pipe-laying. When a feasible pipeline route is confined to a narrow strip, great care has to be taken that the lateral position of the pipeline is within the boundaries of that strip. Continuous observation and measuring of the coordinates of the as-laid position of the pipeline during construction is therefore needed to check that deviations from the pre-selected route are kept within allowable limits.

(3) Post-lay inspection. After completion of pipe-laying operations direct visual inspection of the pipeline to check the exterior condition and the possible occurrence of scouring leading to unsupported spans is of great advantage. Conditions which could result in failure of the pipeline can be detected and timely and remedial work carried out to prevent serious incidents.

The success with which submarines can be utilized to fulfill the aforementioned tasks depends a great deal on the accuracy with which they can be located.

Clear observations of underwater objects and superb television or photographic records become quite useless when they cannot be related precisely to geographical coordinates.

NAVIGATION SYSTEMS

Electro-magnetic radiation is strongly attenuated in seawater. Therefore, positioning systems based on transmission and reception of radio signals as successfully applied for surface ships cannot be used for submersibles. Sound on the other hand is very well transmitted by seawater and navigation of submersibles is therefore carried out with sonic equipment. In 1972, when the oil industry began to use submersibles for offshore projects in the North Sea, position checks of the submersible were obtained by measuring its position relative to the support ship (Fig. 1). This vessel was therefore fitted out with a standard sonar system, displaying the echo signal from the submersible as a blip on the sonar screen. The position of the blip relative to the centre of the screen is a measure for the slant range and direction of the submersible relative to the support ship. By projecting these values on to the horizontal plane, relative range and bearing are determined. The water depth information necessary for this operation was read by the pilot of the submersible on his depth gauge. He then relayed these data to the control team on board the support ship via an acoustic communications link between it and the submersible. The geographical coordinates of the support vessel, measured by means of conventional ship's navigation and her headings as shown on the compass, provided all the information necessary to calculate the absolute coordinates of the submersible. The repetitive character of these calculations easily justified the use of a simple computer on board the support ship.

This system, although very simple, also proved to be very inaccurate, even when it was possible to position the support ship within about 2 m. This accuracy can be achieved when operating close to shore which facilitates using a line of sight microwave system. However, subsequent fixes with the

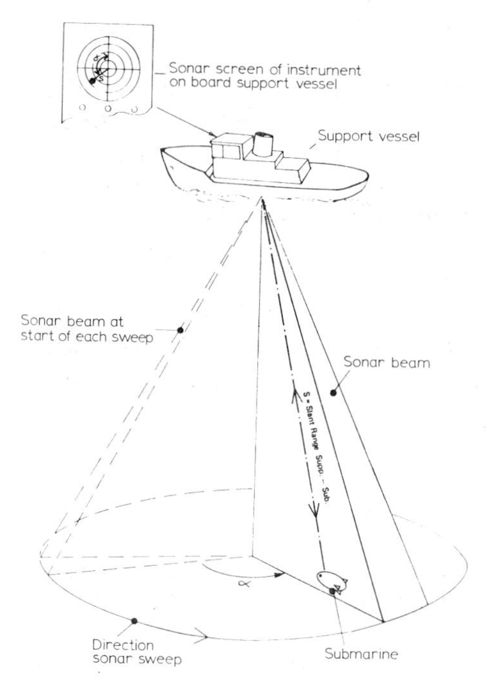

Fig. 1. Fixing submarine with sonar system.

submersible stationary on the seabed could vary as much as 75 m. Apart from the errors inherent in the sonar system, it is difficult to measure the position of the blip accurately on the screen with the support vessel constantly moving about. This results in a continuous shifting of the position of the blip, adding to the overall inaccuracy of the system. In theory the latter problem could be solved by anchoring the support vessel firmly. However, submersible operators would be very reluctant to adhere to such procedure, since it would greatly reduce the possibility of tending to the submersible quickly in case of an emergency. Besides, the mobility of the entire spread would be very much restricted, making the investigation of large areas very tedious.

Much greater accuracy can be achieved with a system of acoustic transponders. These are electronic devices, which upon receiving an acoustic signal of precise, defined frequency, respond by transmitting a signal, also of a pertinent frequency, but different from the interrogation signal and for a limited time only. The frequency of the signals is usually of the order of 10—30 kHz. By interrogating a transponder and measuring the time elapsing between transmission of the interrogation signal and reception of the reply signal, the distance between the point from which the interrogation signal is sent and the transponder can be derived. This is done by multiplying the measured time interval by the value of the sound velocity. With three transponders, each acting and reacting on their own frequencies it is thus possible to position a vehicle relative to these three beacons by interrogating them in turn. The system relies entirely on direct measurement of prime signals. It is therefore not influenced by distorting echo signals from other objects, which occur in sonic systems which measure the time interval between transmission of a signal and the reception of its echo.

Setting up a transponder system begins by launching the transponders from the support ship to the seabed. The coordinates of the support vessel at the time of launching, which are also the approximate coordinates of the transponder, are recorded together with the water depth. Variations in water depth at the launch position and the actual landing spot of the transponder will usually be negligible. Consequently the water depth in which the transponder is located is therefore considered to be the same as that measured at launching. Once all the transponders have been deployed, their geographical coordinates are determined. Should it be possible to measure the position of the surface ship with absolute accuracy, this could be achieved by reading the slant range between each transponder and ship with the latter at three different locations. The locus of the three circles, with as centre point the ship's position, and as radii the horizontal projection of the slant range (obtained from the water depth), would then give the actual coordinates of the transponder in question. However, more measurements are required because of the errors in the ship's navigation system, as well as in the transponder readings.

It is normal practice to determine the position per transponder from some 20 measurements instead of three. The average coordinates are then used for the actual transponder position. The accuracy of the absolute position of the transponders is thus primarily dependent on the accuracy of the ship's navigation system. Close to shore, i.e., at a maximum distance of some 50 km, it is possible to use a microwave system, with which the travel time of a radio signal between the ship and two beacons of known coordinates on shore is measured. With such a system an accuracy can be achieved of the same order as of the transponder system, i.e., about 1 m. The absolute transponder coordinates are then thus known within about 2 m.

Further offshore, present ship navigation systems are far less precise and not more accurate than about 50 m. However, it is usually not essential to know the absolute position of the various structures and pipelines in an oilfield this accurately. Far more important is to insure that all objects, which are to be interconnected, are precisely located relative to each other. This is achieved by relating all further measurements to the transponder grid, which in itself is accurate, although the absolute position of the grid may be far from exact. Once the transponders are fixed, further navigation of the submersible is conducted by measuring her relative distance to the transponders. But now of course there is no need to project the ranges because the submersible is moving in the same plane as the transponders, i.e., just above the seabed.

As mentioned before, the transponder method is based on measuring the time elapsed between transmission and reception of an acoustic signal and multiplying this by the sound velocity. As the latter varies with salinity, temperature and depth, it is important to first determine the exact value under actual operational conditions. The measured ranges can be presented in digital form and stored on magnetic tape for future reference.

PIPELINE ROUTE SURVEY

The following parameters broadly determine the selection of the most favourable route for a submarine pipeline:

(1) Water depth. During construction the pipeline is subject to tensile and bending stresses caused by the weight of the section between the laybarge pontoon and the seabed. The magnitude of these laying stresses depends on the length of the unsupported section which increases with greater water depth. The pipeline is also subject to external hydrostatic pressure of the seawater, and can require additional wall thickness to prevent collapse where the water is deep.

(2) Profile. A sharply undulating topographic profile will cause the pipeline to span between the high points, and stresses due to bending and deflection may be unacceptable. Where cross currents are also present, dynamic

stresses may be introduced by vortex shedding. The laying stresses mentioned under (1) will increase in proportion to the downward slope of the pipeline in the direction of the laybarge because of extra bending at the touch-down point.

(3) Soil conditions. Most pipelines are buried after laying to limit damage from anchors, and trawlboards and to increase their stability against currents. The success of the burial operation depends to a great extent on the nature of the soil at the seabed.

(4) Absence of obstructions. Shipwrecks, lost anchors and other debris, as well as rock pinnacles can obstruct the free passage of the pipeline, causing damage as well as spanning.

(5) Currents. The magnitude of the lift and drag forces resulting from the currents along the seabed, will govern the amount of negative buoyancy required. In case of a suspended line, vortex shedding may occur. The positioning of the laybarge is influenced by surface currents, which are liable to sudden change with tide or wind.

The following equipment is traditionally used in a conventional survey in acquiring the above information:

(a) Echo sounder for measuring water depth and profile.

(b) Seismic subbottom profiler and soil sampler for determining subsoil and soil conditions, respectively.

(c) Side-scan sonar for inspecting the surface of the seabed.

(d) Current meters for measuring magnitude and direction of the currents.

Occasionally it is impossible to avoid rock outcrops on the pipeline route. If the rock surface is sharply corrugated, the pipeline trace has to be levelled to avoid spanning. As an example Figs. 2—5 show the isopach maps of four locations where consderable rock outcrops occur in an area where a pipeline was to be laid. The maps shown in these figures were the result of the interpretation of a conventional survey, and the areas marked A, B, C and D, are characterized as follows:

Area A. (Water depth approximately 50 m.) Sharply corrugated uncovered bedrock, over a length of approximately 300 m. An analysis of the seabed profile along the route indicated that without prior seabed treatment, the pipeline would be suspended over unacceptable lengths.

Area B. (Water depth approximately 90 m.) Uncovered bedrock, but with contours considered to be sufficiently smooth to prevent unacceptable spanning.

Area C. (Water depth approximately 100 m.) Two isolated rock pinnacles within 10 m distance from the route, could possibly impede the free passage of the pipeline.

Area D. (Water depth approximately 60 m.) Two massive rock outcrops, close to each other would leave a passage for a pipeline at one point no wider than 20 m.

All four areas were located close to shore, with the farthest, Area D, about

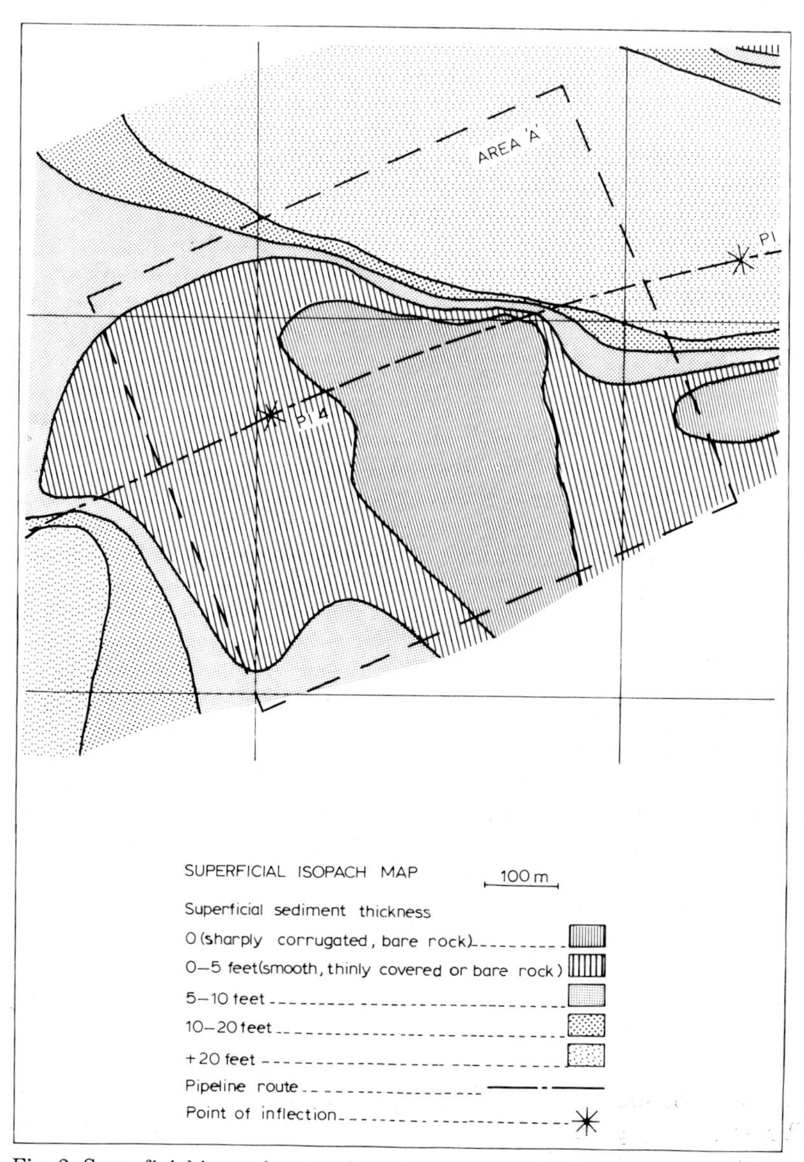

Fig. 2. Superficial isopach map, Area A.

9 km away from the point where the pipeline would reach land. It was thus possible to very accurately, utilizing a microwave positioning system, navigate the survey vessel used to obtain the data from which the isopach map was constructed. Because of the great water depths in which the survey took place together with the movement of the vessel by wave action, it was still impossible to predict limits of the various obstacles with an accuracy greater

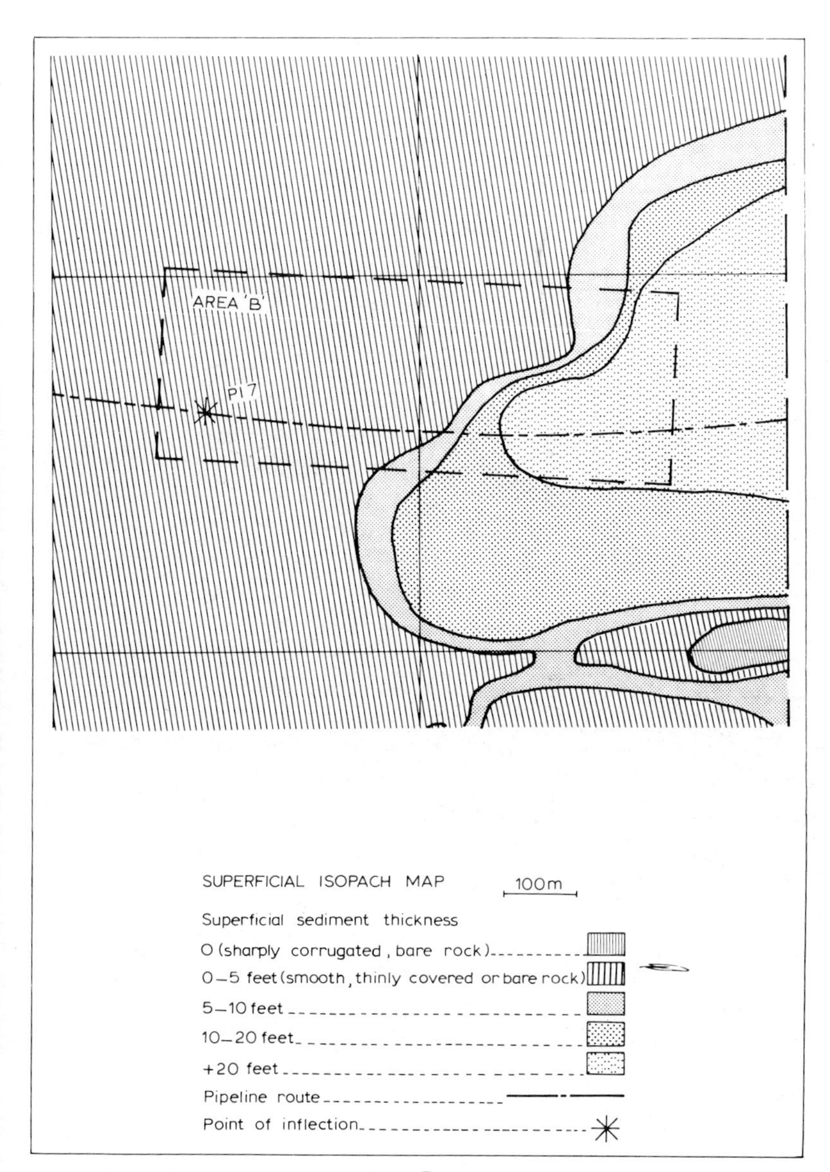

Fig. 3. Superficial isopach map, Area B.

than about 10 m. Furthermore, the complexity of the seabed made precise interpretation of echo sounder, seismic sub-bottom profiler and side-scan sonar records extremely difficult. As the feasibility of the selected pipeline route would depend entirely on the exact location of the various obstacles, direct visual inspection of the seabed with a submersible was conducted. A sonar system was deployed as described in the section describing the navigation

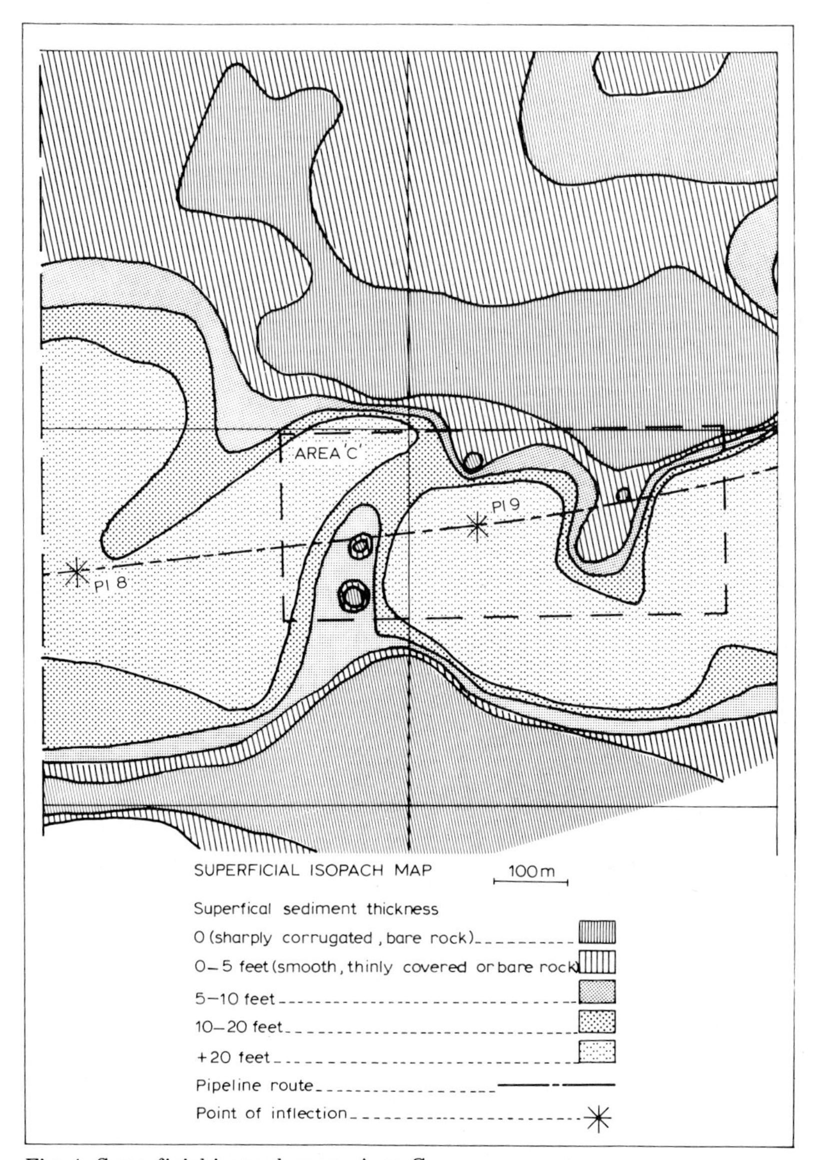

Fig. 4. Superficial isopach map, Area C.

systems and as shown schematically in Fig. 6 for navigating the submersible. A block diagram of the actual components of the system is shown in Fig. 7.

This system proved to be very inaccurate, but better results were obtained when the position of the submersible was determined by the crew from dead reckoning, relative to a pinger and transponder attached to an anchor-weight. These were dropped from the submarine support vessel at slack tide, when it

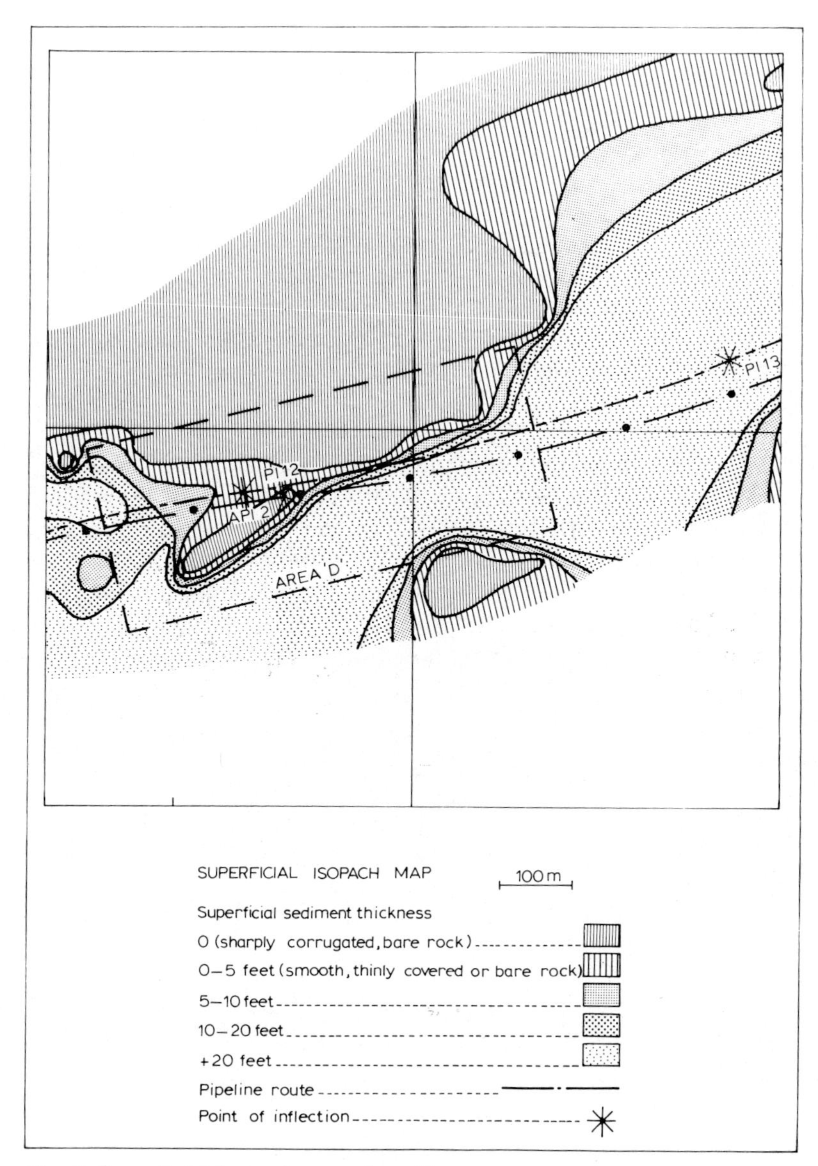

Fig. 5. Superficial isopach map, Area D.

could be assumed that the anchor-weight would drop vertically. Therefore, its landing position would be equal to that of the vessel's at launching, and the coordinates could be marked on a chart which the submersible crew took with them. On the seabed the submersible's position was determined by measuring bearing and range relative to the pinger and transponder, using the gyro compass of the submersible and its sonar.

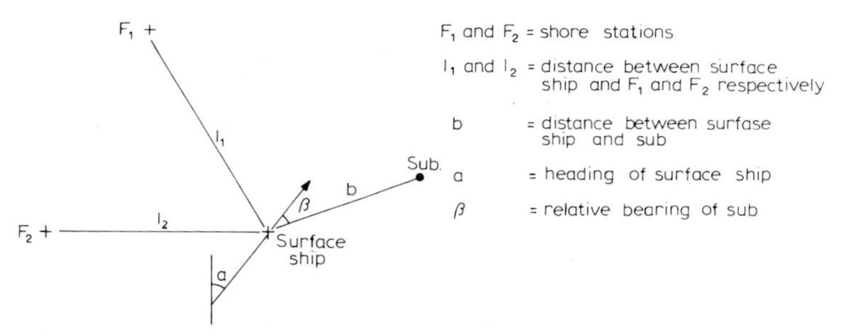

Fig. 6. Schematic diagram of submersible/surface ship navigation system, utilizing sonar system for submersible and microwave system for surface ship.

Each time the position was computed the vehicle was kept stationary by placing it on the seabed. At the same time a position-fix was also taken from the ship, so that a direct comparison could be made between the two systems. A plot sheet (Fig. 8) with the route-tracks as plotted within the submersible and on the surface vessel, shows that although some correlation exists, there is still a large difference. The objective of the submersible's crew had been to steer along the proposed pipeline route. Although this pinger/

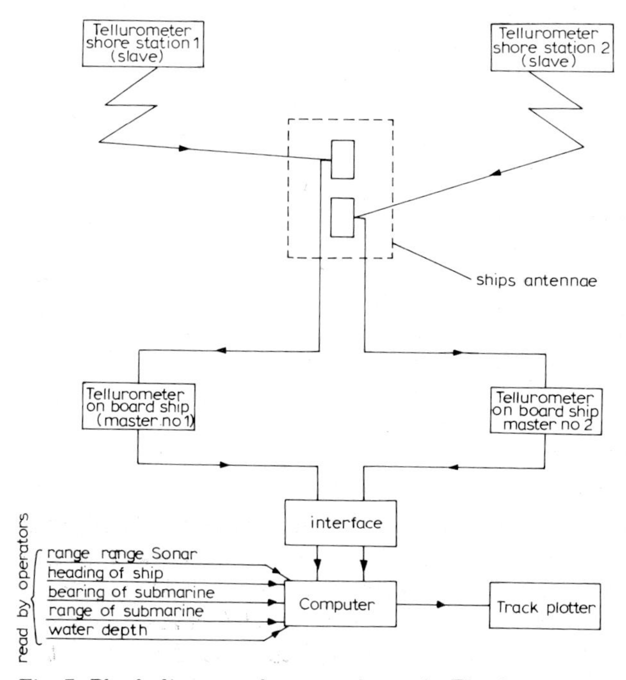

Fig. 7. Block diagram of system shown in Fig. 6.

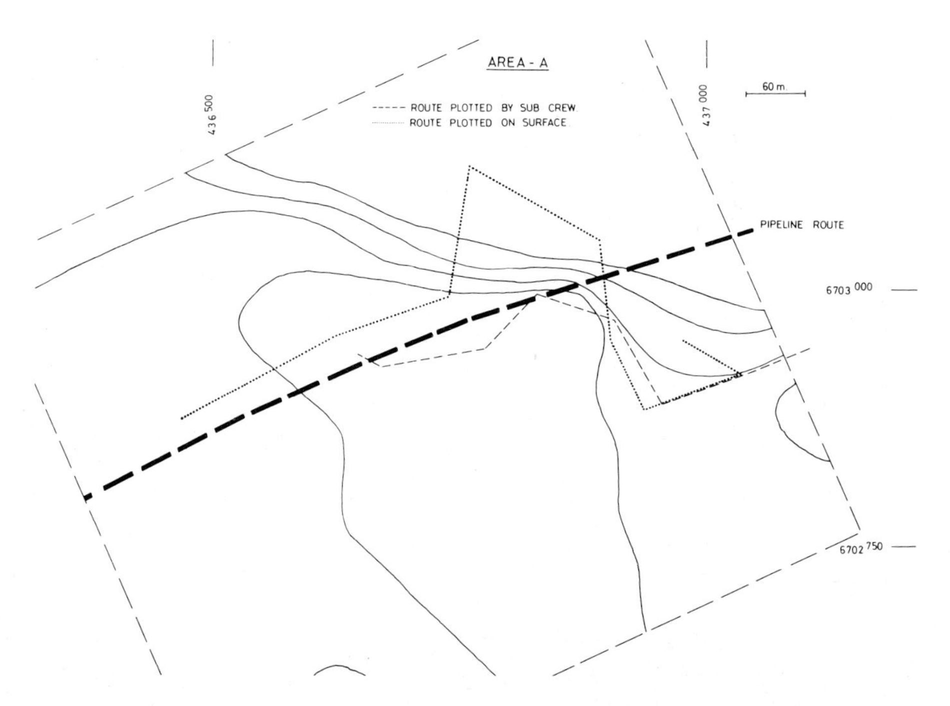

Fig. 8. Comparison of submersible route determined by sonar system and pinger/transponder system.

transponder system seemed to give encouraging results, the actual improvement could not be determined from the limited data obtained. The observations made of the sea floor, showed all features described in the isopach map actually existed. However, the submarine survey had failed entirely to provide further information about the location of the various obstacles. But one firm conclusion could be drawn, namely, the pipeline did indeed cross a patch of sharply corrugated bare rock in Area A. It was decided therefore to build a "causeway" in this area to provide a suitable bearing capability for the pipeline. This was done by dumping crushed rock over a width of 20 m, 10 m on either side of the pipeline route. The average size of the rock was 12.5 cm to ensure sufficient stability against tidal currents. Similar observations confirmed the belief that the rock in area B was smooth enough to need no further treatment. The observations in areas C and D, however, had revealed that there could still be a good possibility for the pipeline being obstructed by rock outcrops.

Dumping in area A was achieved by bringing split-hopper barges alongside a workbarge anchored over the pipeline route. The same microwave positioning system used for the previous surveys was also used for positioning this barge. Divers, operating from a bell-diving system on board, inspected the

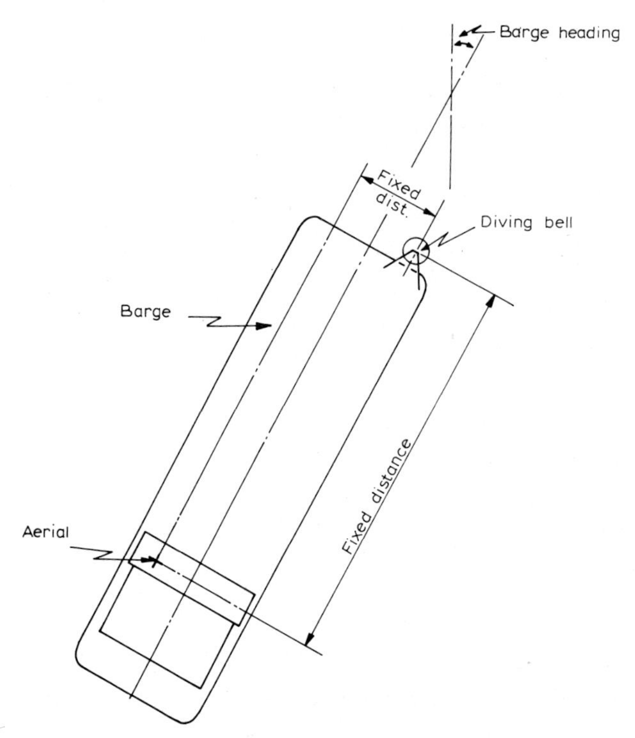

Fig. 9. Lay-out for stone dumping support barge.

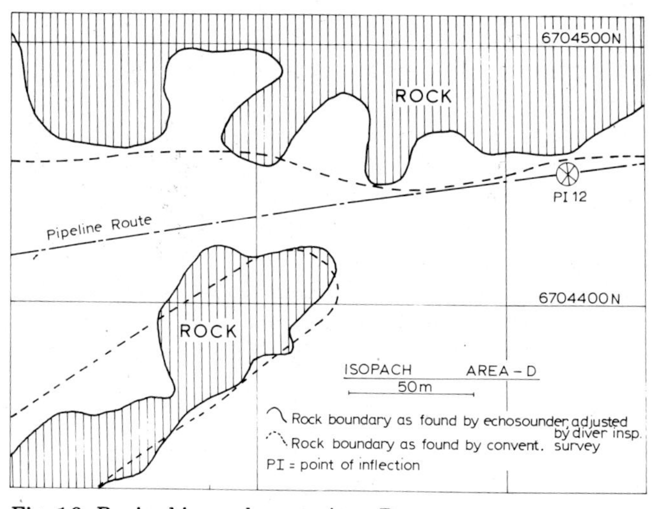

Fig. 10. Revised isopach map, Area D.

results of the operations after each dump. The dispersal of the material dropped was determined by winching the barge sideways until the divers reported to be on the edge of the stone path.

Diving operations were conducted only during slack water. The suspension wire of the diving bell would thus be in a vertical position. As the launching place of the bell relative to the barge microwave aerial was fixed, the actual bell position could easily be calculated. The barge heading also required for this computation was read from the gyro compass on the barge (Fig. 9). With this arrangement it was also possible to verify more accurately the isopach conditions in areas C and D. The revised isopach map of area D is shown superimposed on the initial interpretation in Fig. 10. The results of this bell-survey showed that a 20-m wide path existed between the outcrops in area D. In area C is was still necessary to cover the side of one of the pinnacles, but with only 20% of the crushed rock that was initially estimated.

PIPE-LAYING

A submarine pipeline is constructed by continuously welding on a new joint to the part already completed on board of a laybarge. After completion of each weld the barge is winched forward on its anchors. Tugs are used for the necessary repositioning of the anchors. From the laybarge the pipeline is guided towards the seabed over a pontoon, or in pipeliners' jargon, "stinger". A typical configuration of a pipeline laid in a depth of 130 m is shown in Fig. 11.

The usual method of moving the laybarge is by winching it forward in

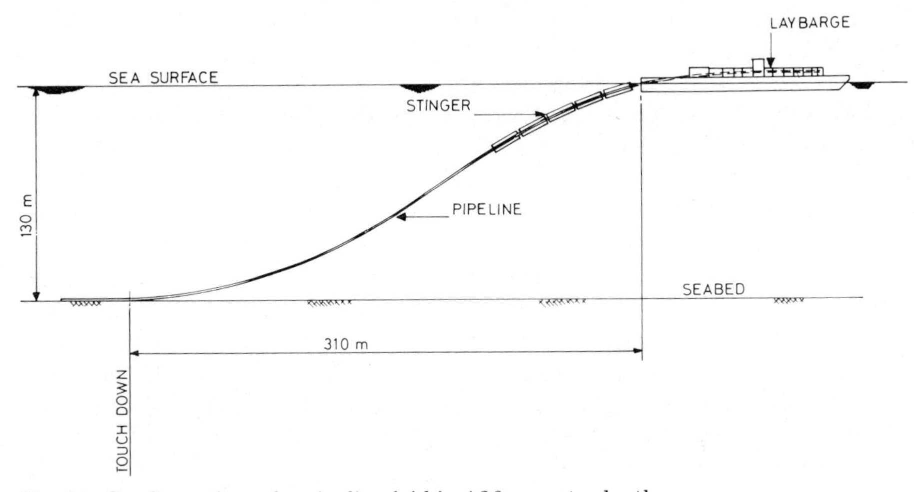

Fig. 11. Configuration of a pipeline laid in 130 m waterdepth.

such a way that it is kept on the pre-selected route. Because of the long distance between barge and the touch-down point, the actual pipeline position may deviate from the planned route. Greater deviations will occur in deeper water and increase when the planned route is curved, or when laying takes place in cross currents. Usually these deviations are not significant, but as described in the previous paragraph, where allowable lateral deviations are to be kept within about 10 m, the traditional system would be unacceptable. It is then that the submersible, if equipped with suitable navigation equipment, again plays an important part. For the laying of the pipeline through the areas A, C and D, a system was applied as shown in Fig. 12. The direction in which the barge was moved after welding of a new joint was now based on the coordinates of the touch-down point. The latter was determined by the submersible which was navigated in relation to three transponders placed on the seabed. The coordinates of the submersible were again determined in relation to two shore stations, by the method described in the section describing the navigation systems. The submersible employed was the same as used for the route survey work conducted the year before. This time however, the submersible could easily relocate the various constraints and check and follow the causeway in area A.

Each time the next pipe joint was welded on, which would take at least ten minutes, the submarine moved forward to the touch-down point. The crew in the submersible then determined the coordinates, which were radioed to surveyors on board the barge. Thereafter they would move out of the way to avoid getting trapped underneath the suspended pipeline during the next barge move. The surveyors plotted the pipeline position and laid out the new course, so that deviations from the actual route were corrected during the next move of the barge. As each pipejoint welded on was only 12 m long, it proved to be relatively easy to keep the pipeline on its predetermined course and within the 10-m limits.

This empirical system therefore proved to be quite satisfactory and it was unnecessary to provide the barge surveyors with any pre-computed correction tables as had been considered previously. Producing such tables would have been extremely difficult because of the many variables influencing the calculations. These include stiffness and friction of the pipe on the sea floor, curvature of the route and value of the current, which not only varied in time, but also along the pipeline route.

The submersible employed for this task needed a 6—8 hr. battery charge after each 6—8 hr. dive. Furthermore submersible operations were not possible when tidal currents exceeded 1 knot. To avoid discontinuing pipelaying when the submersible was not available, position checks of the pipeline were made during such periods by a survey vessel. It zig-zagged behind the barge across the pipeline and thus picked up the pipeline as a parabole on the echo sounder, with the parabole maximum appearing when the echo sounder was

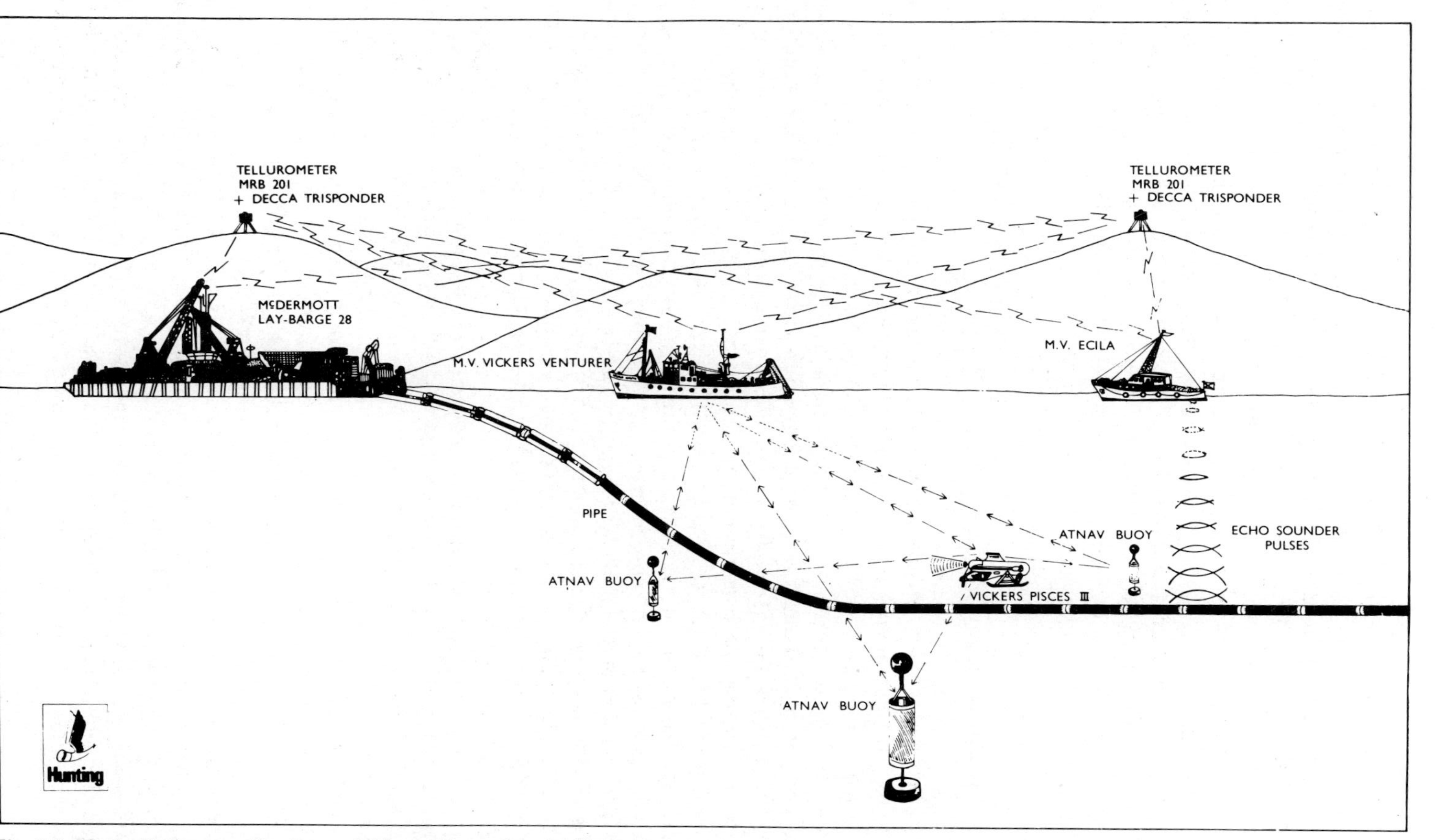

Fig. 12. Methods for monitoring as-laid position of a pipeline.

directly above the pipeline. The survey vessel was also used to plot the pipeline previously measured by the submersible. By correlating the plots obtained by both methods sufficient corrections for the survey vessel records could be established, to enable continuation of laying activities when the submersible could not be used. Obviously as soon as the submersible became operational again, it was used to further check the pipeline position. The total system worked out quite satisfactorily, but great care had to be taken to insure that the microwave positioning system on board all craft involved kept functioning properly. Continuous cross reference of these systems, however, was possible by bringing two vessels alongside one other and comparing recorded coordinates.

POST-LAY INSPECTION

The first application of submersible inspection in pipeline work, was visual inspection of the pipeline on the seabed. When out-fitted with an underwater TV-camera, video tapes of the pipeline can be made. These can also contain the spoken commentary of an observer inside the submersible. The video tapes can be played back by engineers on a video-tape recorder, enabling them to process the information and study at leisure any phenomenon of particular interest, such as defects in the outer coating of the pipeline, etc.

It is normal pipeline practice to identify each weld by painting a number on the exterior of the pipe adjacent to the weld. This enables tallying personnel to keep a record of the number of pipe-joints made, and also to trace back the welders responsible when faulty welds are detected. When the pipeline is on the seabed, underwater inspection can thus directly be related to this numbering system. It also makes it possible to measure the extent of any possible spans. Because each of the pipes comprising a pipeline is of equal length, an immediate record of possible spanning can be made, by counting the consecutive number of unsupported pipejoints in a span.

The height of the pipeline above the seabed will be given by the observer, advising in his commentary that the pipeline is "15 cm off the bottom, 1 m above the bottom", etc., as he moves along. Fig. 13 shows a sample of a pipeline sheet, as derived from the video-tape information. During the inspection, the position of the submersible is also continuously recorded. Therefore, when it becomes necessary to conduct any extra remedial work, the work location can easily be found.

Apart from straightforward inspection, video tapes of pipeline connections made underwater using underwater welding, flanged connections or special couplings are also of great interest.

Following completion of the construction, it is common practice to trench the pipeline into the seabed. This reduces the risk of damage by fishing gear and adds to the stability of the line. Trenching is usually achieved by

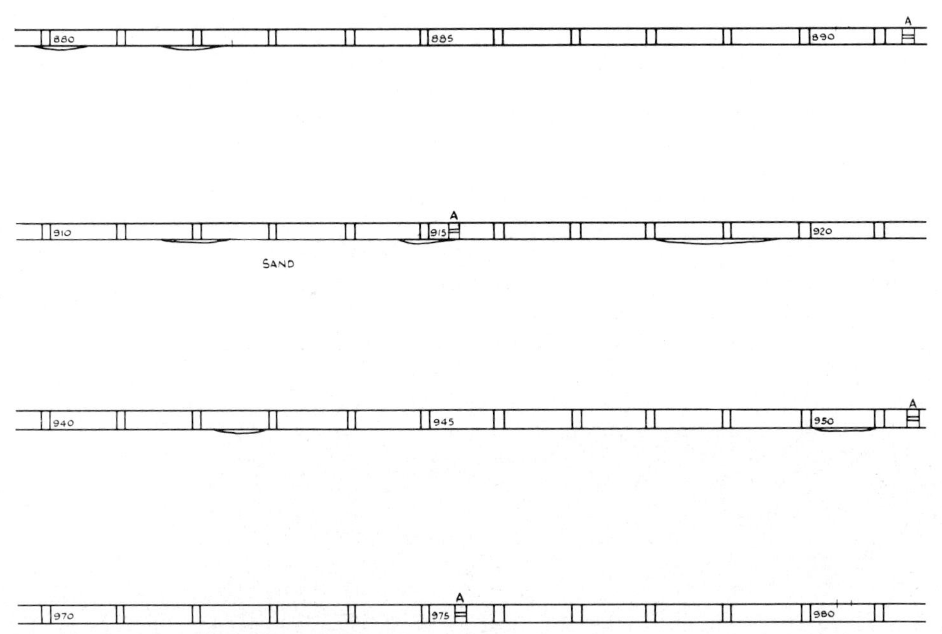

Fig. 13. Pipeline inspection record.

jetting the soil from underneath the pipeline with high pressure waterjets mounted on a sledge pulled along the pipeline. Backfilling of the trench relies on natural sedimentation. Once this has occurred, the pipeline is completely covered and hence no longer visible. It may take a long time, up to some years, to achieve full backfill, and in the meantime the pipeline will only be partially buried. Migrating sanddunes sometimes present in areas of high currents and a sandy sea floor may also cause a pipeline previously buried to become uncovered or even suspended.

Checks on the state of coverage can be made by a surface vessel in a way similar to that described in the section on pipe-laying operations. The sound-beams of the echo sounder are completely reflected by the seabed surface. Therefore a seismic sub-bottom profiler, which transmits at lower frequency to achieve greater penetration, must be used to obtain a reflection from the pipe when covered. The top of the parabolic reflection in relation to the sea-bed indicates if the pipeline is resting on the seabed, or is covered or suspended. The height of cover or suspension can be scaled off the record (Fig. 14). The survey provides information at only those points, where the survey vessel crossed the pipeline. A buried or unburied pipeline, can also be detected with a magnetometer, but no information on possible spanning or burial is obtained with this system because the instrument does not record the seabed level, which the seismic sub-bottom profiler can.

With a submersible, equipped with one of the above instruments, a similar

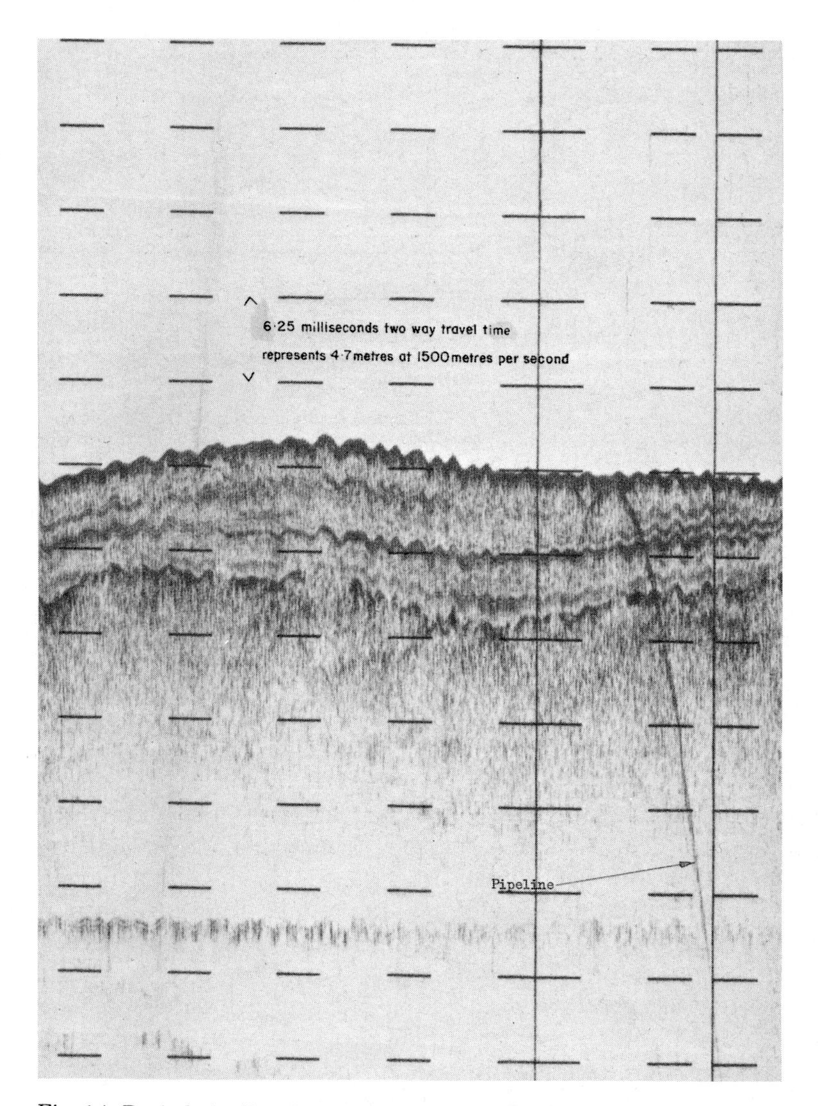

Fig. 14. Buried pipeline detected with a subbottom profiler.

detection of the buried section of the pipeline can be made. The submersible has the added advantage, that wherever the line is not fully covered, it can be followed by the pilot, thus resulting in a continuous record of such sections.

FUTURE DEVELOPMENTS

The really big issues governing the future developments of submersibles in pipeline or indeed all offshore operations, apart from further refinements in

TV facilities, including colour pictures, and still photography, extended to stereophotography, are as follows:

(1) Extended manipulator capabilities

Most submersibles available are equipped with hydraulically operated manipulators enabling the performance of simple tasks. Small objects of the order of 250 kg, can be lifted from the seabed, valves can be operated, etc. When it comes to inspection, however, it would be very useful to have a tool available with which the coating of the pipeline could be pierced to bare metal, so that the electric potential of the steel relative to a reference cell can be measured. This would provide a means of determining if the cathodic protection system of the pipeline, or other underwater steel structures, is functioning properly. Such checks can be made now only by divers, but in water-depths over 50 m, this requires mixed gas or even saturation diving, which is very costly and time consuming. It would therefore be very useful if such work could be included in overall submersible underwater inspections.

(2) Lock-out submersibles

Deep-water pipeline connections to platform risers require saturation divers. These operate from a diving bell to which they are connected with an umbilical cable supplying them with breathing gas, hot water for heating their suits and providing their communication link with the surface. The length of the umbilical cable limits the range of the diver to some 20 m away from the bell. The diving bell itself is operated from a support ship, which if operating in northern North Sea waters has to be of substantial size to cope with the weather. Hence, when diving operations are conducted close to a platform the diving tender also has to be anchored very close to it. This presents inherent dangers when the ship's mooring system, either by conventional anchorage, or dynamic positioning, would fail. The preferred method would be a system whereby the support ship is at safe distance, and the divers are launched in a submersible with a saturation compartment. The submersible could then travel to the work-site, after which the divers leave it and do the work. These so-called lock-out submersibles already exist, but because of their limited capacity for supplying divers with heating and breathing gas, diving time at present is limited to about 3 hrs. Thereafter the submarine has to be recovered and the batteries recharged. Availability of a long-life, light, small, high-power battery would therefore result in a breakthrough in the application of this type of submersible.

(3) Launching/recovering systems

The traditional way of launching and recovering of submersibles is by lifting them from the deck of the support vessel and then lowering them into the sea, either over the side, or more usually, over the stern with help of a A-frame. This method is very vulnerable to weather conditions. Even with the large, 80 m long support vessels in use, excessive ship's movement limits operations in general to significant wave heights of 3 m. Even if the support vessel would be stable enough, working in more severe conditions is still prohibited, when surface-divers are used for connecting and disconnecting lifting and tow-wires. As a result, much weather-down-time is encountered. Weather data collected over the past two-and-one-half years in the North Sea indicate that if work is to stop at a significant wave height of 3 m and can be resumed at 2.4 m, only 214 workdays per year are available.

Great improvement would thus be achieved, if a support ship would be available of the semi-submersible type extending below the surface well beyond the turbulent air—water interface, at a depth of about 30 m. If a system could then be devised whereby the vehicle could leave the submersible underwater and re-enter after its operation without diver assistance, great reduction in weather-down-time would be achieved. Developments in this direction are certainly needed to extend the effectiveness of mini-submersibles in North Sea operations.

(4) Submarine pipe-lay barges

What has been said above about weather influence also applies to pipelaying operations. With day rates for pipe-lay spreads of the order of $200,000 a day, decreasing weather-down-time is even more important. At present this is tried by building large, semi-submersible laybarges at great cost, up to 100 million dollars. Even then, weather remains an important influence. Supply boats for example bringing loads of pipes, cannot come alongside and anchor-handling tugs have serious problems in repositioning anchors, when the weather gets too rough.

In addition to the weather, the performance of laybarges is also strongly influenced by the water depth in which they operate. The length of pipeline suspended between barge and seabed, increases with increasing water depth and concurrently the stresses in the pipe. To keep these at an acceptable level, and to prevent buckling, large tension machines, providing a buckle counteracting tension force, are required. Therefore, a logical step would be transferring the pipeline manufacturing process from the sea surface to the seabed. Submarine pipe-lay barges, supplied by submarine pipeline carriers could be used to construct the pipeline. However, the cost and technical problems for such techniques are now so formidable, that it will be a long time, if ever, before surface pipe-laying equipment will be replaced by submarine pipe-laying equipment. (However, a prototype of the proposed system is described in Chapter 3.) At the moment therefore, the role of submersibles in the construction of submarine pipelines will probably be limited to surveys and inspections.

CHAPTER 13

THE USE OF MANNED SUBMERSIBLES IN THE STUDY OF OCEAN WASTE DISPOSAL

Harold D. Palmer

INTRODUCTION

The urban environment has for centuries been concentrated in the coastal areas of the world. This results from the economics of commerce and industry, and coastal sites are now becoming so crowded that offshore facilities for ports, power generation and even housing are in the planning and design stages. There are many problems associated with the continued growth of these urban—industrial centers, and one of the most serious is that of disposal of the unwanted by-products generated by industry and man's activities in the coastal zone. Unfortunately, to many, the proximity of the ocean to these centers of activity has led them to consider the continental margin as an ideal and limitless disposal site for municipal refuse and industrial wastes.

By 1972 the general lack of control over the nature and magnitude of waste disposal at sea was recognized as a definitive environmental hazard. This situation prompted federal regulation in the form of two Public Laws: PL 92—500, The Federal Water Pollution Control Act Amendments of 1972, and more specifically, The Marine Protection, Research and Sanctuaries Act of 1972 (PL 92—532). These regulations included both the establishment of criteria for evaluating applications for ocean disposal and a rationale for management of disposal sites. PL 92—532 included appendices describing mandatory baseline and monitoring surveys, and in some cases the requirements have been met through the use of submersibles in and near designated ocean dumping grounds.

SUBMERSIBLES AS TOOLS FOR OCEAN DISPOSAL STUDY

As one would infer, the economics of delivering wastes to the marine environment are primarily affected by the distance of the discharge or release site from a port or coastal facility. Thus, those responsible for waste disposal prefer sites as close to shore as possible. Indeed, some smaller municipal sewage-treatment plants discharge into shallow waters only tens of meters offshore. In such cases, SCUBA techniques are appropriate for line inspection and maintenance, but for sites or installations deeper than several

TABLE I

Advantages (+) and disadvantages (—) of employing divers and/or submersibles in seafloor surveys and inspections (Palmer, 1972)

Diver +	Diver —	Submersible +	Submersible —
Retains manual dexterity	cold, fatigue, limited working time	one-atmosphere dry environment	complex support logistics
Maneuverability in confined space	environmental hazards	power for accessories	physical containment
Retention of tactile senses	poor to no communication with partner and surface	instant communication and recording of data	restricted maneuverability
Free ascent	low payload	high payload complex tasks	

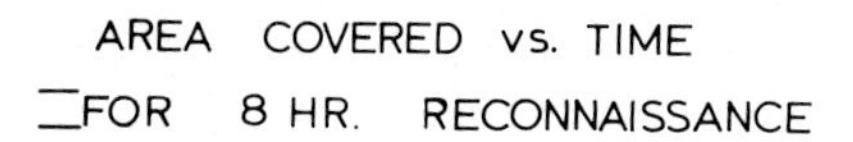

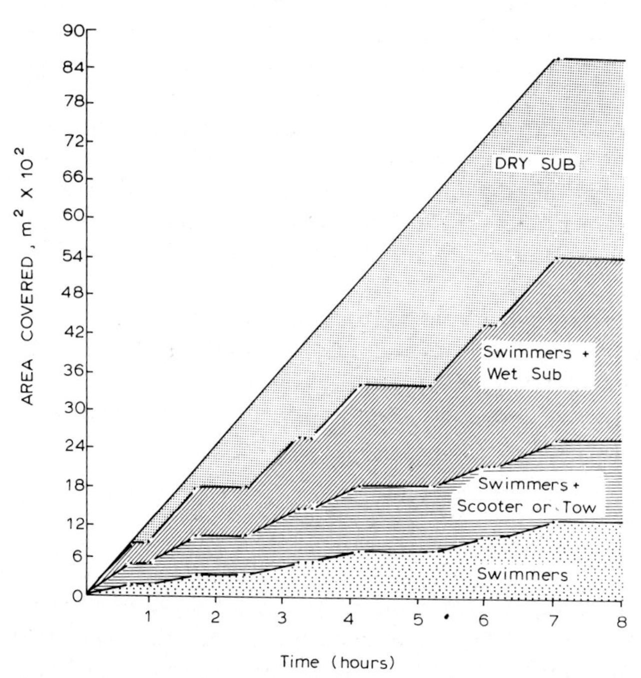

Fig. 1. A comparison of swimmers employing SCUBA and various means of propulsion with performance of a dry submersible reveals that the most efficient means of covering a large area is by the dry sub. In this analysis, we assume the SCUBA survey is conducted by two teams of two divers. The "steps" in the coverage curves are decompression times, for the example shown here a working depth of 20 m was employed.

tens of meters, it may prove more economical to employ a small submersible for periodic surveys of pipe integrity and discharge effects.

For outfall lines, it is common to survey only the trunk line and diffuser segments. For waste discharged from barges, or other vessels releasing their cargo into the open sea while underway, the assessment of effects requires a regional reconnaissance. This in turn demands extensive surveys of the sea floor in and around the designated point of introduction. Here SCUBA is of marginal use, and in deeper sites on the continental shelf and beyond it becomes impractical. Table I supplies a partial listing of the advantages and

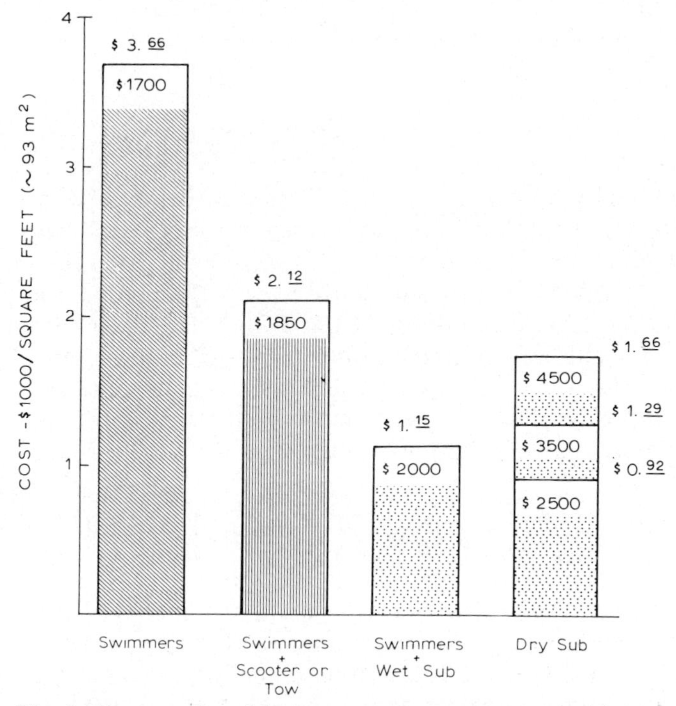

Fig. 2. Four modes of bottom survey are compared with average costing figures for these services. The ordinate in this bar graph gives the cost of the survey in dollars per 1,000 square feet (10^3 ft^2 ~ 93 m^2). Relative heights are inversely proportional to the efficiency, or cost-effectiveness, of each mode. The dollar figure *inside* the upper portion of the bar graph is the daily rate for the various services. The dollar figure *above* the bar is the cost in dollars per 1,000 square feet (the ordinate values). For dry submersibles, these costs appear to the right of the appropriate bar level. From this graph, it appears that the most efficient mode of bottom survey for under about $3,000 per day is the dry submersible. See text for discussion.

disadvantages of employing divers vs. submersibles. For extended missions at depths as shallow as 20 m Figs. 1 and 2 clearly demonstrate that a dry submersible offers the most economical approach to surveying (Palmer, 1972).

A third approach to waste-disposal assessment has been introduced through the use of remote vehicles which operate on a tether, or umbilical cable which permits control of the unmanned device from the surface. These useful tools are discussed elsewhere in this volume (see Chapters 2—4) but it should be noted that they are now in use for routine outfall surveys where only a narrow corridor must be examined. Large, complex units, such as the CURV vehicle, described in Chapters 2 and 4, have been used to investigate deep sites (approximately 1,000 m) off San Francisco where more than 47,000 drums of radioactive wastes were dumped prior to 1970.

To illustrate the utility of manned submersibles in waste-disposal studies, the remainder of this chapter comprises case histories describing various technical methods and results of operations at specific dumpsites.

CASE HISTORIES: DREDGE SPOIL AND MUNICIPAL/INDUSTRIAL WASTES

By weight, dredge spoil is by far the major class of material disposed of at sea. According to a recent report by the National Academy of Science (Gorsline, 1976) nearly 40 million tons per year of mud, sand, shell and gravel dredged from river beds, marginal channels and harbors were dumped at sea. This is five times the tonnage of all other dumped waste materials combined. Of this amount 34% was considered "polluted" on the basis of chlorine content, oil and grease, measures of anoxicity such as biological and chemical oxygen demand (BOD and COD), iron, silica and other elements and components. At least 90% of this total was dumped on the Atlantic shelf, primarily in the New York Bight area.

New York Bight

Ocean disposal off the New York—New Jersey coast dates back to 1888, and as a result of great accumulations of materials (Pararas-Carayannis, 1973; Williams and Duane, 1974; Williams, 1975) serious study of this area was begun by the National Oceanic and Atmospheric Administration (NOAA), in its Marine EcoSystems Analysis (MESA) Program. As a part of this study, a Perry Oceanographics PC-8 submersible was used for seven days during the fall of 1971 to perform a series of dives to determine the distribution of sewage sludge, dredge spoil and "cellar dirt" (construction debris). Although the passenger's view from this craft is normally excellent, the turbid nature of the bottom waters and loose surficial sediments reduced visibility to zero. Other dives conducted some five miles seaward of the main dumping area provided little improvement over previous dives. In this site, visibility was restricted to about 0.3 m by a "jellylike brownish-orange sediment and suspended matter" similar to blackish ooze "seen" in the inshore dive (Burks, 1971). The orange material may have been floccules of ferric hydroxide, a precipitate formed from the dumping of iron-rich acid wastes in a nearby dumpsite.

In the summer of 1972, the submersible DEEPSTAR 2000 described in Chapter 3 was used in a series of divers in the New York Bight dumping grounds. Similar to the PC-8 dives the previous year, visibility was reduced by black flocculated material in the sewage sludge disposal site. Only three species of macrofauna were observed by biologists from the National Marine Fisheries Service (Anonymous, 1973). Diving studies were extended to include a proposed disposal site east of the existing dump site, and the richness of the benthos at the site prompted recommendation that it not be considered as an alternative waste disposal area. Similar conclusions were reached as a result of dives with NEKTON GAMMA in the New York Bight during the summer of 1974. Nine dives in various areas proposed as new dumping sites disclosed a clean bottom supporting a wide variety of benthos and infauna. This led scientists from NOAA to recommend that these areas not be further considered as potential waste-disposal sites.

Although Long Island Sound is not within the region referred to as the New York Bight, a novel and important experiment in that area during the summer of 1974 will be described. In a joint venture by NOAA, the Atomic Energy Administration (now the Energy Research and Development Administration), the U.S. Geological Survey (USGS), and the National Aeronautic and Space Administration (NASA), in-situ X-ray fluorescence and neutron activation instruments were mounted in the BEAVER MARK IV submersible. These devices consist of an energy source (isotopes) and a detector which records radiation emitted from sea-floor sediments exposed to neutron or electron bombardment. These back-radiation "signatures" can be used to quantify the presence of heavy metals in the upper few millimeters of sediment down to a concentration of 20 to 50 parts per million (Vadus, personal communication, 1976).

Because of the complexity of the equipment and its relatively large power requirement, few in-situ studies of this nature have been performed with submersibles. In 1970, a test was made using a prototype neutron activation system aboard the DEEPSTAR 2000, but the progress in developing compact systems with low power demand has been slow.

Middle Atlantic Bight

Two rectangular dumpsites, each 172 km^2 in area, have been designated by the Environmental Protection Agency (EPA) as interim disposal sites for untreated sewage and sludge from the cities of Camden and Philadelphia and for iron-rich acid wastes from the du Pont de Nemours and Company. These mid-shelf sites lie some 60—72 km off the Delaware—Maryland boundary. Water depths in these locations range from 32 to 56 m. The quantity of acid wastes released is approximately 20 million gallons per month, the maximum amount authorized under the present permit. A special barge of one-million gallon capacity is used to transport and discharge this material, and a typical release at 6 knots leaves a track 11—14 km in length. Disposal of this material, which has a pH of 0.01, has continued since July, 1968 (Champ, 1973). Under the current permit, dumping must stop in 1978.

Sludge dumping by the City of Philadelphia has contributed approximately 54 million kg of solids per year since June, 1973. More recently, the City of Camden was authorized to dump untreated municipal sewage at a rate of 15 million gallons per year. This permit has been cancelled, and raw-sewage dumping must cease in November, 1976. However, the sludge, consisting of 3—10% solids, will continue to be discharged in this area until the permit expires in 1981.

As a result of the dispersion of effluent from acid waste and sludge barges, solids and precipitates are spread over a broad area, and inspection of the sea floor to assess possible impacts must be carried out by submersible. In the summer of 1974, and again in 1975, NEKTON GAMMA was employed on surveys sponsored by both NOAA and the EPA. Prior to these efforts, several oceanographic studies had been performed, including a television scan of selected areas (Palmer and Lear, 1973; Lear and Palmer, 1974), but no direct observation of the areas had been attempted.

In 1974, fourteen dives within the two dumpsites and one control area provided biologists and geologists with a first-hand inspection of the sea floor. Video-tape, still and motion picture film and complete voice tape recording permitted reconstruction of each dive. These provided the basis for comparison between dumpsites and the control area some 25 km north of the acid dump site. Specific objectives of the dives were mainly of a reconnaissance nature to: (1) determine where sewage sludge and acid waste might be accumulating; (2) measure current flow close to the bottom; (3) determine the nature of micro-topography; and (4) assess the general distribution and nature of the bottom fauna.

The sea floor in the sludge disposal area was characterized by a grayish to brownish thin and discontinuous veneer of "tufted" materials which projected several millimeters above the sediment surface. Grazing animals left a cleared path through this material (Fig. 3), and shallow excavations with the submersible's manipulator failed to reveal similar materials in the subsurface. It appears that the gray matter is a surface deposit, and it was also noted atop scallops and other biota (Fig. 3) suggesting that it is a precipitate. However, the origin and nature of this material have not been determined, and it cannot be attributed to dumping practices unless further studies reveal a connection to the sludge barged to the site. Although cursory analyses of the

Fig. 3. Characteristic sea-floor appearance in the sludge disposal area, Middle Atlantic Bight. The "tufted" material carpeting the sediment surface lies as a thin, discontinuous veneer over much of the area. Grazing organisms, such as the hermit crab (*Pagurus pollicaris*), appear to clear a path through this material, some of which is seen atop the shell which is approximately 5 cm in diameter. Depth approximately 52 m.

gray "flocs" forming the surface veneer have shown locally high concentrations of lead and cadmium within the sludge dumpsite (2700 and 44 μg/kg dissolved solids, respectively), the highest values for these elements occur well outside the site boundaries (Cox et al., 1975). Since these metals are generally associated with sludge, their abundance in shelf sediments which should not have received particulate matter as a result of dumping poses a question of origin. Although these data are inconclusive, they suggest a source related to estuarine discharge from Delaware Bay.

The sea floor in the acid dumpsite was found to be similar in appearance and morphology to that in the sludge area. Microtopography consisted of broad and subdued megaripples and, frequently the crests were occupied by numerous sand dollars (*Echinarachnius parma*), whereas the troughs contained gray to brown tufted material and abundant shell debris (Fig. 4). Over much of the area surveyed, the sediment was characterized by a veneer of yellowish to brownish orange granular material. This substance was distinctly different in color and appearance from the regional substrate, but its distribution with depth could not be resolved.

Fig. 4. Subdued microrelief in the acid waste disposal site, Middle Atlantic Bight. The numerous sand dollars (*Echinarachnius parma*) lie atop the crest of a low megaripple which trends from the lower center to the upper left. Orange-tinted sediment is exposed on the crests of these features, while the intervening troughs contain brownish to gray "floccules" similar to that seen in the sludge site. The wavelength of these features is about 1 m, with heights being generally less than 15 cm. Numerous dead clam shells (*Arctica islandica*) were commonly seen in a concave-up attitude suggesting that weak currents prevail in this site. Sand dollars are approximately 5—8 cm in diameter; water depth about 44 m.

The control site 25 km northeast of the acid site appeared to have characteristics similar to both the acid and sludge disposal areas. Microtopography consisted of subdued megaripples, and there were no features attributable to sediment transport by currents. No flocculated materials or discolored granular sediments were encountered, and benthic organisms were essentially the same as in the designated dumpsites. The conclusion reached after extensive analyses of the data from both the 1974 and 1975 dive series was that no visible adverse impact on the sea floor or benthic life in the area could be attributed to dumping practices. This series of dives represents the most thorough investigation of a shelf dumpsite by scientists employing submersibles. Although these sites were not visited during the 1976 summer season of diving operations, the results, to be published by the EPA, will describe the detailed findings from the previous two years of investigation.

Southern California Bight

Another center of urban-industrial concentration lies along the coast of southern California between Santa Barbara and the Mexican border. For decades, solid waste disposal at sea has been commonplace in the channel between the mainland and Santa Catalina Island. Since 1931, one site, at the southeastern end of the insular shelf, has been a dumpsite for refuse from ships berthed in Los Angeles Harbor (Brown and Shenton, 1973).

In June of 1972, several dives with the submersible DEEP QUEST were carried out in this area. The longest sea-floor traverse, 10.4 km, was performed in the refuse dumping ground southeast of Santa Catalina Island. Surprisingly little trash was found in the area between 650 and 800 m. Although a few cans, bottles and paper products were observed, the paucity of materials suggests that much of the refuse dumped here has been transported out of the designated dumpsite, or that the release of materials took place outside (shoreward) of the disposal area (Brown and Shenton, 1973).

Another DEEP QUEST dive was performed in the deep Santa Cruz Basin where drums containing radioactive waste materials have been dumped. In addition to the drums, a great variety of outdated explosives and munitions have been dumped in this area by the U.S. Navy. During the 3 km traverse, at least eight sites were found where ammunition boxes, some still strapped to pallets, were scattered over the sea floor. Most of the boxes appeared to be buried, but some were broken open, exposing live ammunition (Fig. 6). No significant deterioration of the munitions was noted.

A third dive with DEEP QUEST was performed in the San Pedro Basin at a site designated for the disposal of a variety of industrial wastes. This location, midway between the mainland and Santa Catalina Island, has been in use since 1947. Chemical fluids containing cyanide, beryllium and magnesium, and wastes containing laboratory acids, film processing chemicals, plating liquids as well as aluminum chloride wastes have been released here, both on the surface and in sealed and weighted containers (Brown and Shenton, 1973). Other wastes have included three million, 42-gallon barrels, of industrial fluids and solids, and three million tons of drilling muds and cuttings from oil wells drilled in the Long Beach Harbor area.

The sea floor in this region is nearly devoid of life. During the 3 km traverse, a loose brownish flocculated material was commonly found on top of a dark gray to black sediment on which numerous dead organisms were observed. Widespread mortality in this area and the lack of rapid decay of organic material (fish and squid carcasses) are attributed to anoxic conditions which preclude high populations and prevent the breakdown of tissue in dead organisms lying on the sea floor. Oxygen concentrations measured by LaFond and others (1967) were uniformly low, 0.02 to 0.10 ml/l, a fact attributed to the lack of circulation in deeper basin waters. His series of dives with DEEPSTAR 4000 (see Chapter 4, fig. 5) in 1966 revealed barely measureable currents (0.05 cm/sec) on the bottom, and the barren sea floor is apparently due to restricted circulation and lack of oxygen exchange in deeper waters rather than to specific dumping activities. Neither DEEP QUEST nor DEEPSTAR 4000 encountered any drums or evidence of drilling muds and in view of the enormous quantities of both, Brown and Shenton (op. cit.) speculate that perhaps dumping was conducted outside of the designated area.

Research in solid waste packaging

It has been estimated (Staples, 1975) that in a typical U.S. household in an urban area the per capita production of household wastes may reach 2.5 kg per day. As more landfill sites become filled to capacity, ocean disposal of compacted trash has received increasing attention (see Loder, 1975). The question of deterioration of the compressed trash bales in sea water must be assessed since many components of the shredded materials are either

Fig. 5. A test rack of baled solid waste is shown (inverted) immediately after placement on the sea floor at a depth of 180 m. This study utilized compressed shredded household wastes (paper, wood, glass, plastics, garbage, metal) within a plastic mesh covering. Square panels on the lower pipe (top of rack when righted) are ceramic and wood settling plates for assessing colonization rates of sessile organisms. (Photo courtesy Ted Loder.)

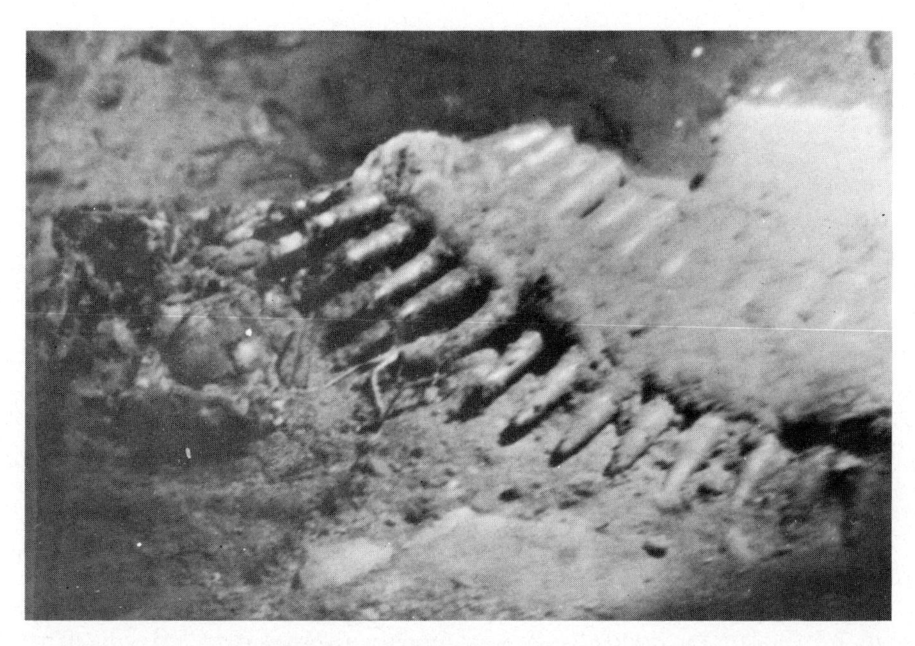

Fig. 6. Live 50-caliber machine-gun ammunition on the bottom of Santa Cruz Basin off southern California. Disposal of outdated munitions at sea is commonplace. Depth 1915 m. (Photo courtesy Brown and Shenton, 1973.)

positively buoyant or barely negative and thus subject to dispersal if bale containment is removed.

In order to assess the integrity of test bales, Loder (personal communication, 1976) performed a series of dives in the vicinity of Veatch Canyon off the New England coast. This program, funded by Sea Grant and utilizing a "NEKTON" submersible, included the placement of one test rack near the shelf break at a depth of 180 m. This unit, shown in Fig. 5, was similar to another placed at a depth of 250 m in the head of Veatch Canyon. Unfortunately, the rack shown in the figure was recovered by a trawler three weeks after emplacement, and the unit deployed in the canyon head has vanished — either swept down the canyon by slumping activity or snared by a trawler and discarded. In 1975 a search of the canyon with ALVIN (see Chapter 4, fig. 2) failed to yield either the test rack or clues to its fate.

RADIOACTIVE WASTES

Containerized radioactive by-products resulting from a variety of nuclear activities and processes were dumped at designated sites in both the Atlantic and Pacific Oceans during the period 1946—1970. Most of the material was disposed of in 55-gallon metal drums which were capped with concrete plugs at both ends. Concern over possible deterioration and the subsequent release of the irradiated materials has prompted two studies in sites where significant amounts of these substances were dumped.

Atlantic Ocean site

According to a recent press release (Morris, 1976), more than 28,000 sealed drums were dumped at a site nearly 200 km east of the Maryland shoreline. EPA scientists employing ALVIN have inspected this area, and they reported some surficial damage, such as corrosion and blistering to the containers, but none appeared to have ruptured. Nevertheless, cesium in significant amounts was found in the sediment near several drums. Water depths at this site are approximately 3000 m, and currents in the region are not well mapped. Further studies with ALVIN in the summer of 1976 will include deployment of current meters and coring of bottom sediments as several aspects of a continuing study in this area.

Pacific Ocean site

There are two major depositories of radioactive wastes off California. The northern dumping ground near the Farallon Islands off San Francisco is in

Fig. 7. Four drums containing radioactive wastes are shown on the sea floor in Santa Cruz Basin, depth 1910 m. Note the lack of scour or any indication of current activity. (Photo courtesy Brown and Shenton, 1973.)

Fig. 8. Paired drums containing radioactive wastes, Santa Cruz Basin. These surfaces form a hard substrate for sessile organisms such as the large sponge shown here. Depth 1910 m. (Photo courtesy Brown and Shenton, 1973.)

water depths less than 1000 m. This site has been studied by the CURV III vehicle (described in Chapter 4) which located, photographed and sampled the sealed drums at this site. The southern site lies in the Santa Cruz Basin within the Southern California Bight. This basin, the deepest on the continental borderland, was the site selected for disposal of more than 3000 steel drums between 1953 and 1962. The dump site, with depths ranging from 1825 to 1960 m, was examined in a dive with DEEP QUEST in 1972. During this dive, 10 drums were located and, with the exception of a coating of marine life (Figs. 7, 8) and some rust on exposed surfaces, the drums appeared unaffected by at least a decade of submergence.

During the inspection of the drums, the manipulator of the DEEP QUEST was employed in an attempt to move the paired drums shown in Fig. 8. The wall of the drum was punctured in the process and the observers noted that the drum appeared to be empty! In as much as the container appeared sound prior to the puncture, this observation is quite puzzling. Perhaps the drum contained fluid, but the rupture, near one end of the drum, was in a location where the concrete cap should have been found. No explanation is offered by the observers reporting on this dive.

PIPE DISCHARGES

Installations which process municipal wastewaters are frequently located near the water's edge. When these facilities are located on coasts, pipes are generally employed to transport the effluent to an offshore site, where the materials are discharged through a diffuser which enhances mixing and dispersion. It is most desirable to discharge the effluent at the maximum depth possible, thus ensuring the maximum diffusion and suppression of buoyant plumes if release can be achieved below the local thermocline. Off many coasts this is economically unfeasible since the modest bathymetric gradient of most shelf surfaces would require a prohibitive length of pipe. As a result, most outfalls are well within SCUBA depth and for short-term inspection swimming the line and "bounce" diving on diffusers is more economical than employment of a manned submersible. However, the proximity of deep water to some sites has led to installations well beyond conventional SCUBA techniques, and submersibles have provided the necessary tools to inspect pipe integrity and effects of disposal.

PICSES IV has been employed in the study of wastewater effects in Howe Sound, British Columbia (Levings and McDaniels, 1973) and another submersible was deployed to investigate outfalls off Oxnard, California (unpublished report, Environmental Quality Analysists, Inc. & Marine Biological Consultants, Inc., 1975).

Southern California Bight — Santa Monica Bay

The Hyperion treatment plant of the Los Angeles County Sanitation District maintains two wastewater discharge pipes extending 8 km and 11 km into Santa Monica Bay. The shorter of the two has two diffusers in a Y-pattern, each having a length of 1.2 km and discharging into waters 60 m deep. The longer pipe carries about 19,000 m^3/day of sludge and effluent into the head of Santa Monica submarine canyon where, at a depth of 100 m, the sludge is released from the open end of the pipe.

In 1966, a small submersible was used to inspect the entire length of the sludge pipe, and in 1968, 11 years after the pipe was placed in service, a NEKTON submersible continued the inspection down the canyon to a depth of 250 m (Anonymous, 1971). The canyon has been described as "narrow" below the discharge, and some solid materials carried by the effluent were observed on the bottom. They varied in thickness from 30 cm near the pipe terminous to 10 cm down-canyon. For a distance of about 1 km down-canyon (to a depth of about 170 m) the bottom was entirely barren of all life save the few species of benthos and infauna capable of surviving in the waste solids. This blanket of waste solids appeared to be intermittent, because bottom sampling in the canyon from surface ships often shows no accumulations in this same area. The lack of significant accumulations and the suspected occurrence of slumps in canyon heads suggest that periodic flushing of natural and artificial sediments from the canyon heads may remove these materials and prevent a build-up of sludge.

More recently, the submersible SNOOPER was employed in trunkline and diffuser inspections along the Hyperion lines. These studies, reported by Allen et al. (1976), are by far the most complete studies of marine organisms associated with ocean outfalls. Because the pipes provide a hard substrate in an otherwise soft bottom habitat, they are populated by a profuse sessile community. Differences in marine life between the two pipes is related to the greater surface area of the shorter (8 km) pipe (diameter 3.6 m) over that of the longer, yet smaller (0.6 m diameter) sludge line. Although the largest sea anemones were found concentrated near the ends of both pipes (Fig. 9) and the greatest concentrations of larger fish were seen in the same sites (Fig. 10); it cannot yet be concluded that effluent is responsible for these or other observations of species distributions.

Fig. 9. Plumose anemones (*Metridium senile*) affixed to the outer segments of the Hyperion sewer outfall. These sessile organisms stand about 60 cm high, and are shown here in a view at a depth of about 100 m. (Photo courtesy Harry Pecorelli.)

Fig. 10. A diffuser port at the southern leg of the Hyperion outfall provides a protected habitat for small rockfish (*Sebastes* sp.). A plumose anemone similar to those in Fig. 9 is attached on the upper surface of the port. Depth 66 m. (Photo courtesy Harry Pecorelli.)

SUMMARY

By law, those who dispose of materials in the sea are required to assess the impact and effect of the effluent on the environment. These requirements can be met by sampling from surface ships, or employing bottom photography or tethered unmanned vehicles bearing television cameras and lights. The sophistication of these vehicles and their relatively low cost (and liability) in comparison with manned submersibles have made them increasingly popular. However, for detailed sampling and study of the regional effects of the discharge of solids and liquids, the manned submersible remains the most efficient, but not necessarily the most economical, means of data collection.

Consideration of the mission to inspect drums of radioactive materials in the Santa Cruz Basin will emphasize the desirability of placing trained observers on the sea floor. This mission evolved as the 12-hour dive continued, with the crew selecting target options, photographing objects from various attitudes, collecting samples and a variety of physical and geological data, and employing the manipulator for moving objects. Another significant aspect of dumpsite sampling techniques was brought into focus through the observations obtained during the studies in the Middle Atlantic Bight.

It became apparent that the physical properties of the sediments and the distribution of epifauna and infauna on and within the substrate were quite localized in distinct provinces. This is evident in Fig. 4, where shell debris are seen to be concentrated in ripple troughs and living sand dollars are restricted to ripple crests. Considering the fact that many bottom surveys in dumpsites are performed with remote techniques which involve lowering of sampling devices from the surface, conclusions reached on a statistical spread of sample station locations may be biased by small-scale local variations as observed and recorded from the submersible.

Where visibility permits, there can be no more efficient method of data collection than placing the trained observer on the sea floor. The range of tools available to him permit the rapid and controlled collection of data necessary for determination of the effects of ocean dumping. Future assessment of dumpsite areas should consider these useful craft as basic tools in the management of ocean disposal techniques and practices.

ACKNOWLEDGEMENTS

Ted Loder of the University of New Hampshire kindly provided information on tests conducted with baled wastes. William Garber, Dept. of Public Works, City of Los Angeles supplied unpublished reports on diving operations on the Hyperion outfall lines, while Harry Pecorelli of Aqua-Con, Redondo Beach, California, furnished photographs of life associated with these studies. Robert Brown, Lockheed Ocean Laboratories, San Diego, provided the unpublished Plessy Environmental Systems report on the DEEP QUEST dives off southern California. Field work in the Middle Atlantic Bight was conducted in the company of David Folger, U.S. Geological Survey, Woods Hole, who kindly provided detailed summaries of all dive observations. Joseph Vadus, Manned Undersea Science and Technology (MUST) office of NOAA furnished data in press from the MUST annual report for 1975. Rich Slater, Northern Colorado University, Greeley, contributed through personal communications not cited herein.

REFERENCES

Allen, M.J., Pecorelli, H. and Word, J., 1976. Marine organisms associated with outfall pipes in Santa Monica Bay, California. J. Water Pollut. Control Fed., in press.

Anonymous, 1971. Environmental effects of liquid wastewater discharge upon Santa Monica Bay. Board of Sanitation, Dept. of Public Works, City of Los Angeles, Calif., 64 pp. (unpublished).

Anonymous, 1973. Manned Undersea Science and Technology, Fiscal Year 1972 Report. National Oceanic and Atmospheric Administration Report, Rockville, Md.
Anonymous, 1975. Manned Undersea Science and Technology, Fiscal Year Report. National Oceanic and Atmospheric Administration Report, Rockville, Md., 51 pp.
Brown, R.P. and Shenton, E.H., 1973. Submersible inspection of deep ocean waste disposal sites off Southern California. Unpublished Report to Manned Undersea Science and Technology Office, NOAA, by Plessey Environmental Systems, San Diego, Calif., 133 pp.
Burks, E.C., 1971. Scientists in a mini-sub seek effect of dumping off the city. In: New York Times, 30 Sept., 1971.
Champ, M.A. (Editor), 1973. Operations SAMS. Sludge Acid Monitoring Survey. The Center for Earth Resources and Environmental Studies, The American University. Publ. No. 1, 169 pp.
Cox, G.V., Townsend, S.A. and Wickramarante, P., 1975. Monitoring Sludge Disposal — Mid-Atlantic Shelf. In: Proc. Civil Engineering in the Oceans, III. Am. Soc. Civ. Eng., 2: 1314—1331.
Gorsline, D.S. (Editor), 1976. Disposal in the Marine Environment; An Oceanographic Assessment. Ocean Disposal Study Steering Comm. Report, National Res. Council, National Academy Sciences, 76 pp.
LaFond, E.C., Good, D.E., Olson, J.R., Jackson, D.L., Cairns, J.L. and Hanson, P.A., 1967. Oceanographic research (sea floor interface — chemical and physical properties). In: NEL Deep Submergence Log No. 3, U.S. Navy Electronic Lab. Report, Feb., 1967, 17—45, 105 pp.
Lear, D.W. and Palmer, H.D., 1974. Ocean Disposal: Middle Atlantic Bight. In: Proc. 10th Annu. Conf., Mar. Technol. Soc., pp. 115—126.
Levings, C.D. and McDaniels, N., 1973. Biological observations from the submersible Pices IV near Britannia Beach, Howe Sound, British Columbia. Fish. Res. Board Can., TR 409.
Loder, T.C., 1975. Effects of baled solid waste disposal in the marine environment — A descriptive model. In: Marine Chemistry in the Coastal Environment. Am. Chem. Soc. Symp. Ser., 18: 467—486.
Loder, T.C., Culliney, J. and Turner, R., 1974. Unpublished reports. Univ. New Hampshire.
Morris, W., 1976. News release. The Washington Post, 2 June, 1976, p. A16.
Palmer, H.D., 1972. Manned submersibles for coastal zone studies — extravagance or economy? In: Proc. Coastal Zone Management Conf. Mar. Technol. Soc., pp. 27—40.
Palmer, H.D. and Lear, D.W., 1973. Environmental Survey of an Interim Ocean Dumpsite, Middle Atlantic Bight. EPA Report 903/9-73-001-A, 170 pp.
Pararas-Carayannis, G., 1973. Ocean dumping in the New York Bight: An assessment of environmental studies. Tech. Mem. No. 39, U.S. Army Corps of Engineers, Coastal Eng. Res. Center, Ft. Belvoir, 159 pp.
Staples, G.M. III, 1975. The solid waste problem. Baltimore Engineer, April, 1975, pp. 7—8.
Williams, S.J., 1975. Anthropogenic filling of the Hudson River (shelf) channel. Geology, 3: 597—600.
Williams, S.J. and Duane, D.B., 1974. Geomorphology and sediments of the inner New York Bight continental shelf. Tech. Memo 45, U.S. Army Corps of Engineers, Coastal Eng. Res. Center, Ft. Belvoir, 81 pp.

CHAPTER 14

THE ASHERAH — A PIONEER IN SEARCH OF THE PAST

George F. Bass and Donald M. Rosencrantz

INTRODUCTION

The story of seafaring is much older than the story of civilized man. Before he had domesticated plants or animals for food, or had invented pottery or learned to smelt metals; and before he had settled in villages, man had learned to cross the Aegean in some type of water craft. Nearly 10,000 years ago he was bringing obsidian from the island of Melos to the Greek mainland, where it was fashioned into blades and scrapers of types excavated in Mesolithic levels of the Franchthi Cave. These levels also produced bones of large fish which presumably were caught at sea.

Since those early beginnings, water craft have been useful for exploration, fishing, pleasure, warfare, religious rites, and, perhaps most important for the archaeologist, transportation of cargo. Water transport remained throughout antiquity the most economical method of carrying goods. Thus everything that man made, from tiny pieces of jewelry to the stone components for entire temples and churches, was at one time or another carried by ship. Many of the ships were lost.

If only one vessel a year had sunk during the ten millennia for which we have evidence of seafaring, there would be 10,000 shipwrecks. But in most years hundreds or thousands of vessels have gone to the bottom of the sea, providing an almost unlimited storeroom of material remains from man's past. Once on the seabed these remains, especially if covered by sand or mud, are often far better preserved than they would have been if left on land.

Although underwater archaeology is a relatively new field, its future importance is clear. The sea has already yielded (1) most of the original life-sized bronze statues in existence from Classical Greek times, (2) the largest hoard of metal utensils from the Bronze Age, (3) the best dated and largest collections of Byzantine pottery, (4) unique sets of weights and glassware, and (5) iron tools. Every shipwreck excavated seems to add a new superlative to the growing list.

Not only are shipboard artifacts well preserved, partly because they are protected from the destruction of man, himself, but they are often quite precisely dated by coins carried on board, making each shipwreck a "time capsule" of known date, a thing of inestimable value to the archaeologist.

The wooden hulls of ancient ships are scarcely less interesting than the cargoes they carried, and the development of ship construction has become the subject of a specialized field of study in recent years. Already we have learned how modern "skeleton-first" construction developed from the older "shell-first" method. Further, new techniques of conservation have made possible at Kyrenia, Cyprus, the display of the only Classical Greek ship raised from the sea and restored. Thus one day scholars and laymen can visit and inspect ships of all periods of antiquity as easily as they now study architectural remains on land.

Not all underwater finds were excavated scientifically. Most of the Classical bronze statues mentioned were found and raised by sponge divers, or caught in fishermen's nets. Even after the development of SCUBA allowed the first attempts at proper underwater excavation in the 1950's, work remained little more than salvage, with bits and pieces of wood and pottery brought to the surface without their context having been recorded. However, in no case did diving teams include archaeologists, those responsible for interpreting the remains and publishing their results.

The staffing of excavation teams with archaeologists and archaeology students more than any technical developments after the invention of SCUBA turned underwater archaeology into a respected branch of science. Such teams, starting at the University Museum of the University of Pennsylvania in 1960, knowing what was required in field archaeology, were able to develop methods to make underwater excavations accurate, efficient, and, at the same time, safer.

The scientific exploitation of an ancient shipwreck begins with its location, and most discoveries have until now been made by chance. Before and during excavation of the wreck, accurate plans and records must be made to show the spacial relationships between all objects, including hull fragments, before they are disturbed. Once uncovered and recorded, remains must be raised safely to the surface where they are cleaned and conserved by techniques appropriate to their material. Restoration, further recording in drawings and photographs, historical interpretation, and publication follow.

The conservation of a shipwreck can prove to be more complex and expensive than its excavation, and research and publication take years longer than actual excavation. However, it is obviously the first stages of excavating a site that will concern us in this chapter, namely, discovery, mapping, excavating, and raising of remains.

RATIONALE FOR A SUBMERSIBLE IN ARCHAEOLOGY

The decision by Bass to have a submersible, the ASHERAH, built for the University Museum of the University of Pennsylvania in 1964 was based mainly on economics. Ironically, our most expensive piece of equipment was the result of the relative poverty of our program. At the same time, however, we were anxious to go deeper than we could with compressed-air SCUBA, especially as a Classical bronze statue had been netted from a presumed wreck at a depth approaching 90 m off the Turkish coast in 1963. The ability to dive deeper with a submersible is obvious, but less obvious is the economic factor.

Because it produces nothing of potential economic value, archaeology is more limited in funds than many marine sciences. Even the multi-million-dollar appraisals of bronze statues pulled from the sea are misleading for the statues are, if excavated legally, simply the automatic property of the countries in whose waters they are found. Thus archaeology's only reward is knowledge, and it depends almost solely on altruistic patronage similar to that which supports music, art museums and the more esoteric scholarly pursuits. Underwater archaeology has a slight advantage over that on land in that occasional grants and contracts for the development of new techniques or pieces of equipment have been obtained.

At the same time, however, underwater archaeology has such an undeniably romantic appeal to many people that lesser funds are often needed than by other marine sciences for similar work. Our own excavations receive more applications from skilled volunteers than can be accepted. These include not only archaeologists and archaeology students, but also doctors, engineers, geologists, professional divers, photogrammetrists, draftsmen, seamen, photographers, mechanics, and architects, most of whom dive.

All of these factors must be considered in planning any new program. A commercial concern with salaried divers, for example, might find it economical to use saturation diving on a site, whereas we could double the time on the bottom on that site simply by accepting a dozen extra volunteers, without large expenditures for saturation-diving equipment. On the other hand, it is sometimes easier to raise funds for a more complex method of excavation than is necessary, if the development of required equipment might be funded by an interested military or industrial agency.

An example of the economics of underwater archaeology is provided by the excavation of a seventh-century Byzantine shipwreck at Yassi Ada in Turkey. The wreck, lying at an average depth of 36 m, was excavated in the four summers of 1961—1964 at a total cost of about $100,000. Our diving logs reveal that we spent 1,243 hours on the wreck in 3,533 individual dives, thus costing approximately $90 per man-hour on the seabed.

In order to lessen these costs, it was never possible to reduce such fixed and necessary annual expenditures as (1) fuel, (2) replacements for diving equipment, (3) insurance, (4) maintenance of diving barge and machinery, (5) hire of local support boats, and (6) storage. Thus considerable savings might be made only by reducing personnel on a project for, although salaries for professional advisers and staff seldom appeared in our budgets, costs of personnel remained a major expense of working in the Mediterranean. Fares from and to the United States or Europe, along with food and shelter, accounted for more than a third of each excavation campaign's $25,000. But, because of limited individual bottom time, a reduction to fewer than fifteen to twenty divers on a shipwreck 35 to 45 m deep would not help, for it is ineffective to mount an expedition to the Mediterranean, with its fixed costs, unless daily bottom time of divers is substantial.

A properly designed submersible, it seemed, might allow two men to accomplish more in a day than could be done by a dozen or more divers, and at the same time allow them to work deeper. What would the submersible be required to do? For an answer we must return to the 1,243 hours spent on the Byzantine shipwreck. If about 160 hours spent making a documentary film are discounted, our logs show that removal of sand had taken 694 hours, or 64% of the total. Of this time, 224 hours were spent air-lifting and 470 hours sweeping sand away by hand. Making plans by photography and drawing consumed 204 hours, or 19% of our time. Removal of pottery, anchors and other finds took 115 hours, or 11% of the time, and bringing up hull remains 41 hours, or 4%. The remaining 2%, or 24 hours, was used to anchor the diving barge and similar non-archaeological tasks.

Removal of sand or mud overburden was the most time-consuming job on that excavation, as on most others. Since a diver can usually control a polyvinyl chloride (PVC) air-lift 5 m high and 10 cm in diameter with one hand (freeing the other to fan sand into its lower end), an extremely simple attachment on a submersible could also hold such an air-lift (perhaps with a small water jet to take the place of the fanning hand). Thus, if the submersible were sufficiently maneuverable and delicately ballasted, and if it afforded its occupants a clear view of the PVC pipe, it could be used for air-lifting for hours at a time. Probably such work should be limited to the upper, sterile layers of overburden where fragile artifacts or wood fragments would most likely not be encountered.

Mapping a site is the second most time-consuming task. We had developed in 1963 a diver-operated method of mapping in three dimensions by stereophotogrammetry, based on normal aerial-survey techniques; and thus surmised that a similar method could be devised for a submersible that would imitate the flight of an airplane.

Removal of sometimes heavy or bulky artifacts from a wreck might present a more serious problem to our hypothetical submersible because it would probably have a low payload. Little engineering would be required, however, to devise a system whereby jars and other ancient cargo would be raised high enough to be dropped into a basket attached to a lifting balloon that could be inflated as required to transport the artifacts to the surface.

A submersible was not seen as taking the place of divers on an excavation, but simply as saving their time on less delicate jobs. Because divers would still be used, the submersible could serve another purpose: centralized control. Unlike on land, where the director can personally supervise his excavators, the underwater archaeologist is often limited in time to only minutes per day on his site. It must be remembered that in 1963, when construction of ASHERAH was ordered, saturation diving was not yet commonplace. A submersible offering good visibility and a sub-to-diver communications system, however, could allow a director to watch and supervise rotating teams of divers for uninterrupted hours, as has been done from a diving bell.

Lastly, it was thought that a submersible could be used for locating new sites, as it would provide searchers longer hours, at greater depths, than they might obtain with normal SCUBA.

SUBMERSIBLE CHARACTERISTICS

Design requirements

The design requirements for a submersible to be used in archaeological research, based on the premises given above, were:

(1) Low cost, based on economic reality.

(2) Reliability and minimal logistic support, based on its intended use in remote areas, and in countries where customs formalities might delay the arrival of essential spare parts for months.

(3) Minimal surface support; no locally available boat would have the capability of lifting a submersible from the water.

(4) Ability to be operated and maintained by archaeologists and others without formal training as engineers.

(5) Good visibility.

(6) High degree of maneuverability.

(7) Two-man crew, for safety and efficiency.

(8) Operating depth of at least 100 m, based on two presumed wrecks at that depth that had yielded bronze statues to nets off the Turkish coast.

Design and construction

The decision to have a submersible designed and built by the Electric Boat Division of General Dynamics was based, partly on the advice of a noted ocean engineer not connected with any of the companies considered, and partly on the fact that Rosencrantz, who understood the needs of underwater archaeology from working with Bass earlier, was about to take a position as an engineer with Electric Boat. Although the estimated cost was higher than could be afforded by our archaeological program, enough of that cost was absorbed by Electric Boat's research and development budget to allow a sales price of $50,000.

The following functional requirements were established:

— Operating depth 180 m.

— Two-man crew.
— Life support: 12 hours minimum.
— Speed: 0—3 knots.
— High maneuverability — forward, reverse, straight up and down, spin on axis, hover if possible.
— Maintain even keel, good stability in pitch and roll.
— Entrance and egress with submersible in water.
— Forward visibility needed as well as to port and starboard.

Construction began in early 1964, and the submersible was launched on 28 May of that year, named ASHERAH after a Phoenician sea goddess. Statistics were:

Length 5.2 m
Beam 2.6 m
Height 2.3 m
Weight (dry) 3,810 kg
Collapse depth 365 m
Life support (maximum) 48 man-hours
Speed (maximum) 2 knots
Payload 45 kg and operators

Fig. 1 is a sectional elevation view showing the major components.

The steel pressure hull is a spherical shell 1.5 m in diameter with a hatch at its top, six view ports, and a number of mechanical, piping, and electrical penetrations to accommodate the ballast, propulsion, and sensor systems. Streaming aft from the pressure hull is a free-flooding teardrop-shaped fiberglass fairing which provides a reasonable hydrodynamic shape. The end of

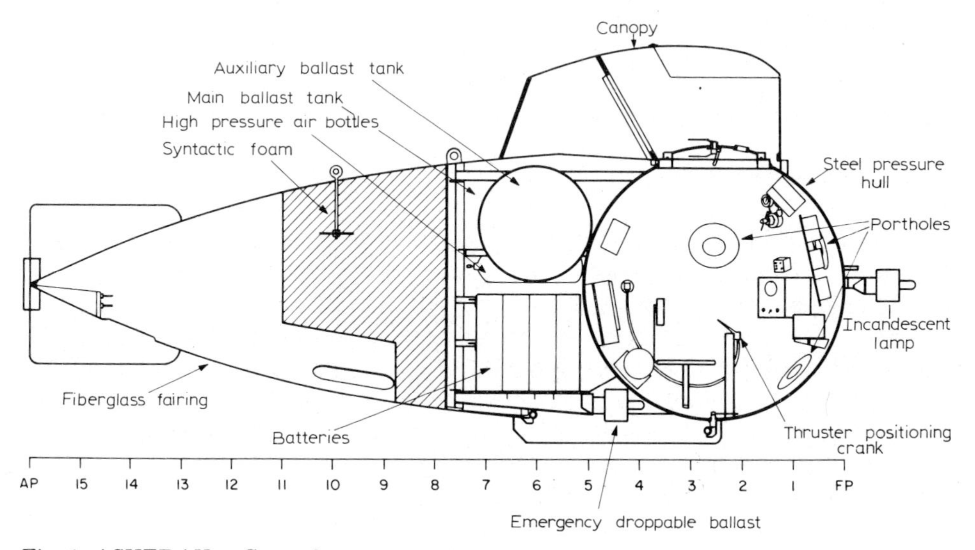

Fig. 1. ASHERAH — General arrangements.

the fairing is fitted with fins to provide stability at the higher speeds. Within the fiberglass fairing are the batteries, high-pressure air bank, main ballast tank, variable ballast tank, and syntactic foam. A free flooding canopy is provided on top of ASHERAH, giving her about 0.6 m of freeboard when she is surfaced.

A portion of the teardrop-shaped fairing is sealed to the pressure hull and to a bulkhead located approximately midships. This volume, open at the bottom to the sea, serves as the main ballast tank to give surface buoyancy. The variable ballast tank is a spherical pressure vessel 0.66 m in diameter. When ASHERAH dives, a vent to the main ballast tank is opened, allowing it to flood; then the variable ballast tank is partially flooded until neutral buoyancy is attained. Neutrally buoyant, the submersible is propelled horizontally and vertically by means of its thrusters. Positive buoyancy is regained by blowing the ballast tanks with high-pressure air which is provided by a bank of 4 SCUBA tanks filled to 158 kg/cm^2 (or, in an emergency, by dropping ASHERAH's heavy skids). Additional fixed buoyancy comes from the use of syntactic foam which is incompressible and has a density of 0.48 g/cm^3.

Fig. 2. ASHERAH.

Power for propulsion and other services comes from lead—acid batteries with a capacity of 23 kilowatt-hours. These are located outboard and have oil on top of the electrolyte for pressure compensation.

The propulsion system consists of two thrusters mounted port and starboard along a common athwartship axis (Fig. 2). Each thruster is independently rotatable about this axis by a pair of manually controlled cranks inside the pressure hull. Thruster speed is varied by silicon-controlled rectifier circuitry. The operator can make the boat go forward, reverse, spin, or move straight up and down by suitable adjustments of thruster orientation and speed.

Life support is accomplished by manually replenishing used-up oxygen from a high-pressure oxygen supply tank. Carbon dioxide is removed by circulating the air through a cannister of absorbent. Equipment is supplied to monitor the O_2 and CO_2 content periodically.

Sensors used for basic navigation are a depth gauge, fathometer and magnesyn compass. An underwater telephone similar to a UQC but using a base frequency of 25 kHz is used for subsurface communications, and a Citizen's Band radio is used on the surface. Two 800 watt incandescent lamps provide lighting.

USE AT SEA

Use in 1964

ASHERAH's use in Turkey was for little more than sea trials. Political considerations led to the training of a Turkish archaeologist, Yuksel Egdemir, as an operator; he usually dived in the submersible with Rosencrantz, William Beran (another engineer from Electric Boat), or Bass.

Ability to operate with a minimum of surface support proved an important factor in ASHERAH's design. Not only could no local boat lift her from the water, but the closest land-based crane capable of lifting her weight was many miles from the area of use. Thus, the submersible was simply towed and moored like a small dinghy. When a cracked viewport needed replacement the repair was made after a dozen men had pulled the vessel up onto a beach as far as possible.

The trials showed that the main problem areas were:

(1) Limited navigational capability.

(2) Poor viewing capability. The positions of the portholes made it very difficult for the operator and observer to place their faces near the forward looking portholes (during the design stages, the archaeologists, who wanted a centrally located downward-looking porthole, had lost their argument with the engineers who said it was not needed).

(3) A tiring control system; the use of hand cranks to control thruster

motor attitude proved most inconvenient, and the effort required to use them totally exhausted the operator after two hours.

(4) An unreliable speed-control system (redesigned and perfected during the following winter).

No attempts were made to search for the deep shipwrecks believed to lie off the coast, but an initial program of mapping stereophotogrammetrically was successful.

A true test of ASHERAH's worth, therefore, did not come until her second use in Turkey, in 1967.

Use in 1967

Visibility was increased by both an optical view port and a closed-circuit television system. Other improvements included a reliable speed control system, an improved underwater telephone, and a perfected stereophotographic system, and ASHERAH returned to Turkey for the summer of 1967. Rosencrantz was again in charge of the submersible, aided by pilot Egdemir.

The two areas in which ASHERAH proved most effective, mapping and searching, were those funded largely by the U.S. Navy for development and evaluation. This suggests that ASHERAH might have been similarly successful in other missions had adequate funds been available for them. Evaluations of the various tasks originally conceived for an archaeological submersible are as follows:

Centralized control

In theory this remains a valid concept, but in practice it was not. Many of the excavators in Turkey were archaeologists or archaeology students, and most of the staff were experienced in and knowledgeable about their work. Photographs taken of each excavation area were available to the director on a daily basis. Thus taking ASHERAH onto the site simply to allow the archaeological director to watch the work in progress involved undue effort, such as: (1) batteries had to be charged, (2) high-pressure air and oxygen tanks required filling or replacement, as did CO_2 absorbent, (3) divers were required to enter the water to loosen or attach mooring lines, (4) the operator and observer had to be rowed to the submersible and back, (5) telephone and radio operators were on duty on the surface, and (6) a surface boat was on standby in case ASHERAH were caught in a current and swept out to sea, which happened once. Also, the route to the wreck site was hazardous because of mooring lines from the diving barge above, lines to various air lifts, and other lines which required skillful dodging.

Bass had hoped, probably naively, that taking ASHERAH to the site each day would have been no more complex than driving a car to work, with occasional refueling and maintenance.

Air-lifting and artifact retrieval

The attachment of an external manipulator, for which funds had been found, proved once more far more complex than Bass had anticipated. ASHERAH would have needed to be largely dismantled for pressure testing had new hull penetrations been made for the manipulator. Thus, no attempts were made to retrieve artifacts from the shipwreck being excavated.

Lack of time more than anything else prohibited us from attaching a PVC air lift, with surface-supplied air, to the nose of ASHERAH for testing. Bass, at least, remains convinced that this concept remains valid and on the seabed it was possible to inch the submarine around the site, with no greater weight on the wreck than that of a diver. Unfortunately ASHERAH was sold, for reasons given below, before this, perhaps her most useful function, was tested.

Mapping

Aerial photogrammetric mapping is the standard technique used for the production of topographic maps of land features. The technique is conducted from an airplane flying over an area at approximately constant altitude and taking a series of overlapping photographs. The overlap coverage between adjacent photographs is about 60%.

We reasoned that ASHERAH with her good pitch and roll stability and high degree of maneuverability could be used as an "underwater airplane" to get similar photographic coverage underwater.

With aerial mapping a single camera is used, and pictures are taken from an altitude of 3.7 km with a separation of 2.1 km between photo stations. Underwater requirements dictated that ASHERAH would "fly" at an altitude of 4.0 m with a camera station separation of 1.6 m.

Analysis of stereo-photographs to produce topographic maps requires knowledge of the precise orientation (positional and angular) of the camera each time a picture is taken. For aerial photography, this is done by using auxiliary sensors as well as surveying selected points on the ground after the pictures are taken.

The underwater mapping situation requires that no direct survey measurements can be made. Therefore all orientation data must be taken using auxiliary sensors.

Most of the orientation data needed for photogrammetric analysis is used to establish relative orientation between adjacent overlapping camera stations. A camera station separation of only 1.6 m underwater allows the use of two synchronously operating cameras to be mounted on a rigid connecting frame. This configuration makes relative orientation constant, and measurable before mounting the system on ASHERAH, and greatly reduces the need for auxiliary sensors.

Two sensors were deemed necessary to provide additional data. A tilt sensor was used to determine the angle of the camera axes with respect to vertical each time a picture was taken. An altitude sensor was also used to determine the system depth with respect to sea level.

During our 1964 season, we experimented with a stereo-camera system utilizing 2 FB-1 aerial reconnaissance cameras which had been modified for underwater use. (Karius et al., 1965) The results were marginal but promising, and pointed the way towards what equipment improvements were necessary.

Office of Naval Research (ONR) funding was made available in 1967 to procure and test a proper system. (Bass and Rosencrantz, 1968; Rosencrantz, 1975) A pair of cameras, Model PC-750, were bought from Hydro Products of San Diego. These used 70-mm film and were equipped with lenses designed by E. Leitz specifically for underwater use. The effective focal length underwater was 60 mm with a field of view of 50° × 50°.

The tilt sensor was an electrolytic bubble level made by General Precision. The altitude sensor was a differential pressure gauge made by the Whittaker Corporation. The gauge was used in conjunction with an adjustable constant-pressure reference so that it could be zeroed for any altitude. This would be done at the beginning of a "flight". Vertical deviation from the zero level was indicated as the submersible moved above and below the zero datum level set earlier. All sensors had electrical outputs and were fed to calibrated panel meters. A data recording camera (a motorized Nikon F 35-mm camera) took a picture of the meters and a frame counter each time the stereocamera was fired.

A series of pictures were taken during these "flights" over the hull remains of a fourth century Roman shipwreck that was undergoing excavation. (Fig. 3) Nine stereo pairs were analyzed by a professional photogrammetrist, Rudolf Karius. The results of his work are shown in Fig. 4. It shows the true-plan position of remaining wooden frames and planking from the hull. A number of spot heights and contours were also plotted to give adequate topographic detail for archaeological purposes. Fig. 5 is a photo mosaic of the same area which was made by enlarging and fitting prints from the same stereo pairs used to make the plan.

The results were adequate for archaeological records. An area of 6 × 12 m was mapped from an altitude of 4.5—6 m with a position error of 3.8 cm. This error can be reduced with further system improvements.

Searching

In the mid-1950's the larger-than-life bronze head of a veiled woman was netted by a sponge dragger off Turkey's southern coast. In 1963, the statue of a Negro youth was similarly netted off the southwest coast. Discussions with boat captains who had found the two pieces revealed that they had

Fig. 3. A photo-"flight" over a Roman shipwreck. (Copyright National Geographic Society.)

both been pulling their nets in about 85 m of water. The discovery areas were revisited, to try and estimate the uncertainty of the positions where they were found. The search area was 6.4 km^2 for the first statue and 3.2 km^2 for the second.

During the 1964 season with the ASHERAH, it became painfully evident that she could not be used very effectively for search because of her lack of adequate navigational equipment.

However, even if she were equipped with the best equipment available today, it is still doubtful how effective she could have been for searching the above areas visually. Simple algebra shows that with a search speed of 3

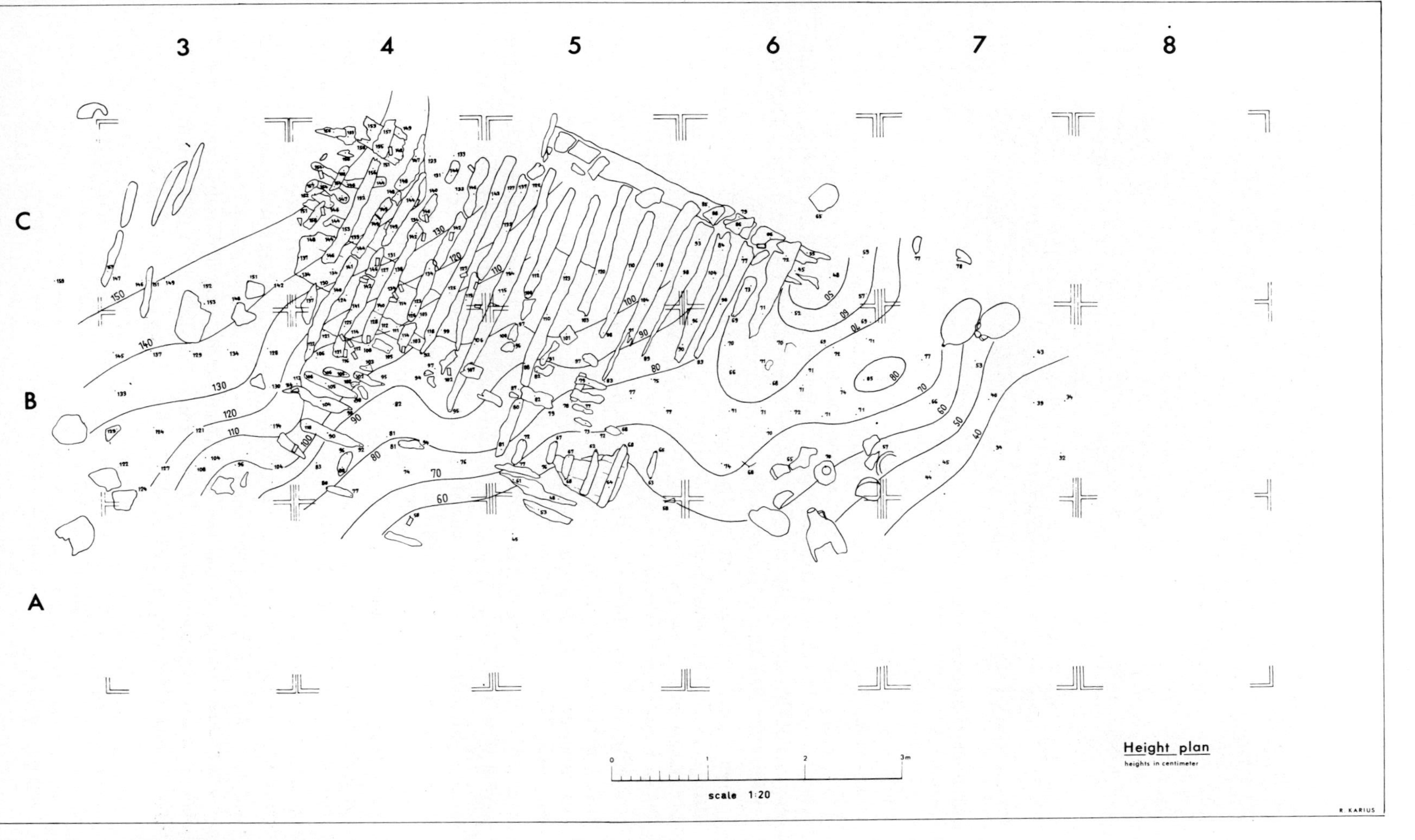

Fig. 4. Topographic map of hull wood.

Fig. 5. Photomosaic of hull wood.

knots and a 6-m search-swath width, it would be impractical to cover the above areas even with perfect navigational capability. The work done during 1964, and some further experiments with underwater TV and a manned towed capsule called a towvane, indicated that the search areas were remarkably smooth, sandy, and free of debris and rock outcroppings. This was terrain that was ideally suited for a side-scanning sonar search.

The ONR funding provided the means for evaluating two side-scan sonars during 1967. One was from the Marine Physical Lab at Scripps Institute of Oceanography, and the other had just been designed by Klein for EG&G. These were towed in the two search areas. Accurate navigation of the sonar tow ship was accomplished by using multiple transit stations ashore linked with citizens band radio. Twenty six possible "hits" were found in the 6.4 km^2 area and only one was found in the 3.2 km^2 area. The ASHERAH was then used to visually check out the "hit" in the smaller area. Fig. 6 shows what this "hit" looked like on the sonar readout. Using data from the initial search, the sonar was towed back and forth along different bearings close to the "hit" until it was apparent the transducer had gone right over the top. When the best fixes were obtained, a buoy was dropped right on the spot. ASHERAH was deployed, and descended vertically trying to follow the buoy rope to the bottom. When she reached bottom, there was a layer of turbid water that limited visibility to about 0.5 m. Also, there was no sign of any shipwreck. An attempt was made to run a square-sided spiral search pattern by making runs for increasingly long periods of time and changing compass heading by 90° at the end of each period. After four hours, nothing was seen, and ASHERAH surfaced.

The sonar fixing technique was tried again, the following day, and a second buoy dropped, about 30 m from the first. This time when ASHERAH followed the rope down, she landed in the middle of a Classical Greek

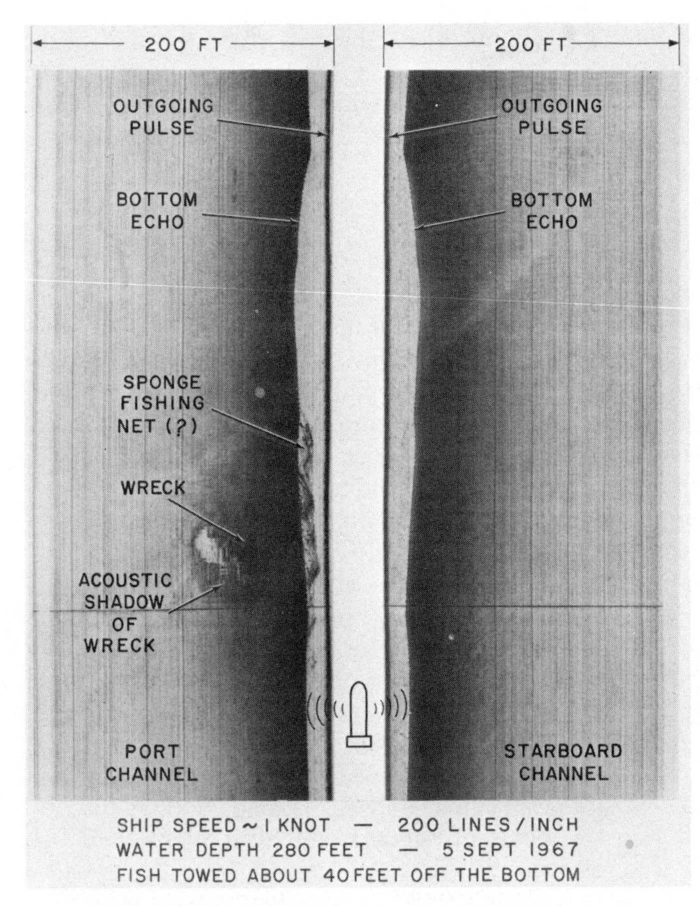

Fig. 6. Side-scan sonar record of Greek shipwreck.

shipwreck. The visibility had cleared up overnight, and it was possible to see 2.5—3 m. ASHERAH was kept at slightly negative buoyancy so she could move about the wreck and set down for a long look at objects.

The following year (1968) underwater TV was brought back to Turkey, and the wreck inspected again. It was then taken to the larger search area and the 26 "hits" from the previous year were examined. The sea state ranged from 2 to 4, and the search team had a difficult time trying to make out just what it was they were seeing. Two out of the 26 hits looked like they might be wrecks, but need further investigation.

Searching operations generally include three phases: detection, localization, and classification. ASHERAH proved to be valuable only for the last phase because of her lack of navigational equipment and the limited optical visibility of the environment. She did well in the last phase for the same reasons that other people buy submersibles. She was free swimming and decoupled from the surface effects of a tether. She carried people with eyes that are far superior to TV cameras for viewing, and could, like a swimmer, move around the target at will to examine different things of interest.

CONCLUSIONS

In 1967, the University of Pennsylvania decided that, in spite of signed waivers by her operators, ASHERAH should be insured for $5,000,000 liability. Insurance was obtained through the University's insurance department, and a bill presented to the Turkish underwater project upon the return of its staff to the U.S. Because the insurance premium for only a dozen or so dives by the submersible equalled the entire cost of a small excavation on land, Bass concluded that such annual sums could not be justified by the use of ASHERAH and recommended that the university sell her, which it did.

Thus, the value of a submersible to archaeology remains a question. As presented at the beginning of this chapter, the main reason for building a submarine was economic. The cost of shipping, operating and maintaining ASHERAH in 1964 and 1967 was small, in fact covered by off-season rentals to other agencies for her crew and even the crew of a surplus U.S. Navy ship loaned to support her in 1964 remained volunteers. The high insurance premium, however, approaching nearly $10,000 a year, prohibited us from realizing ASHERAH's full potential just as she was found to be an effective tool in at least some of the tasks for which she was planned. Had we discovered priceless bronze statues, the resultant publicity might have enriched our program to the extent that we could have continued to operate the submarine. Since the sale of ASHERAH in 1969, a single bronze statue netted in the Adriatic has been valued at $4,000,000, a pair of bronzes found off southern Italy was appraised at $1,000,000; and the amateur diver who found them rewarded by the Italian Government with $250,000. A group of bronzes found in the Straits of Messina were illegally raised and sold to antiquities dealers, and another classical bronze found near or on the Negro Youth site was spirited out of Turkey by Turkish dealers. One day, hopefully, archaeologists rather than those interested only in financial profit may find such examples of art. A submersible may or may not play a role in their location, and in the excavations of the ships that carried them.

REFERENCES

Bass, G.F. and Rosencrantz, D.M., 1968. A Diversified Program for the Study of Shallow Water Searching and Mapping Techniques. National Technical Information Service, U.S. Department of Commerce, Washington, D.C., AD 686 487, 130 pp.

Karius, R., Merifield, P.M. and Rosencrantz, D.M., 1965. Stereo-mapping of underwater terrain from a submarine. In: MTS/ASLO Ocean Science and Ocean Engineering, 1965, 2. Marine Technology Society, Washington, D.C., pp. 1167—1177.
Rosencrantz, D.M., 1975. Underwater Photography and Photogrammetry. In: E. Harp Jr. (Editor), Photography in Archeological Research. University of New Mexico Press, Albuquerque, N.M., pp. 265—309.

Appendix

APPENDIX

SUMMARY OF CHARACTERISTICS, STATUS AND USE OF UNDERSEA VEHICLES WORLDWIDE

J.R. Vadus

INTRODUCTION

The undersea vehicle has evolved over the last decade from merely demonstrating its scientific and technological potential into a very useful means for conducting a wide variety of undersea technological tasks and research activities. Unlike the earlier system the new vehicles are designed and built to not only reliably achieve specific objectives, but to do so in an economical manner. This is necessary if they are to maintain a competitive position when cost-effective comparisons are made with other methods. Most of the information presented is available in considerable detail in published reports by Vadus (1975, 1976).

UNDERSEA VEHICLE/STATISTICS

The major characteristics of 137 undersea vehicles are listed and grouped by countries in Tables I and II. Ninety-one of this total are manned and 46 unmanned. Their statistics are summarized as well as averaged in Table III. Eighty-three of the manned vehicles are either operational or ready for use and eight are under construction, with most of the latter scheduled for completion before the end of 1977. Forty-one of those in the unmanned category are operational or ready for use and five are under construction. It is estimated that between 5 and 10% of the systems classified as operational may be considered marginal with respect to their level of readiness, because of the additional time that would be required for mobilization, and crew training. Wet submersibles operated by divers and those designed to operate in depths of less than 183 m are excluded from the summary.

The ownership by country of these vehicles is analyzed in Table III. The United States is first with 52, with France second (25) and the Soviet Union third (19). One-half of the undersea vehicles listed in Table I were built in the United States. It also owns and operates about 50% of the world's unmanned vehicles and the U.S. Naval Undersea Center, is their major developer.

In averaging the characteristics of all these vehicles in Table III it was necessary to exclude several systems, such as the large bathyscaphs to avoid

TABLE I

Major characteristics of manned undersea vehicles worldwide

Vehicle	Operator	Crew	Depth (m)	Length (m)	Weight (kg)	Payload (kg)
Australia						
PLATYPUS 1c	Univ. of Sydney	2	300	4.6	5000	
Canada						
AQUARIUS I	HYCO Subsea	2	330	4.4	5000	400
AQUARIUS IIc	MARTECH	2	330	4.4	5000	400
AQUARIUS III	MARTECH	2	330	4.4	5000	400
SEA OTTER	CANDIVE	2	330	4.3	2860	250
SDL-1*	Canadian Armed Forces	5	600	6.1	13636	1163
AUG. PICCARD	Horton Maritime	4	670	28.5	166300	9090
PISCES VI	HYCO Subsea	3	2000	5.8	11090	863
PISCES IV	Environment Canada	3	2000	5.8	10954	861
PISCES V	HYCO Subsea	3	2000	5.8	11090	863
France						
GLOBULE	COMEX	2	200	2.8	2434	
PC8B	InterSub	2	240	5.9	5000	227
SHELF DIVER *	French Navy	4	240	7.2	7727	545
PC1201	InterSub	2	300	6.9	8180	454
PC1202 *	InterSub	5	300	9.6	15000	681
PC1203	COMEX	2	300	6.9	8180	454
PC1204	InterSub	2	300	6.9	8180	454
MOANA I	COMEX	3	430	4.2	9090	
MOANA IIIc	COMEX	3	430	4.2	9090	
MOANA IVc	COMEX	3	430	4.2	9090	
MOANA Vc	COMEX	3	430	4.2	9090	
SP-350	COF	2	410	3.0	3818	136
SP-500 (2)	COF	1	490	3.2	2409	45
GRIFFON	French Navy	3	600	7.5	13363	200
DEEPSTAR 2000	G.O. Int'l		610	6.3	7045	454

PC 16	InterSub	4	910	7.8	15000	272
DEEPSTAR 4000	COMEX	3	1212	5.5	8180	227
CYANA	CNEXO	3	2980	5.8	8000	200
ARCHIMEDE	CNEXO	3	10900	20.9	55450	2727
Germany (F.R.G.)						
MERMAID IV *c		3	300		12727	
Italy						
PC-5C	SubSea Oil	2	360	6.7	4545	340
PC-8C	SubSea Oil	2	360	7.0	5454	500
Japan						
UZUSHIO	Nippon Kokan	2	155	5.5	4755	
HAKUYO	Japan Ocean Sys.	3	300	6.4	6000	150
SHINKAI	Japan Maritime	4	600	15.3	90900	1818
Netherlands						
SKADOC 1000*	Skadoc Sub Sys.	3	330	5.5	3000	
Poland						
DELFIN-2*	Geological Inst.	2	195			1672
Soviet Union						
TRITON c (Amphibious URV)	Giprorybflot Institute					
GVIDON	Research Inst. of Fish. and Oceanog.	3	250	5.0	3900	
ATLANT II	Atlantic Inst. of Fisheries	2	330	5.0	2954	
AQUARIUS	Acad. of Science	3	390			
PISCES VII	Acad. of Science	3	450	5.8	10909	
TINRO II	Pacific Fish. Lab	2	450		3636	
ARGUS	Acad. of Science	3	450		10181	
OSA-3-600 I	Research Inst. of Fish. and Oceanog.	3	600			
OSA-3-600 II	Research Inst. of Fish. and Oceanog.	3	600			

(cont.)

TABLE I (cont.)

Vehicle	Operator	Crew	Depth (m)	Length (m)	Weight (kg)	Payload (kg)
SEVER I	Research Inst. of Fish and Oceanog.	1	600			
PISCES XI	Acad. of Science	3	2000	5.8	10954	6818
SEVER II	Polar Inst. of Fish and Oceanog.	2	2000	11.0	29545	
Sweden						
URF c	Royal Swedish Navy	5	450	13.7	50000	2000
Taiwan						
BURKHOLDER I	Kuofeng Ocean Develop. Corp.	2	300	6.0	9090	400
United Kingdom						
MERMAID III*	P and O Subsea	5	260	6.4	12727	
VOL-L1* and L2*	Vickers Oceanics	4	360	10.0	12727	909
PC-9	P and O Subsea	4	400	7.9	10227	250
PISCES I	Vickers Oceanics	2	450	4.9	2250	725
LEO I	P and O Subsea	3	600	5.8	12045	818
TAURUS*	P and O Subsea	4	660	10.5	24090	1818
PISCES II	Vickers Oceanics	3	720	5.8	10909	860
PISCES VIII	Vickers Oceanics	3	900	5.8	10909	680
PISCES III	Vickers Oceanics	3	900	5.8	10909	860
PISCES X	Vickers Oceanics	3	900	5.8	10909	860

United States						
SEA RANGER	Verne Engr. Corp.	4	180	5.2	8363	1000
NEMO	SW Research Inst.	2	180	1.9	909	386
SEA EXPLORER	Sea Line Inc.	2	180	4.6	1636	135
PRV-2*	Pierce Subs Inc.	3	180	5.8	7045	455
NEKTON ALPHA	Gen. Oceanographics	2	300	4.6	2030	135
NEKTON BETA	Gen. Oceanographics	2	300	4.6	2030	209
NEKTON GAMMA	Gen. Oceanographics	2	300	4.6	2030	209
JOHNSON SEA LINK *	Harbor Br. Found.	4	300	7.0	9545	545
SNOOPER	Undersea Graphics	2	300	4.6	2045	91
GUPPY	SunShip and Drydock	2	300	3.4	2770	180
OPSUB	Ocean Systems	2	300	5.5	4727	180
SEA RAY	Sub. R and D Corp	2	300	6.1	4545	160
MERMAID II	Int'l U.W. Contractors	2	300	5.2	4545	455
NEMO I	Seaborne Ventures	3	300	3.1	9090	545
DIAPHUS	Texas A and M Univ.	2	360	4.0	4545	102
PC-14c-2	Army Missile Command	2	360	4.0	4545	102
STAR II	Deepwater Explor. Ltd.	2	360	5.2	4545	227
DEEP VIEW	SW Research Inst.	2	450	4.9	5455	227
JOHNSON SEA LINK*	Harbor Br. Found.	4	300	7.0	9545	545
BEAVER MK IV*	Int'l U.W. Contractors	5	620	7.6	15455	909
DSRV-1	U.S. Navy	4	1500	15.3	34090	1954
DSRV-2	U.S. Navy	4	1500	15.3	34090	1954
SEA CLIFF	U.S. Navy	3	1950	7.9	19090	318
TURTLE	U.S. Navy	3	1950	7.9	19090	318
DEEP QUEST	Lockheed	4	2400	12.2	52270	3181
ALVIN	Woods Hole Oceanog. Inst.	3	3600	7.0	14545	681
TRIESTE II	U.S. Navy	3	6000	24.0	81818	909
DOWB	Friendship S.A.	3	1960	5.2	9090	477

c = construction.
* = diver lockout.

TABLE II
Major characteristics of unmanned undersea vehicles worldwide

Vehicle	Operator	Depth (m)	Length (m)	Weight (kg)	Payload (kg)
Canada					
BATFISH	Bedford Inst.	216	1.3	70	
TROV	Canada Center Inland Waters	360	1.6	513	
France					
PAP	Societe ECA	200	2.7	800	
TELENAUTE I	Institute Francais Petrol	300	4.1	1000	
TELENAUTE II	Institute Francais Petrol	300	4.1	1000	
ERIC	French Navy	1000	4.5	2000	
TROIKA	French Navy	2150	4.2	909	7272
Japan					
OCEAN SPACE ROBOT	Mitsubishi Ind.	240	4.5	1604	
Germany (F.R.G.)					
IBAK	IBAK	5950			
MANKA c (test model)	GKSS		3.7		
Norway					
CABLE CONTROLLED VEHICLE	Royal Norwegian Navy	340			
Soviet Union					
MANTA (2 Units)	Acad. of Science	300			
GIDROPLAN c	Acad. of Science	300			
KAYMAN		600		363	
SKORPENA	Research Inst. Fish. and Oceanog.	1000	3.3	454	
KRAB-1	Acad. of Science	3000	2.7	454	
KRAB-2	Acad. of Science	3000			
United Kingdom					
TROV-01	Underground Location Services	360	2.1	909	
CONSUB	BAC	600	2.4	800	132
SEXTON	MATSU				
CUTLET	Ministry of Defence				
United States					
SOLARIS	Naval Torpedo Sta.	450			
ELEC. SNOOPY	Naval Undersea Ctr.	450	1.0	68	
ELEC. SNOOPY II	Naval Facilities Engr. Center	450	1.1	136	20

TABLE II (continued)

Vehicle	Operator	Depth (m)	Length (m)	Weight (kg)	Payload (kg)
CORD	Harbor Br. Found	450	1.8	350	22
RECON II	Perry Oceanog.	450	1.1	205	
SCAT	Naval Undersea Ctr.	600	1.8	181	36
DEEP DRONE	SupSalvage	600		2727	
RUFAS II	Miss. State Univ.	700	3.3	454	
CURV II	Naval Undersea Ctr.	750	4.5	1568	181
CURV IIB	Naval Torpedo Sta.	750	4.5	1363	181
SCARAB (2)	A.T. and T. Co.	1800		2727	
RC 225	HYDRO Products	2000			
CURV III	Naval Undersea Ctr.	2100	4.5	2045	909
RUM/ORB	Scripps Inst. of Oceanog.	2400	3.8	10909	
SEA PROBE	Ocean Search Inc.	3600		181818	
TELEPROBE	Naval Oceanog. Off.	600	2.4	1590	727
DEEP TOW	Scripps Inst. of Oceanog.	600	1.6	147	
MIZAR FISH	Naval Research Lab	600	2.6	818	
RUWS	Naval Undersea Ctr.	400	3.1	1954	454
WORK VEHICLE c	HYDROTECH	1200	1.3	50000	18181
VERTICAL c TRANSPORT VEH.	HYDROTECH	1200	1.8	59090	45454
UDOSS [1]	Jet Prop. Lab	6000	3.0	1363	
ROBOT VEHICLE	M.I.T.		2.4	114	

[1] Design stage.

skewing the statistics of the more typical systems. The average depth capability of the world's manned submersibles is now 800 m. This represents an increase of about 10% greater than in 1974; and this trend will continue because of the need to operate at greater depths. The average weight of manned vehicles has increased from 8600 kg a year ago, to about 10,000 kg, or about 15%. The average payload was calculated to be about 600 kg for manned vehicles and a very low average value of 100 kg for the unmanned vehicles. This difference can be attributed to the fact that most of the latter are instrumented for a specific mission and do not provide additional payload space. For the manned vehicles, the average life support is calculated to be about 120 man-hours or 40 hours per man. This figure is considered low and many believe that 72 hours per man should be the minimum requirements for safety in the event of disablement and the resulting need to await search and rescue. However, in some missions that require working in rougher waters, further offshore and at greater depths, provisions for additional emergency life-support capability should be provided.

TABLE III

Summary statistics on undersea vehicles

		Manned	Unmanned
Status			
Worldwide — operational or ready		83	41
Worldwide — under construction		8	5
Worldwide	Total	91	46
Average characteristics			
Design depth (m)		800	2000
Weight (kg)		11818	730
Payload (kg)		590	—
Crew size		3	—
Life support (man-hours)		85	—
Ownership by country			
United States		28	24
France		20	5
Soviet Union		12	7
United Kingdom		11	4
Canada		9	2
Japan		3	1
Italy		2	0
Germany (F.R.G.)		1	2
Netherlands		1	0
Poland		1	0
Australia		1	0
Sweden		1	0
Taiwan		1	0
Norway		0	1

VEHICLE UTILIZATION

Within the last year, there has been over a 30% increase worldwide in available undersea vehicles, primarily in support of offshore development activities, especially for the oil industry. The summation of data on manned vehicles listed in Figs. 1, 2, and Tables IV and V, reveals that inspection, mainly of pipelines and cables, was the leading mission category, followed by cable burial. A list of the leading activities sampled on a worldwide, dive-day basis, in descending order are:

Inspection (pipeline, cable, etc.) — 50%.
Cable burial — 18%.
Engineering, salvage, etc. — 12%.

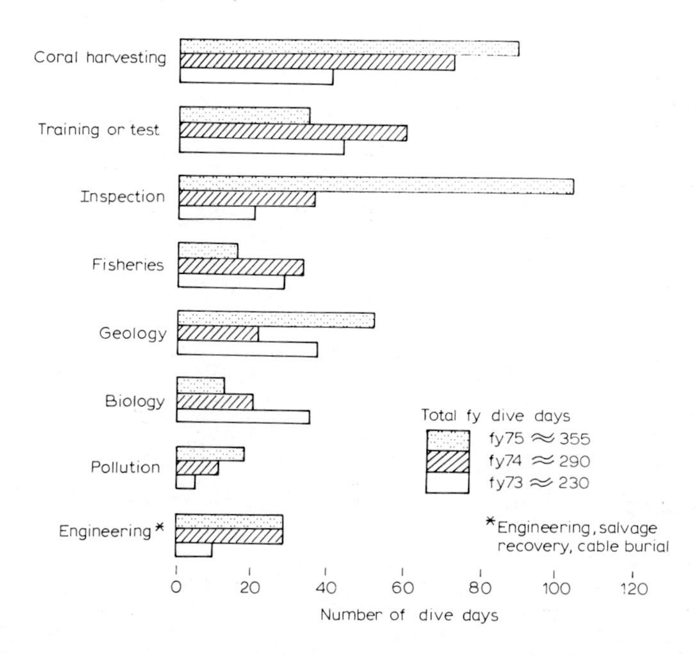

Fig. 1. Civilian manned undersea vehicle utilization in the U.S. during Fiscal Years 1973, 1974 and 1975.

placed in descending order, although there are only small differences between them:

Coral harvesting.
Geological.
Biological, fisheries.
Pollution, ocean dumping.

There are no data in this summary on unmanned vehicle activities, although unmanned vehicles have been busy, but in general not as much as manned systems. An example of one noteworthy mission, carried out for

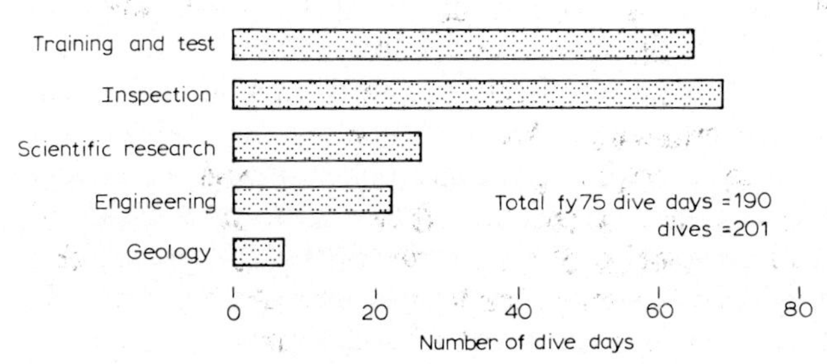

Fig. 2. U.S. Navy's manned undersea vehicle utilization in Fiscal Year 1975.

TABLE IV

Utilization of the ALVIN submersible

	Year 74	Totals 75	Cumulative totals to 1 Jan. 1976
Total number of dives	60	58	604
Total dives, test and training	7	4	136
Total mission dives	53	54	468
Mission Categories:			
Orientation	0	0	60
Biology	13	37	122
Geology	32	9	142
Search and recovery	2	3	45
Equipment inspection	6	0	28
Navigation experiments	0	0	24
Other science and engineering	0	5	47
Total time submerged (hrs)			2,236
Average time for dive (hrs)			4.3

several weeks in the summers of 1974 and 1975, was conducted by the U.S. Environmental Protection Agency, using the CURV III unmanned vehicle to survey, photograph, and sample around a radioactive dumpsite near the Farralon Islands, off the coast of California. Data concerning the integrity of the radioactive waste containers and the fate of any leaking pollutants is of worldwide interest in establishing a policy for future dumping. These activities are discussed in Chapter 13.

Although many new undersea vehicles were built in the United States, those available for use in the U.S. have changed negligibly. The use of underwater vehicles in the U.S. over the last three fiscal years, is illustrated in Fig. 1. The number of total dive-days in FY 1975 decreased by about 15%, from FY 1974. Inspection, mainly of pipelines and cables, was the leading U.S. mission, and this coincides with worldwide activities. Coral harvesting, of jewelry-quality, pink and black coral at 300 m depth is represented only by the STAR II's activities off the east coast of Oahu, in the Hawaiian Islands; but has been increasing steadily over the last three years.

Fisheries and biology mission have exhibited slight decreases each year, whereas geology missions have increased. Most of the biology efforts are attributed to ALVIN operations in studying the deep-ocean food chain, and also the deep-benthic fish and other organisms. Other uses have included studies on: (1) the under-utilized species of crab at the 680—1000-m depths; (2) the habitation and migration of deep-water lobster and shrimp; and (3) the deployment and effectiveness of line arrays of lobster traps.

TABLE V

Utilization of undersea vehicles as sampled on a worldwide basis excluding the U.S. (July 1974 through December 1975)

Vehicle	Mission category	Mission location	Dives	Dive days	Average depth (m) or range
Canada					
SDL-1	Test	Nova Scotia	16	15	75
SDL-1	Training	Nova Scotia	68	36	75
SDL-1	Inspection	Nova Scotia	10	10	300
PISCES V	Cable Burial	Nova Scotia	27	20	1450
AQUARIUS I	Survey oil Barge	Prince Ed. Isl.	6	6	76
AQUARIUS I	Guideline Replacement for Well Head	Prince Ed. Isl.	14	15	75
AQUARIUS I	Cable Burial	Block Isl. U.S.A.	15	15	75
AQUARIUS I	Cable Inspection	Nova Scotia	6	6	75
France					
CYANA	Test & Training	Mediterranean	28	28	30–2700
CYANA	Geology (FAMOUS Proj.).	Azores	15	15	3000
CYANA	Pipeline Insp	Mediterranean	3	1	400
CYANA	Pipeline & Cable Inspection	Sicily	44	36	100–600
PC8B	(Offshore Support	North Sea	400	212	150
PC1202	Activities-Mainly	North Sea	231	90	200
PC1202	Pipeline Survey)	North Sea	79	50	200
Japan					
HAKUYO	Pipeline Insp	Aga. Niigata	230	15	30–81
HAKUYO	Fisheries	Shizuoka	204	17	30–200
HAKUYO	Fisheries	Kanagawa	44	1	65
HAKUYO	Biology	Sagami Bay	84	3	115–134
HAKUYO	Equipment Emplacement	Wakayama	49	4	147–250
HAKUYO	Cable Inspection	Ibaragi	9	2	83–167
HAKUYO	Salvage	Kagoshima	4	3	125
United Kingdom					
PISCES I	Navy Missions	W. Scotland	292	312	40–200
PISCES II	Pipeline Work	North Sea	150	151	40–200
PISCES II	Cable Burial	Bay Biscay	26	32	40–200
PISCES III	Pipeline Work	North Sea	30	30	15–120
PISCES III	Cable Burial	Bay Biscay	107	107	15–120
PISCES III	Platform Survey	North Sea	130	128	15–120
PISCES V	Pipeline Work	North Sea	77	77	30–160
PISCES V	Cable Burial	North Sea	59	59	30–160
PISCES VIII	Pipeline Work	North Sea	5	5	30–140
PISCES VIII	Cable Burial	Bay Biscay	46	46	30–140
VOL-L1	Pipeline Work	North Sea	26	26	3–160
VOL-L2	Trials	North Sea	20	20	10–60

As noted in Table IV, ALVIN has made over 600 dives, of which about 22% involved test and training, and the balance of the missions were primarily oriented to geology and biology. ALVIN has spent an equivalent total of almost 100 continuous days under the sea, and has developed a steadily increasing average time for dives, which is now 4.3 hours.

Federal use during the last three years of American Bureau of Shipping (ABS)-classed civilian-operated manned vehicles was less than 10% of the total available submersible time. However, the U.S. Navy's undersea vehicle utilization in FY 1975 involved about 190 dive-days, mainly for deep undersea inspection missions, training and testing, as categorized in Fig. 2. The PC-14C-2, owned by the Army's Ballistic Missile Command, has the special mission of recovering missile debris entering the splashdown area of the Kwajalein Missile Range.

Statistical data are not available but it is reported that the Soviet undersea vehicles are mainly involved in fisheries research. The OSA-3-600, owned and operated by the National Institute of Sea Fisheries and Oceanography, has been used in fisheries research; for example, to hover over a school of fish and transmit data on the extent, location, and speed of movement of the school. It is also capable of taking core samples from the ocean bottom for later analysis by petroleum scientists. The unmanned tethered vehicle, SKORPENA is also operated by this institute, and is reportedly utilized in oceanographic and biological research on illuminescence and bioluminescence. The SEVER 2, operated by the Polar Institute of Fish and Oceanography, is reportedly operating in the North Atlantic, looking for schools of fish, studying the sea bottom, and selecting areas for trawl fishing. In the Black Sea, most of the Soviet activities originate from their base at Gelendzhik. Information on Soviet undersea vehicle activities is presented by Boylan (1975) and Anonymous (1975).

CLASSIFICATION, CERTIFICATION AND STANDARDIZATION

An important consideration in vehicle development, ownership, and operation is to have the vehicle system designed, built and tested in accordance with a classification code. This provides an added degree of confidence regarding performance and personnel safety. Insurance companies often consider this as one of the criteria in establishing underwriting coverage. There are nine classification organizations worldwide:

(1) American Bureau of Shipping.
(2) Bureau Veritas.
(3) Det Norske Veritas.
(4) Germanischer Lloyd.
(5) Lloyd's Register of Shipping.
(6) Nippon Kaiji Kyokai.
(7) Polish Register of Shipping.

The following categories, representing the balance of the missions, are
(8) Registro Italiano Navale.
(9) USSR Register of Shipping.

Some major guidelines and classification data provided by four classification organizations are listed in Table VI, together with the U.S. Navy's certification requirements. In reviewing the classification manuals, slight variations between each agency were noted, and occasionally an item was listed as a guideline and not as a requirement. The classification process relies heavily on design review and observation of tests by an inspector. Table VI also summarizes requirements for U.S. Navy certification in the same categories used for the classification organizations. Navy certification requirements are far more stringent than are the guidelines for classification. In addition to review of design, material and construction, the U.S. Navy certification process requires the system to be tested to verify its performance within the scope of its intended use.

Based on the comparison of classification data, there appears to be a need for some degree of standardization pertaining to personnel safety, including emergency features and means to aid in search and rescue. For example in the event of disablement on the bottom, it would be desirable to provide the crew with a minimum number of hours of life support per man, e.g., 72 hours, under normal operating conditions; and some greater number based on distance offshore, depth, expected sea state, and weather conditions. In order to communicate and signal location during disablement, it is desirable to standardize on frequencies for underwater telephones and emergency acoustic beacons. Although the support ship can probably make contact, other rescue forces brought to the scene may not be so equipped.

Once located, the next step is to recover the submersible and it would be very desirable for each one to have a standard hooking arrangement located at an established lift point. A report by Talkington (1975) provides the following list of items considered mandatory as self-help rescue features for undersea vehicles:

Acoustic beacon on a standard distress frequency (37 kHz).

External standard lift points.

Acoustic communications on a standard underwater telephone (8—11 kHz).

Minimum operator qualifications.

Filling of dive plan with a potential rescue unit.

Passenger pre-dive briefing.

The Marine Technology Society's Undersea Vehicles Safety Standards Subcommittee (Anonymous, 1976) has developed plans to formulate submersible safety standards. The objectives are to improve safety in vehicle operation, and in rescue response capabilities.

The plan involves establishing standards in three areas:

Personnel qualifications and training.

TABLE VI

Some major guidelines provided by classification organizations and corresponding U.S. Navy certification requirements

	American Bureau of Shipping	Lloyd's Register of Shipping	Det Norske Veritas	Germanischer Lloyd							U.S. Navy (Certification)
				Depth (m)							
				50	100	200	267	400	500	600	
Pressure hull											
Design collapse depth / operating depth	1.5	1.43		2.5	2.4	2.2	2.0	1.9	1.8	1.7	1.5 Acceptable
Test depth / operating depth	1.2	1.5	1.3	1.5	1.4	1.3	1.2	1.2	1.2	1.2	1.10–1.25
Design criteria and documentation	membrane=2/3 yield combined=3/4 yield equations and drawings specified	equations and drawings specified	any recognized design code, equation and drawings specified	0.83 yield equations and drawings specified							membrane=2/3 yield combined=3/4 yield equations and drawings specified
Weld inspection	100% radiograph		100% radiograph	100% radiograph							see NavShips 0900–000–1000
Piping penetrations	hull stop valves required	hull stop valves required	hull stop valves required	hull stop valves required							desirable but no requirement
Electrical penetrations	1.5 of operating depth		no plastic penetrators	must be approved							capable of maintaining seal if sheared off (1.5 of operating depth is acceptable)

Life support and environmental control					
O_2 limits	18.4% to 23.6%		161 ± 20 mm. Hg		normal 18···21% set points 17···23%
CO_2 limits	1%		<7.6 mm. Hg		0.5···2.0%
CO_2 scrubber duration	mission <24 hrs mission + 72 hrs mission >24 hrs mission + 3 days + 2 hrs/day	5 times mission	mission +72 hrs	48 hrs minimum	depends on stated mission time with 50% safety factor
Standard man consumption	0.036 m^3 O_2/hr		26 1 O_2/hr 22 1 CO_2/hr	25 1 O_2/hr 20 1 CO_2/hr	0.973 m^3 O_2/hr for planning
Temperature control					
Humidity control	30—90%				discretion of designer
Instrumentation	O_2, CO_2 R.H. temperature pressure	gas analysis R.H. temperature pressure	(2) CO_2 (2) O_2 (±10 mm. Hg) barometer (±3%)	O_2, CO_2 pressure	O_2, CO_2 R.H. temperature pressure recommended
Emergency life-support equipment	1.5 × surfacing time	O_2 breathing system	1.5 × surfacing time		1.5 × surfacing time
Communications (considered)					
U.W. telephone	yes	yes	yes		recommended
Radio (with distress frequency)	yes	yes	yes		recommended
Pinger	yes	yes	yes		recommended

Operational plans and procedures.
Emergency equipment.

The results of this effort will be documented in an MTS book "Recommended Safety Standards for Undersea Vehicles", to be published by the end of 1977. This will be a third in the series of books prepared by this Subcommittee; the other two are entitled "Safety and Operational Guidelines for Undersea Vehicles". (Anonymous, 1974).

MANNED SUBMERSIBLE ACCIDENTS

There have been eight major submersible accidents, taking the lives of eight persons, within the last seven years, which were reported to have occurred during underwater operations. Details on the last two accidents are now available, and a brief description of the information reported is as follows: In September 1975, the STAR II submersible, attached to its Launch and Retrieval Transport (LRT) was involved in a fatal accident while in the process of being launched from its LRT, off the west coast of Oahu, Hawaii. It was reported that two of the divers, supporting the submerged launching of the STAR II, lost their lives trying to free it as the LRT continued to sink and safe diver depths for air breathing were exceeded. The third diver barely made it back to the surface. In June 1976 it was reported that a MOANA vehicle, while involved in a demonstration dive off the coast of Marseille, France with three persons on board, sunk as a result of water entering the vehicle's main entrance hatch. One of the three persons on board lost his life and the other two managed to escape.

A good reference source, pertaining to submersible safety through accident analysis, is Appendix IV of Book II, "Safety and Operational Guidelines for Undersea Vehicles. (Anonymous, 1974). Busby's (1976) "Manned Submersibles" also contains a chapter "Emergency Devices and Procedures" and another "Emergency Incidents and the Potential for Rescue."

CONCLUSIONS

Undersea vehicles, within the last five years, have proven to be a significant tool in ocean research and development, and their number and use is steadily increasing. The offshore industry is the principal user, and there are many other mission applications that will require more extensive usage. The latest designs feature fully integrated systems (vehicle, ship, handling gear, and logistics and maintenance support) to ensure an effective high utilization rate.

More specificity and standardization is needed by the classification societies in the vital areas pertaining to improved safety, search and rescue. Safety standards in areas of crew qualifications, operating procedures, and emergency equipment, should be developed by the user community but not to the extent of hindering innovation in the design and effective utilization of vehicles.

ACKNOWLEDGEMENTS

The author would like to thank the many submersible builders, owners, and operators worldwide, who furnished data on the design and utilization of their vehicles.

REFERENCES

Anonymous, 1974. Safety and Operational Guidelines for Undersea Vehicles, I and II. Marine Technology Society, Washington, D.C.

Anonymous, 1975. Soviet Ocean Activities: A Preliminary Survey. Library of Congress Report, 1975, prepared for the U.S. Senate Committee on Commerce.

Anonymous, 1976. MTS Undersea Vehicles Safety Standards Subcommittee, 1976. Chairman: J. Pritzlaff; Steering Committee: F. Busby, L. Shumaker and H. Talkington. Marine Technology Society, Washington, D.C.

Boylan, L., 1975. Underwater Activities in the Soviet Union. Informatics, Rockville, Md.

Busby, F., 1976. Manned Submersibles. U.S. Government Printing Office, Washington, D.C.

Talkington, H., 1975. Self-Help Rescue Capability for Submersibles. Naval Undersea Center, San Diego, Calif.

Vadus, J.R., 1975. International Review of Manned Submersibles and Habitats. U.S. Department of Commerce, Springfield, Va. (Order No. PB 246 428/7WO).

Vadus, J.R., 1976. International Status and Utilization of Undersea Vehicles, 1976. U.S. Department of Commerce. Available from U.S. Government Printing Office, Washington, D.C.

INDEX *

* Italicized pagenumbers refer to illustrations.